System Dynamics for Engineering Students

System Dynamics for Engineering Students
Concepts and Applications

Nicolae Lobontiu

University of Alaska Anchorage

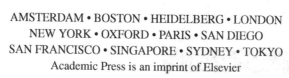

AMSTERDAM • BOSTON • HEIDELBERG • LONDON
NEW YORK • OXFORD • PARIS • SAN DIEGO
SAN FRANCISCO • SINGAPORE • SYDNEY • TOKYO
Academic Press is an imprint of Elsevier

ELSEVIER

Academic Press is an imprint of Elsevier
30 Corporate Drive, Suite 400, Burlington, MA 01803, USA
The Boulevard, Langford Lane, Kidlington, Oxford, OX5 1GB, UK

MATLAB and Simulink are registered trademarks of The MathWorks, Inc. See www.mathworks.com/trademarks for a list of additional trademarks. The MathWorks Publisher Logo identifies books that contain MATLAB® and Simulink® content. Used with permission. The MathWorks does not warrant the accuracy of the text or exercises in this book. This book's use or discussion of MATLAB® and Simulink® software or related products does not constitute endorsement or sponsorship by The MathWorks of a particular use of the MATLAB® and Simulink® software or related products.

For MATLAB® and Simulink® product information, or information on other related products, please contact: The MathWorks, Inc., 3 Apple Hill Drive, Natick, MA, 01760-2098 USA; Tel: 508-647-7000; Fax: 508-647-7001; E-mail: info@mathworks.com; Web: www.mathworks.com (A full listing of editorial standards, *The MathWorks Book Program Book Development, Production, and Promotion Guidelines,* is available from The Book Program Team at bookprogram@mathworks.com.)

Library of Congress Cataloging-in-Publication Data
A catalog record for this book is available from the Library of Congress.

British Library Cataloguing-in-Publication Data
A catalogue record for this book is available from the British Library.

ISBN: 978-0-240-81128-4 (Case bound)
ISBN: 978-0-12-381990-1 (Case bound with on-line testing)

For information on all Academic Press publications, visit our
Web site: www.elsevierdirect.com

Printed in the United States of America
Transferred to Digital Printing, 2014

Typeset by: diacriTech, Chennai, India

To all readers coming across this book,
with friendly consideration.

"Pro captu lectoris habent sua fata libelli." *(It is on the reader's understanding that the fate of books depends).*

Terentianus Maurus, Latin writer
De litteris, De syllabis, De metris

Contents

Foreword

This text is a modern treatment of system dynamics and its relation to traditional mechanical engineering problems as well as modern microscale devices and machines. It provides an excellent course of study for students who want to grasp the fundamentals of dynamic systems and it covers a significant amount of material also taught in engineering modeling, systems dynamics, and vibrations, all combined in a dense form. The book is designed as a text for juniors and seniors in aerospace, mechanical, electrical, biomedical, and civil engineering. It is useful for understanding the design and development of micro- and macroscale structures, electric and fluidic systems with an introduction to transduction, and numerous simulations using MATLAB and SIMULINK.

The creation of machines is essentially what much of engineering is all about. Critical to almost all machines imaginable is a transient response, which is fundamental to their functionality and needs to be our primary concern in their design. This might be in the form of changing voltage levels in a sensor, the deflection of a spring supported mass, or the flow of fluid through a device. The phenomena which govern dynamics are not simply its mechanical components but often involve the dynamics of transducers as well, which are often electro-mechanical or fluidic based. This text discusses traditional electro-magnetic type actuators, but also ventures into electrostatics which are the dominant form of actuators in microelectromechanical systems (MEMs).

This book presents an opportunity for introducing dynamic systems to scientists and engineers who are concerned with the engineering of machines both at the micro- and macroscopic scale. Mechanism and movement are considered from the types of springs and joints that are critical to micro-machined, lithographic based devices to traditional models of macroscale electrical, fluidic, and electromechanical systems. The examples discussed and the problems at the end of each chapter have applicability at both scales. In essence this is a more modern treatment of dynamical systems, presenting views of modeling and substructures more consistent with the variety of problems that many engineers will face in the future. Any university with a substantive interest in microscale engineering would do well to consider a course that covers the material herein. Finally, this text lays the foundation and framework for the development of controllers applied to these dynamical systems.

Professor Ephrahim Garcia
Sibley School of Mechanical and Aerospace Engineering
Cornell University
Ithaca, New York

Preface

Engineering system dynamics is a discipline that focuses on deriving mathematical models based on simplified physical representations of actual systems, such as mechanical, electrical, fluid, or thermal, and on solving the mathematical models (most often consisting of differential equations). The resulting solution (which reflects the system response or behavior) is utilized in design or analysis before producing and testing the actual system. Because dynamic systems are characterized by similar mathematical models, a unitary approach can be used to characterize individual systems pertaining to different fields as well as to consider the interaction of systems from multiple fields as in coupled-field problems.

This book was designed to be utilized as a one-semester system dynamics text for upper-level undergraduate students with emphasis on mechanical, aerospace, or electrical engineering. Comprising important components from these areas, the material should also serve cross-listed courses (mechanical-electrical) at a similar study level. In addition to the printed chapters, the book contains an equal number of chapter extensions that have been assembled into a companion website section; this makes it useful as an introductory text for more advanced courses, such as vibrations, controls, instrumentation, or mechatronics. The book can also be useful in graduate coursework or in individual study as reference material. The material contained in this book most probably exceeds the time allotted for a one-semester course lecture, and therefore topical selection becomes necessary, based on particular instruction emphasis and teaching preferences.

While the book maintains its focus on the classical approach to system dynamics, a new feature of this text is the introduction of examples from compliant mechanisms and micro- and nano-electromechanical systems (MEMS/NEMS), which are recognized as increasingly important application areas. As demonstrated in the book, and for the relatively simple examples that have been selected here, this inclusion can really be treated within the regular system dynamics lumped-parameter (pointlike) modeling; therefore, the students not so familiar with these topics should face no major comprehension difficulties. Another central point of this book is proposing a chapter on coupled-field (or multiple-field) systems, whereby interactions between the mechanical, electrical, fluid, and thermal fields occur and generate means for actuation or sensing applications, such as in piezoelectric, electromagnetomechanical, or electrothermomechanical applications.

Another key objective was to assemble a text that is structured, balanced, cohesive, and providing a fluent and logical sequence of topics along the following lines:

1. It starts from simple objects (the components), proceeds to the objects' assembly (the individual system), and arrives at the system interaction level (coupled-field systems).

2. It uses modeling and solution techniques that are familiar from other disciplines, such as physics or ordinary differential equations, and subsequently introduces new modeling and solution procedures.

3. It provides a rather even coverage (space) to each book chapter.

4. While various chapter structures are possible in a system dynamics text, this book proposes a sequence that was intended to be systematic and consistent with the logical structure and progression of the presented material.

As such, the book begins with an introductory Chapter 1, which offers an overview of the main aspects of a system dynamics course for engineering students. The next four chapters—Chapters 2, 3, 4, and 5—are dedicated, in order, to mechanical (Chapters 2 and 3), electrical (Chapter 4), and fluid and thermal (Chapter 5) system modeling. They contain basic information on components, systems, and the principal physical and mathematical tools that make it possible to model a dynamic system and determine its solution.

Once the main engineering dynamic systems have been studied, Chapter 6 presents the Laplace transform technique, a mathematical tool that allows simplifying the differential equation solution process for any of the individual systems. This chapter is directly connected to the next segment of the book, containing Chapters 7, 8, and 9. Chapter 7 introduces the transfer function approach, which facilitates finding the time-domain response (solution) of a dynamic system after the corresponding unknowns have been determined in the Laplace domain. The complex impedance, which is actually a transfer function connecting the Laplace-transformed input and output of a specific system element, is also introduced and thoroughly treated in this chapter. Chapter 8 studies the state space modeling and solution approach, which is also related to the Laplace transform of Chapter 6 and the transfer function of Chapter 7. Chapter 9 discusses modeling system dynamics in the frequency domain by means of the sinusoidal (harmonic) transfer function.

Chapter 10 analyzes coupled-field (or multiple-field) dynamic systems, which are combinations of mechanical, electrical, magnetic, piezoelectric, fluid, or thermal systems. In this chapter, dynamic models are formulated and solved by means of the procedures studied in previous chapters. Because of the partial and natural overlap between system dynamics and controls, the majority of textbooks on either of these two areas contain coverage of material from the adjoining domain. Consistent with this approach, the companion website contains one chapter, Chapter 11, on introductory controls, where basic time-domain and frequency-domain topics are addressed.

The book also includes four appendices: Appendix A presents the solutions to linear differential equations with constant coefficients, Appendix B is a review of matrix algebra, Appendix C contains basic MATLAB® commands that have been used throughout this text, and Appendix D gives a summary of equations for calculating deformations, strains, and stresses of deformable mechanical components.

The book introduces several topics that are new to engineering system dynamics, as highlighted here:

Chapter 3, Mechanical Systems II

- Lumped-parameter inertia properties of basic compliant (flexible) members.
- Lumped-parameter dynamic modeling of simple compliant mechanical microsystems.
- Mass detection in MEMS by the resonance shift method.

Chapter 4, Electrical Systems

- Capacitive sensing and actuation in MEMS.

Chapter 5, Fluid and Thermal Systems

- Comprehensive coverage of liquid, pneumatic, and thermal systems.
- Natural response of fluid systems.

Chapters 3, 4, and 5

- Notion of degrees of freedom (DOFs) for defining the system configuration of dynamic systems.
- Application of the energy method to calculate the natural frequencies of single- and multiple-DOF conservative systems.
- Utilization of the vector-matrix method to calculate the eigenvalues either analytically or using MATLAB®.

Chapter 6, Laplace Transform

- Linear ordinary differential equations with time-varying coefficients.
- Laplace transformation of vector-matrix differential equations.
- Use of the convolution theorem to solve integral and integral-differential equations.
- Time-domain system identification.

Chapter 7, Transfer Function Approach

- Extension of the single-input, single-output (SISO) transfer function approach to multiple-input, multiple-output (MIMO) systems by means of the transfer function matrix.
- Application of the transfer function approach to solve the forced and the free responses with nonzero initial conditions.
- Systematic introduction and comprehensive application of the complex impedance approach to electrical, mechanical, and fluid and thermal systems.
- MATLAB® conversion between zero-pole-gain (zpk) and transfer function (tf) models.

Chapter 8, State Space Approach

- Treatment of the descriptor state equation.
- Application of the state space approach to solve the forced and free responses with nonzero initial conditions.
- MATLAB® conversion between state space (ss) models and zpk or tf models.

Chapter 9, Frequency-Domain Approach

- State space approach and the frequency domain.
- MATLAB® conversion from zpk, tf, or ss models to frequency response data (frd) models.
- Steady-state response of cascading unloading systems.
- Mechanical and electrical filters.

Chapter 10, Coupled-Field Systems

- Formulation of the coupled-field (multiple-field) problem.
- Principles and applications of sensing and actuation.
- Strain gauge and Wheatstone bridge circuits for measuring mechanical deformation.
- Applications of electromagnetomechanical system dynamics.
- Principles and applications of piezoelectric coupling with mechanical deformable systems.
- Nonlinear electrothermomechanical coupling.

Within this printed book's space limitations, attention has been directed at generating a balanced coverage of minimally necessary theory presentation, solved examples, and end-of-chapter proposed problems. Whenever possible, examples are solved analytically, using hand calculation, so that any mathematical software can be used in conjunction with any model developed here. The book is not constructed on MATLAB®, but it uses this software to determine numerical solutions and to solve symbolically mathematical models too involved to be obtained by hand. It would be difficult to overlook the built-in capabilities of MATLAB®'s tool boxes (really programs within the main program, such as the ones designed for symbolic calculation or controls), which many times use one-line codes to solve complex system dynamics problems and which have been used in this text. Equally appealing solutions to system dynamics problems are the ones provided by Simulink®, a graphical user interface program built atop MATLAB®, and applications are included in almost all the chapters of solved and proposed exercises that can be approached by Simulink®.

Through a companion website, the book comprises more ancillary support material, including companion book chapters with extensions to the printed book (with more advanced topics, details of the printed book material, and additional solved examples, this section could be of interest and assistance to both the instructor and the student). The sign ⬤ is used in the printed book to signal associated material on the companion website. The companion website chapters address the following topics:

Chapter 3, Mechanical Systems II

- Details on lumped-parameter stiffness and inertia properties of basic compliant (flexible) members.
- Additional springs for macro and micro system applications.
- Pulley systems.

Chapter 4, Electrical Systems

- Equivalent resistance method.
- Transformer elements and electrical circuits.
- Operational amplifier circuits as analog computers.

Chapter 5, Fluid and Thermal Systems

- Capacitance of compressible pipes.

Chapter 6, Laplace Transform

- Thorough presentation of the partial-fraction expansion.
- Application of the Laplace transform method to calculating natural frequencies.
- Method of integrating factor and the Laplace transform.

Chapter 7, Transfer Function Approach

- System identification from time response.
- Cascading loading systems.
- Mutual inductance impedance.
- Impedance node analysis.

Chapter 8, State Space Approach

- State space modeling of MIMO systems with input time derivative.
- Calculation of natural frequencies and determination of modes.
- Matrix exponential method.

Chapter 10, Coupled-Field Systems

- Three-dimensional piezoelectricity.
- Energy coupling in piezoelectric elements.
- Time stepping algorithms for the solution of coupled-field nonlinear differential equations.

Whenever possible, alternative solution methods have been provided in the text to enable using the algorithm that best suits various individual approaches to the same problem. Examples include Newton's second law and the energy method for the free response of systems, which have been used in Chapters 3, 4, and 5, or the mesh analysis and the node analysis methods for electrical systems in Chapter 4.

The ancillary material also comprises an instructor's manual, an image bank of figures from the book, MATLAB® code for the book's solved examples, PowerPoint lecture slides, and a longer project whereby the material introduced in the chapter sequence is applied progressively. After publication and as a result of specific requirements or suggestions expressed by instructors who adopted the text and feedback from students, additional problems resulting from this interaction will be provided on the website, as well as corrections of the unwanted but possible errors.

To make distinction between variables, small-cap symbols are generally used for the time domain (such as f for force, m for moment, or v for voltage), whereas capital symbols denote Laplace transforms (such as F for force, M for moment, or V for voltage). With regard to matrix notation, the probably old-fashioned symbols { } for vectors and [ ] for matrices are used here, which can be replicated easily on the board.

Several solved examples and end-of-chapter problems in this book resulted from exercises that I have used and tested in class over the last years while teaching courses such as system dynamics, controls, or instrumentation, and I am grateful to the students who contributed to enhancing the scope and quality of the original variants. I am indebted to the anonymous academic reviewers who critically checked this project at its initial (proposal) phase, as well as at two intermediate stages. They have made valid suggestions for improvement of this text, which were well taken and applied to this current version. I appreciate the valuable suggestions by Mr. Tzuliang Loh from the MathWorks Inc. on improving the presentation of the MATLAB® material in this book. I am very thankful to Steven Merken, Associate Acquisition Editor at Elsevier Science & Technology Books, whose quality and timely assistance have been instrumental in converting this project from its embryonic to its current stage.

In closing, I would like again to acknowledge and thank the unwavering support of my wife, Simona, and my daughters, Diana and Ioana—they definitely made this project possible. As always, my thoughts and profound gratitude for everything they gave me go to my parents.

Resources That Accompany This Book

System dynamics instructors and students will find additional resources at the book's Web site: www.booksite.academicpress.com/lobontiu

Available to All

Bonus Online Chapter For courses that include lectures on controls, Chapter 11, Introduction to Modeling and Design of Feedback Control Systems, is an online chapter available free to instructors and students.

Additional Online Content Linked to specific sections of the book by an identifiable Web icon, extra content includes advanced topics, additional worked examples, and more.

Downloadable MATLAB® Code For the book's solved examples.

For Instructors Only

Instructor's Manual The book itself contains a comprehensive set of exercises. Worked-out solutions to the exercises are available online to instructors who adopt this book.

Image Bank The Image Bank provides adopting instructors with various electronic versions of the figures from the book that may be used in lecture slides and class presentations.

PowerPoint Lecture Slides Use the available set of lecture slides in your own course as provided, or edit and reorganize them to meet your individual course needs.

Instructors should contact their Elsevier textbook sales representative at textbooks@elsevier.com to obtain a password to access the instructor-only resources.

Also Available for Use with This Book

Web-based testing and assessment feature that allows instructors to create online tests and assignments which automatically assess student responses and performance, providing them with immediate feedback. Elsevier's online testing includes a selection of algorithmic questions, giving instructors the ability to create virtually unlimited variations of the same problem. Contact your local sales representative for additional information, or visit www.booksite.academicpress.com/lobontiu to view a demo chapter.

Introduction

This chapter discusses the notion of modeling or simulation of dynamic engineering systems as a process that involves physical modeling of an actual (real) system, mathematical modeling of the resulting physical representation (which generates differential equations), and solution of the mathematical model followed by interpretation of the result (response). Modeling is placed in the context of either analysis or design. The dynamic system mathematical model is studied in connection to its input and output signals such that single-input, single-output (SISO) and multiple-input, multiple-output (MIMO) systems can be formed. Systems are categorized depending on the order of the governing differential equations as zero-, first-, second- or higher-order systems. In addition to the examples usually encountered in system dynamics texts, examples of compliant (or flexible) mechanisms that are incorporated in micro- or nano-electromechanical systems (MEMS or NEMS) are included here. The nature of presentation is mainly descriptive in this chapter, as it attempts to introduce an overview of a few of the concepts that will be covered in more detail in subsequent chapters.

1.1 ENGINEERING SYSTEM DYNAMICS

Engineering system dynamics is a discipline that studies the dynamic behavior of various systems, such as mechanical, electrical, fluid, and thermal, either as isolated entities or in their interaction, the case where they are coupled-field (or multiple-field) systems. One trait specific to this discipline consists in emphasizing that systems belonging to different physical fields are described by similar mathematical models (expressed most often as differential equations); therefore, the same mathematical apparatus can be utilized for analysis or design. This similitude also enables migration between systems in the form of analogies as well as application of a unitary approach to coupled-field problems.

System dynamics relies on previously studied subject matter, such as differential equations, matrix algebra, and physics and the dynamics of systems (mechanical, electrical, and fluid or thermal), which it integrates in probably the first

DOI: 10.1016/B978-0-240-81128-4.00001-5

engineering-oriented material in the undergraduate course work. Engineering system dynamics is concerned with physically and mathematically modeling dynamic systems, which means deriving the differential equations that govern the behavior (response) of these systems, as well as solving the mathematical model and obtaining the system response. In addition to known modeling and solution procedures, such as Newton's second law of motion for mechanical systems or Kirchhoff's laws for electrical systems, the student will learn or reinforce new techniques, such as direct and inverse Laplace transforms, the transfer function, the state space approach, and frequency-domain analysis.

This course teaches the use of simplified physical models for real-world engineering applications to design or analyze a dynamic system. Once an approximate and sufficiently accurate mathematical model has been obtained, one can employ MATLAB®, a software program possessing numerous built-in functions that simplifies solving system dynamics problems. Simulink®, a graphical user interface computing environment that is built atop MATLAB® and which allows using blocks and signals to perform various mathematical operations, can also be used to model and solve engineering system dynamics problems. At the end of this course, the student should feel more confident in approaching an engineering design project from the model-based standpoint, rather than the empirical one; this approach should enable selecting the key physical parameters of an actual system, combining them into a relevant mathematical model and finding the solution (either time response or frequency response). Complementing the classical examples encountered in previous courses (such as the rigid body, the spring, and the damper in mechanical systems), new examples are offered in this course of compliant (flexible) mechanisms and micro- or nano-electromechanical systems. These devices can be modeled using the approach used for regular systems, which is the lumped-parameter procedure (according to which system parameters are pointlike).

In addition to being designed as an introduction to actual engineering course work and as a subject matter that studies various systems through a common prism, engineering system dynamics is also valuable to subsequent courses in the engineering curricula, such as vibrations, controls, instrumentation, and mechatronics.

1.2 MODELING ENGINEERING SYSTEM DYNAMICS

The *modeling* process of *engineering system dynamics* starts by identifying the *fundamental properties* of an actual system. The *minimum set of variables* necessary to fully define the *system configuration* is formed of the *degrees of freedom (DOF)*. Key to this selection is a schematic or diagram, which pictorially identifies the parameters and the variables, such as the *free-body diagram* that corresponds to the dynamics of a point-like body in mechanical systems with forces and moments shown and which plays the role of a *physical model* for the actual system.

It is then necessary to utilize an appropriate *modeling procedure* that will result in the *mathematical model* of the system. Generally, a mathematical model describing

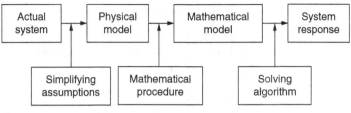

FIGURE 1.1

Flow in a Process Connecting an Actual Dynamic System to Its Response.

the dynamic behavior of an engineering system consists of a differential equation (or a system of differential equations) combining parameters with known functions, unknown functions, and derivatives. The next step involves *solving the mathematical model* through adequate mathematical procedures that deliver the *solution*, that is, expressions (equations) of variables as functions of the system parameters and time (or frequency), and that reflect the *system response* or *behavior*. Figure 1.1 gives a graphical depiction of this process that connects an actual dynamic system under the action of external forcing to its response. There are also situations when interrogation of the system response results in information that is fed back to the actual system at the start of the chain to allow for corrections to be applied, very similar to *feedback control systems*.

1.2.1 Modeling Variants

Various steps can be adopted in transitioning from the actual system to a physical model, then from a physical model to a mathematical one, as sketched in Figure 1.1. Several physical models can be developed, starting from an actual system, depending on the severity of the simplifying assumptions employed. Once a physical model has been selected, several modalities are available to mathematically describe that physical model. The application of different algorithms to the same mathematical model should produce the same result or solution, as the system response is unique.

In the case of a car that runs on even terrain, the car vertical motion has a direct impact on its passengers. A basic physical model is shown schematically in Figure 1.2, which indicates the car mass is lumped at its center of gravity (CG) and the front and rear suspensions are modeled as springs. Because the interest here lies only in the car vertical motion and the terrain is assumed even (perfectly flat), it is safe to consider, as a rough approximation conducing to a first-level physical model, that the impact points between the wheels and the road surface are fixed points. Under these simplifying assumptions, the parameters that define the car's properties are its mass, its mechanical moment of inertia about an axis passing through the CG and perpendicular to the drawing plane, and the stiffness (spring) features of the two suspensions. What is the *minimum number of variables fully describing the state* (or *configuration*) of this simplified system at any moment in time? If we

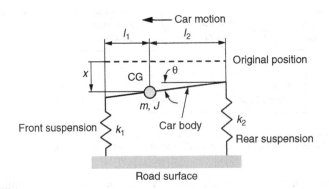

FIGURE 1.2

Simplified Physical Model of a Car That Moves over Even Terrain. Shown Are the Degrees of Freedom of the Center of Gravity and Pitch of the Body.

attach the system motion to the CG, it follows that knowing the vertical motion of the CG (measured by the variable x) and the rotation of the rigid rod (which symbolizes the car body) about a horizontal axis and measured by an angle θ are sufficient to specify the position of the car body at any time moment. Of course, we have used another simplifying assumption, that the rotations and vertical displacements are relatively small and therefore the motions of the suspensions at their joining points with the car (the rod) are purely vertical.

It follows that the *system parameters* are the car mass m and its moment of inertia J, the suspension spring constants (stiffnesses) k_1 and k_2, as well as the distances l_1 and l_2, which position the CG of the car. Generally, all these parameters have known values. The *variables* (*unknowns* or *DOFs*) are x, the vertical motion of the CG, and θ, the rotation of the body car about its CG. The next step is deriving the *mathematical model* corresponding to the identified physical model, and this phase can be achieved using a specific *modeling technique*, such as Newton's second law of motion, the energy method, or the state space representation for this mechanical system—all these modeling techniques are discussed in subsequent chapters. The result, as mentioned previously, consists of a system of two differential equations containing the system parameters m, J, k_1, k_2, l_1, l_2 and the unknowns x, θ together with their time derivatives. Solving for x and θ in terms of initial conditions (for this system, these are the initial displacements when $t = 0$, namely $x(0)$, $\theta(0)$, and the initial velocities $\dot{x}(0)$, $\dot{\theta}(0)$) provides explicitly the functions $x(t)$ and $\theta(t)$, and this constitutes the *system's response*. The *system behavior* can be studied by plotting, for instance, x and θ as functions of time.

More complexity can be added to the simple car physical model of Figure 1.2, for instance by considering the wheels are separate from the mechanical suspension through the tire elasticity and damping. The assumption of an uneven terrain surface can also be introduced. Figure 1.3 is the physical model of the car when all these system properties are taken into account—please note that the masses of wheels and

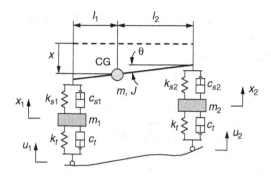

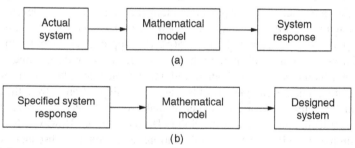

FIGURE 1.3

Simplified Physical Model of a Car Moving over Uneven Terrain, with the Degrees of Freedom of the Suspensions Shown.

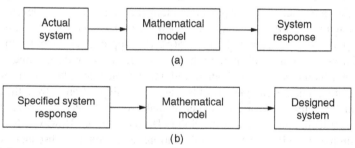

(a)

(b)

FIGURE 1.4

Processes Utilizing Dynamic Models: (a) Analysis; (b) Design (Synthesis).

tires and suspensions are included and combined together (they are denoted by m_1 and m_2 in Figure 1.3) and that the two wheels are considered identical. It can now be seen that two more DOFs are added to the existing ones, so that the system becomes a four-DOF system (they are x, θ, x_1, and x_2), whereas the input is formed by the two displacements applied to the front and rear tires, u_1 and u_2.

Dynamic modeling is involved in two apparently opposite directions: the *analysis* and the *design* (or *synthesis*) of a specific system. Analysis starts from a given system whose parameters are known. The dynamic analysis objective is to establish the response of a system through its mathematical model. Conversely, the design needs to find an actual dynamic system capable of producing a specified performance or response.

In analysis we start from a real-world, well-defined system, which we attempt to characterize through a mathematical model, whereas in design (synthesis), we embark with a set of requirements and use a model to obtain the skeleton of an actual system. Figure 1.4 gives a graphical representation of the two processes.

1.2.2 Dynamical Systems Lumped-Parameter Modeling and Solution

Lumped-Parameter Modeling

It is convenient from the modeling viewpoint to consider that the parameters defining the dynamic behavior of a system are located at well-specified stations, so they can be considered pointlike. The mass of a rigid body, for instance, is considered to be concentrated at the center of mass (gravity) of that body so that the center of mass becomes representative for the whole body, which simplifies the modeling task substantially without diminishing the modeling accuracy. Similar *lumping* considerations can be applied to springs or dampers in the mechanical realm but as well in the electrical domain, where resistances, capacitances, and inductances are considered lumped-parameter system properties.

Also, in some cases, the lumped-parameter modeling can be used for components that have inherently distributed properties. Take the example of a cantilever, such as the one sketched in Figure 1.5(a). Both its *inertia* and *elastic properties* are *distributed*, as they are functions of the position x along the length of the cantilever. Chapter 3 shows how to *transform* the actual *distributed-parameter model* into an *equivalent lumped-parameter model*, as in Figure 1.5(b). That approach provides the tip mass m_e and stiffness k_e that are equivalent to the dynamic response of the original cantilever.

Caution should be exercised when studying complex flexible systems, where the lumping of parameters can yield results that are sensibly different from the expected and actual results, as measured experimentally or simulated by more advanced (numerical) techniques, such as the finite element method. However, for the relatively simple compliant device configurations analyzed in subsequent chapters, lumped-parameter modeling yields results with relatively small errors.

Modeling Methods

Several procedures or methods are available for deriving the mathematical model of a specified lumped-parameter physical model. Some of them are specific to a certain system (such as *Newton's second law of motion*, which is applied to mechanical

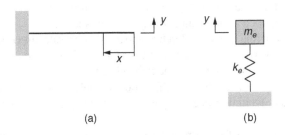

(a)   (b)

FIGURE 1.5

Cantilever Beam: (a) Actual, Distributed-Parameter Inertia and Stiffness; (b) Equivalent, Lumped-Parameter Inertia and Stiffness.

systems; *Kirchhoff's laws*, which are used in electrical systems; or *Bernoulli's law*, which is employed to model fluid systems) but others can be utilized more across the board for all dynamic systems, such as the *energy method*, the *Lagrange's equations*, the *transfer function method*, and the *state space approach*. These methods are detailed in subsequent chapters or in companion website material.

Solutions Methods

Once the mathematical model of a dynamic system has been obtained, which consists of one differential equation or a system of differential equations, the solution can be obtained mainly using two methods. One method is the *direct integration* of the differential equations, and the other method uses the *direct* and *inverse Laplace transforms*. The big advantage of the Laplace method, as will be shown in Chapter 6, consists in the fact that the original, time-defined differential equations are transformed into algebraic equations, whose solution can be found by simpler means. The Laplace-domain solutions are subsequently converted back into the time-domain solutions by means of the inverse Laplace transform. The *transfer function* and the *state-space methods* are also employed to determine the time response in Chapters 7 and 8, respectively.

System Response

Solving for the unknowns of a mathematical model based on differential equations provides the solution. In general, the solution to a differential equation that describes the system behavior is the sum of two parts: One is the *complementary* (or *homogeneous*) *solution*, $y_c(t)$ (which is the solution when no input or excitation is applied to the system) and the other is the *particular solution*, $y_p(t)$ (which is one solution of the equation when a specific forcing or input acts on the system):

$$y(t) = y_c(t) + y_p(t) \tag{1.1}$$

The complementary solution is representative of the *free response* and usually vanishes after a period of time with dissipation present; thus, it is indicative of the *transient response*. The particular solution, on the other hand, persists in the overall solution and, therefore, defines the *forced* or *steady-state response* of the system to a particular type of input.

1.3 COMPONENTS, SYSTEM, INPUT, AND OUTPUT

A *system* in general (and an engineering one in particular in this text) is a combination of various *components*, which together form an entity that can be studied in its entirety. Take for instance a resistor, an inductor, a capacitor, and a voltage source, as shown in Figure 1.6(a); they are individual electrical components that can be combined in the series connection of Figure 1.6(b) to form an electrical system.

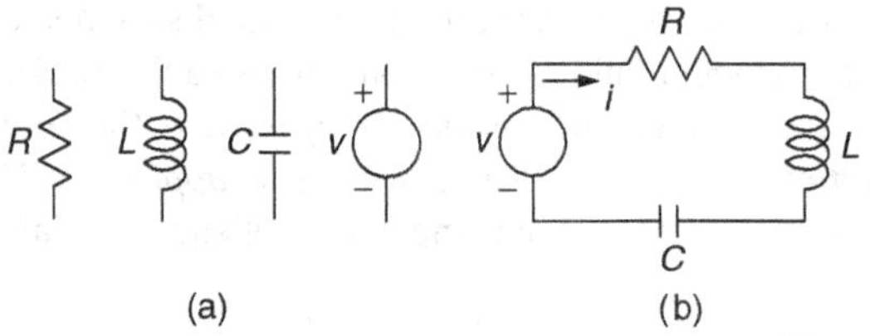

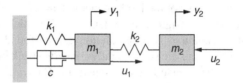

FIGURE 1.6

(a) Individual Electrical Components; (b) Electrical System Formed of These Components.

FIGURE 1.7

MIMO Mechanical System with Linear Motion.

Similarly, mechanical components such as inertia (mass), stiffness, damping, and forcing can be combined in various ways to generate mechanical systems. There are also fluid systems, thermal systems, and systems that combine elements from at least two different fields (or domains) to generate coupled-field (or multiple-field) systems, such as electro-mechanical or thermo-electro-mechanical, to mention just two possibilities.

The response of a dynamical system is generated by external causes, such as *forcing* or *initial conditions*, and it is customary to name the cause that generates the change in the system as *input* whereas the resulting response is known as *output*. A system can have one input and one output, in which case it is a *single-input, single-output system* (SISO), or it can have several inputs or several outputs, consequently known as *multiple-input, multiple-output system* (MIMO).

A SISO example is the single-mesh series-connection electrical circuit of Figure 1.6(b). For this example the input is the voltage v whereas the output is the current i. A MIMO mechanical system is sketched in Figure 1.7, where there are two inputs, the forces f_1 and f_2, and two outputs, the displacements x_1 and x_2. The car models just analyzed are also MIMO systems, as they all have more than one input or output.

The *input signals* (or *forcing functions* and generally denoted by u) that are applied to dynamic systems can be deterministic or random (arbitrary) in nature. *Deterministic* signals are known functions of time whereas *random signals* show no pattern connecting the signal function to its time variable. This text is concerned

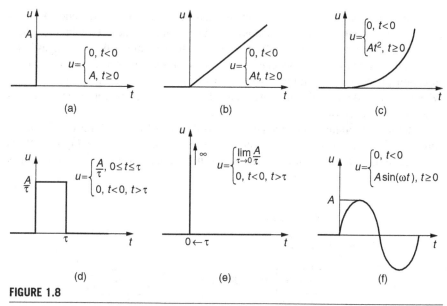

FIGURE 1.8

A Few Input Functions: (a) Step; (b) Ramp; (c) Parabolic; (d) Pulse; (e) Impulse; (f) Sinusoidal.

with deterministic input signals only. Elementary input signals include the *step*, *ramp*, *parabolic*, *sine* (*cosine*), *pulse*, and *impulse* functions; Figure 1.8 plots these functions.

1.4 COMPLIANT MECHANISMS AND MICROELECTROMECHANICAL SYSTEMS

In addition to examples that are somewhat classical for dynamics of engineering systems, this text discusses several applications from the fields of *compliant mechanisms* and *micro-* and *nano-electromechanical systems*, so a brief presentation of these two domains is given here. The effort has been made throughout this book to demonstrate that, under regular circumstances, simple applications from compliant mechanisms and MEMS can be reduced to lumped-parameter (most often) linear systems that are similar to other well-established system dynamics examples.

Compliant (flexible) mechanisms are devices that use the elastic deformation of slender, springlike portions instead of classical rotation or sliding pairs to create, transmit, or sense mechanical motion. The example of Figure 1.9 illustrates the relationship between a classical translation (sliding) joint with regular springs and the corresponding compliant joint formed of *flexure hinges* (slender portions that bend and enable motion transmission). The compliant device of Figure 1.9(a) is constrained to move horizontally because the four identical flexure hinges bend identically (in pairs of two) whenever a mechanical excitation is applied about the direction of motion. The lumped-parameter counterpart is drawn in Figure 1.9(b), where the four identical flexure hinges have been substituted by four identical translation springs, each of stiffness k.

Another compliant mechanism example is the one of Figure 1.10(a), which pictures a piezoelectrically-actuated, displacement-amplification device. Figure 1.10(b) is the schematic representation of the actual mechanism, where the flexure hinges

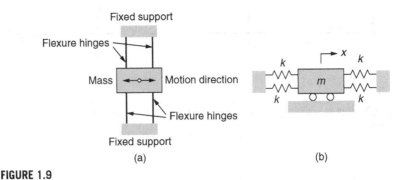

(a)   (b)

FIGURE 1.9

Realizing Translation: (a) Compliant Mechanism with Flexure Hinges; (b) Equivalent Lumped-Parameter Model.

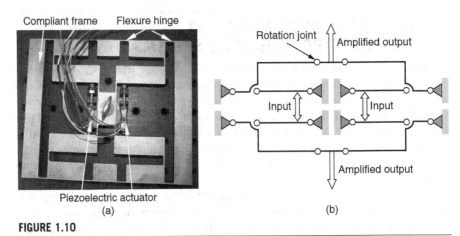

(a)   (b)

FIGURE 1.10

Flexure-Based Planar Compliant Mechanism for Motion Amplification: (a) Photograph of Actual Device; (b) Schematic Representation with Pointlike Rotation Joints.

are replaced by classical pointlike rotation joints. The schematic shows that the input from the two piezoelectric actuators is amplified twice by means of two lever stages. The mechanism is clamped to and offset above the base centrally, as indicated in Figure 1.10(a) and is free to deform and move in a plane parallel to the base plane.

As monolithic (single-piece) devices, compliant mechanisms present several advantages over their classical counterparts, such as lack of assembly, no moving parts, and therefore no losses due to friction between adjacent parts, no need for maintenance, and simplicity of fabrication (although at costs that are higher generally compared to classical manufacturing procedures). Their main drawback is that the range of motion is reduced because of the constraints posed by limited deformations of their compliant joints.

Compliant mechanisms are encountered in both macro-scale applications (with dimensions larger than millimeters) and micro- or nano-scale ones (when the device dimensions are in the micrometer or nanometer range (1 μm = 10^{-6} m, 1 nm = 10^{-9} m), particularly in *microelectromechanical systems*. In many situations, compliant mechanisms are built as single-piece (monolithic) devices with techniques such as *wire electro-discharge machining* (wire EDM), through *water jet* machining, or by microfabrication techniques (for MEMS) such as *surface* or *bulk micromachining*.

MEMS applications, such as sensors, actuators, pumps, motors, accelerometers, gyroscopes, electrical or mechanical filters, electronic or optical switches, GPS devices (to mention just a few), are encountered in the automotive, defense, medical, biology, computing, and communications domains. Figure 1.11 is the microphotograph of a flexure-hinge thermal microactuator whose motion is sensed electrostatically by several pairs of capacitors. The entire device floats over a substrate on which it is attached by four anchors (two are shown in the figure).

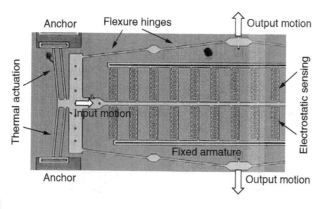

FIGURE 1.11

Top View of Compliant MEMS with Thermal Actuation and Electrostatic Sensing of Motion.

A similar application, where both actuation and sensing are performed electrostatically, is the microcantilever shown in the side view of the three-dimensional rendition of Figure 1.12. Out-of-plane bending of the microcantilever can be used to realize switching in an electrical circuit, for instance.

Another MEMS application is sketched in Figures 1.13(a) and 1.13(b). It represents a *torsional micro-mirror*, which can be used in several applications such as dynamic redirectioning of incoming optical signals. External actuation on the side of and underneath the central plate (the mirror) through attraction/repulsion forces that can be produced electrostatically or magnetically generates partial rotation of the plate about the axis that passes through the two elastic end hinges. The hinges deform in torsion; hence the name *torsional mirror*. Figure 1.11(c) shows the equivalent lumped-parameter model of the actual torsional mirror. Several other MEMS devices are analyzed in subsequent chapters as mechanical, electrical, or coupled-field systems.

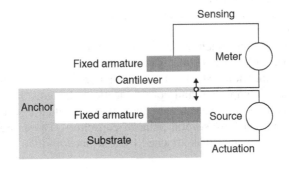

FIGURE 1.12

Side View of a Microcantilever with Electrostatic Actuation and Sensing.

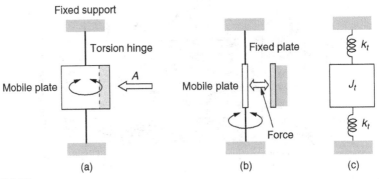

FIGURE 1.13

MEMS Torsional Micromirror: (a) Top View; (b) Side View (from A); (c) Equivalent Lumped-Parameter Model.

1.5 SYSTEM ORDER

As mentioned previously, an engineering system is described by a mathematical model that consists of a differential equation (for SISO systems) or a system of differential equations (the case of MIMO systems). The order of the differential equation(s) gives the order of the system, as is shown next.

For a SISO system, the relationship between the input and the output is described by a differential equation of the type

$$\sum_{i=0}^{n} a_i \frac{d^i y(t)}{dt^i} = bu(t) \tag{1.2}$$

where a_i ($i = 0$ to n) and b are constant factors, and the input function, $u(t)$, can also include derivatives. The maximum derivation order of the output function $y(t)$ in a system of the nth order is n such that, a second-order system for instance is defined by a maximum-order input derivative of 2 and so on.

1.5.1 Zero-Order Systems

A *zero-order system* is defined by the equation

$$a_0 y(t) = bu(t) \tag{1.3}$$

which is also written as

$$y(t) = Ku(t) \tag{1.4}$$

where $K = b/a_0$ is the *constant gain* or *static sensitivity*. The static sensitivity constant reflects the *storage* nature of a zero-order system, and this is illustrated in the following example. The electrostatic MEMS actuator sketched in Figure 1.14(a) can be regarded as a zero-order system in the following circumstances. A force f that

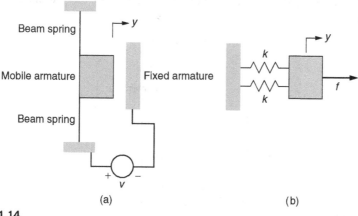

(a)   (b)

FIGURE 1.14

MEMS with Electrostatic Actuation and Beam-Spring Elastic Support: (a) Actual System; (b) Mechanically Equivalent Lumped-Parameter Model.

can be generated electrostatically attracts the mobile armature to the right until it is statically balanced by the elastic reactions of the two side beam springs. If only the mechanical domain is of interest, the lumped-parameter equivalent model of Figure 1.14(b) can be employed to describe the quasistatic behavior of the MEMS. The static equilibrium requires $f = f_e$, the elastic force being produced by two springs as $f_e = 2ky$. As a consequence, the following equation is produced:

$$y = \frac{1}{2k}u \tag{1.5}$$

where the static sensitivity is $K = 1/(2k)$ and the input is $f = u$.

1.5.2 First-Order Systems

First-order systems are described by a differential equation:

$$a_1\frac{dy(t)}{dt} + a_0y(t) = bu(t) \tag{1.6}$$

which can also be written as

$$\tau\frac{dy(t)}{dt} + y(t) = Ku(t) \tag{1.7}$$

where the new constant, τ, is the *time constant* and is defined as

$$\tau = \frac{a_1}{a_0} \tag{1.8}$$

It can be seen that the input-system interaction of a first-order system is described by two constants: the static sensitivity K and the time constant τ—this latter one displays the *dissipative* side of a first-order system. The thermal system sketched in Figure 1.15 is a first-order system, as shown next.

The heat quantity exchanged between the bath and the thermometer during contact is

$$Q = mc\big[\theta(t) - \theta_b\big] \tag{1.9}$$

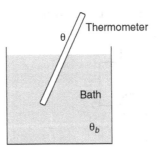

FIGURE 1.15

Bath-Thermometer Thermal System.

where c is the specific heat of the thermometer, m is its mass, θ_b is the bath temperature (assumed to be constant here), and θ is the thermometer temperature. At the same time, it is known that convective heat exchange between the bath and thermometer is governed by the equation

$$\dot{Q} = hA\left[\theta_b - \theta(t)\right] \tag{1.10}$$

where h is the convection heat transfer coefficient and A is the thermometer area in contact with the fluid. Applying the time derivative to Eq. (1.9) and combining the resulting equation with Eq. (1.10) results in

$$\frac{mc}{hA}\frac{d\theta(t)}{dt} + \theta(t) = \theta_b \tag{1.11}$$

which indicates that $\tau = mc/(hA)$ and $K = 1$. The input is the bath temperature θ_b and the output is the thermometer temperature $\theta(t)$. Figure 1.16 displays a typical first-order system response for the particular case where $\tau = 10$ s and $\theta_b = 80°C$.

For a first-order system, the response to a step input can be characterized by the *steady-state response*, $y(\infty)$, the *rise time* (time after which the response gets to 90% of the steady-state response, but other definitions are also applicable), and the *settling time* (time necessary for the response to stay within 2% of the steady-state response values); more details on this topic are given in the website Chapter 11, which studies dynamics of control systems.

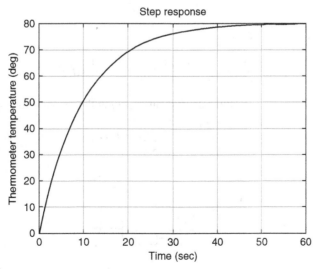

FIGURE 1.16

Thermometer Temperature as a Function of Time (Typical First-Order System Time Response).

1.5.3 Second- and Higher-Order Systems

Second-order systems are defined by the following differential equation:

$$a_2 \frac{d^2 y(t)}{dt^2} + a_1 \frac{dy(t)}{dt} + a_0 y(t) = b u(t) \tag{1.12}$$

Division of Eq. (1.12) by a_0 then rearrangement of the resulting equation yields

$$\frac{d^2 y(t)}{dt^2} + 2\xi \omega_n \frac{dy(t)}{dt} + \omega_n^2 y(t) = \omega_n^2 K u(t) \tag{1.13}$$

where the new constants, the natural frequency ω_n and the damping ratio ξ are defined as

$$\begin{cases} \dfrac{1}{\omega_n^2} = \dfrac{a_2}{a_0} \\[2mm] \dfrac{2\xi}{\omega_n} = \dfrac{a_1}{a_0} \end{cases} \tag{1.14}$$

Let us prove that the mechanical system of Figure 1.17 operates as a second-order system.

The equation of motion is derived by means of Newton's second law of motion as

$$m \frac{d^2 y(t)}{dt^2} = f(t) - c \frac{dy(t)}{dt} - k y(t) \tag{1.15}$$

which can be rearranged as

$$\frac{d^2 y(t)}{dt^2} + \frac{c}{m} \times \frac{dy(t)}{dt} + \frac{k}{m} y(t) = \frac{1}{m} f(t) \tag{1.16}$$

where $f = u$; therefore, the three coefficients defining the second-order system are

$$\omega_n = \sqrt{\frac{k}{m}}; \ \xi = \frac{c}{2\sqrt{mk}}; \ K = \frac{1}{k} \tag{1.17}$$

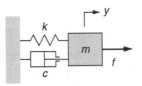

FIGURE 1.17

Mechanical System with Mass, Spring, and Damper.

Based on this mechanical system, Figure 1.18 shows the time response of a typical second-order system for $f = 50\,N$, $\xi = 0.5$, $\omega_n = 100\,rad/s$, and $K = 1$. The main characteristics of the time response of a second-order system to a step input are the *steady-state response*, the *rise time*, the *peak time* (time required for the response to reach its maximum value) and the *peak response* (the maximum response), and the *settling time*; all these parameters are studied in more detail in the website Chapter 11 in the context of controls.

Systems of orders larger than two are also encountered in engineering applications, such as the following example, which results in a *third-order system model*. The electromechanical system of Figure 1.19 consists of a dc (direct-current) motor and load shaft. The dynamic model of a motor consists of equations that describe the mechanical, electrical, and mechanical-electrical (coupled-field) behavior. Essentially, the electromechanical system sketched in Figure 1.19 is formed of a mobile part (the rotor armature), which rotates under the action of a magnetic field produced by the

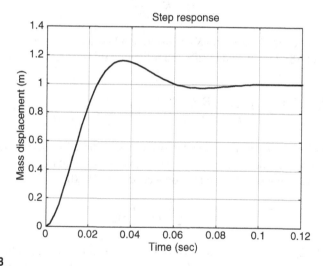

FIGURE 1.18

Mass Displacement as a Function of Time (Typical Second-Order System Time Response).

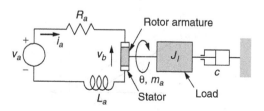

FIGURE 1.19

Schematic of a dc Motor as an Electromechanical System.

electrical circuit of an armature (the stator). The electrical circuit is governed by Kirchhoff's second law, according to which

$$R_a i_a(t) + L_a \frac{di_a(t)}{dt} = v_a(t) - v_b(t) \tag{1.18}$$

where the subscript a indicates the armature and v_b is the back electromotive force (voltage). The mechanical part of the system is governed by the equation

$$J_l \frac{d^2\theta(t)}{dt^2} = m_a(t) - c\frac{d\theta(t)}{dt} \tag{1.19}$$

where m_a is the torque developed due to the stator-rotor interaction. It is also known that the following equations couple the mechanical and electrical fields:

$$\begin{cases} m_a(t) = K_t i_a(t) \\ v_b(t) = K_e \dfrac{d\theta(t)}{dt} \end{cases} \tag{1.20}$$

with K_t (measured in N-m/A) and K_e (measured in V-s/rad) being constants. By combining Eqs. (1.18), (1.19), and (1.20), the following third-order differential equation is produced:

$$\frac{L_a J_l}{K_t} \times \frac{d^3\theta(t)}{dt^3} + \left(\frac{L_a c}{K_t} + \frac{R_a J_l}{K_t}\right)\frac{d^2\theta(t)}{dt^2} + \left(\frac{R_a c}{K_t} + K_e\right)\frac{d\theta(t)}{dt} = v_a(t) \tag{1.21}$$

The order of Eq. (1.21) can be reduced to two by using the substitution $\omega(t) = d\theta(t)/dt$.

When neglecting the armature inductance, that is $L_a = 0$ in Eq. (1.18), the following equation can be written using the connection Eqs. (1.20):

$$m_a = -\frac{K_e K_t}{R_a}\omega + \frac{K_t}{R_a}v_a \tag{1.22}$$

Equation (1.22) indicates a linear relationship between the actuation torque m_a and the angular velocity ω for a specified armature voltage, as shown in Figure 1.20. Any

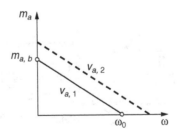

FIGURE 1.20

Actuation Torque versus Shaft Angular Frequency in a dc Motor.

operation point defined by an angular velocity and a delivered torque is located on this characteristic curve.

The torque-angular velocity curve generated by an armature voltage v_{a1} intersects the coordinate axes at ω_0—the *free (no-load) angular velocity*, when basically the dc motor shaft spins freely with no external load acting on it, and at $m_{a,b}$—the *block (stall) torque*, where an external load torque equal to $m_{a,b}$ stalls the shaft rotation altogether. These two parameters are obtained from Eq. (1.22) as

$$\omega_0 = \frac{v_a}{K_e}; \; m_{a,b} = \frac{K_t}{R_a}v_a \qquad (1.23)$$

For a different armature voltage, say $v_{a2} > v_{a1}$, the torque-angular velocity characteristic shifts up but preserves its slope for constant values of K_t, K_e and R_a, as indicated in Eq. (1.22).

1.6 COUPLED-FIELD (MULTIPLE-FIELD) SYSTEMS

The previous example illustrated the interaction between mechanical and electrical elements and systems that resulted in an electro-mechanical system. The corresponding mathematical model is formed of equations pertaining to a single domain or field (either mechanical or electrical) and equations combining elements from both fields. Such a mathematical model is representative of *coupled-field* (or *multiple-field*) *systems*. Another example is presented in a descriptive manner and more details on coupled-field systems are given in Chapter 10. Consider the system of Figure 1.21, which is formed of a piezoelectric (PZT) block with a strain gauge attached to it.

Each of these two subsystems has its own electrical circuit. Piezoelectric materials essentially deform when an external voltage is applied to them due to the *piezoelectric effect*. A voltage applied at the end points of the bloc sketched in Figure 1.21

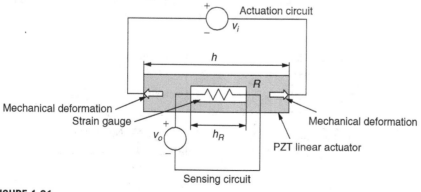

FIGURE 1.21

Coupled-Field System with Mechanical, Electrical, and Piezoelectric Elements.

generates an axial deformation, which is proportional to the applied voltage. Assuming now that the strain gauge (which is a resistor) is affixed longitudinally to the block, the resistor incurs the same axial deformation as the block; therefore, its resistance changes by a quantity proportional to the mechanical deformation. This resistance change can be sensed in the external circuit. To summarize, an equation is obtained that combines electrical, piezoelectric, and mechanical elements; therefore, this system is a coupled-field one.

As illustrated by this example, the piezoelectric block behaves as an *actuator* when supplied with a voltage generating the mechanical motion. The dc motor is another actuator (or *motor*) example, where the armature voltage is the source of shaft angular rotation. More generically, an actuator transforms one form of energy (such as electrical, most often) into mechanical energy. For a piezoelectric block working against a translatory spring, as sketched in Figure 1.22(a), the characteristic curve connecting the delivered force to the tip displacement is similar to that of the dc motor, see Figure 1.22(b). Without a spring, the piezoelectric actuator deforms by a quantity y_0—the free displacement, which (more details to come in Chapter 10) is produced by an actuation voltage v_a as

$$y_0 = dv_a \tag{1.24}$$

where d is a material constant. The maximum force preventing the piezoelectric actuator from deforming is the block force, f_b, which is calculated as

$$f_b = k_{PZT}y_0 = \left(\frac{EA}{h}\right)_{PZT} y_0 \tag{1.25}$$

E is the longitudinal (Young's) modulus of the piezoelectric material, A is the block cross-sectional area, and h is the original (undeformed) length. The linear force-deformation characteristic curve of Figure 1.22(b) has the equation

$$f = my + n \text{ or } f = -\left(\frac{EA}{h}\right)_{PZT} y + \left(\frac{EA}{h}\right)_{PZT} dv_a \tag{1.26}$$

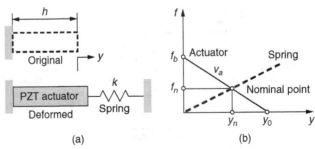

(a)   (b)

FIGURE 1.22

Linear Piezoelectric Actuation: (a) Against Spring; (b) Force-Displacement Characteristics.

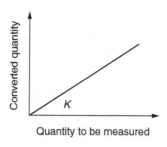

FIGURE 1.23

Sensing Characteristics in a Linear Measurement Process.

which is similar to the dc equation relating the developed torque to the angular velocity—Eq. (1.22). The force-displacement characteristic of the spring is indicated with dotted line in Figure 1.22(b) and, because there is permanent contact between the piezoelectric actuator and the spring, a *nominal* (or *operation*) *point* lies at the intersection of the actuator and spring characteristics. For a specified actuator, a family of parallel lines is obtained for various actuation voltages.

As mentioned in this section, a piezoelectric block can also produce voltage when subjected to mechanical pressure or deformation, a case where it behaves as a *sensor* (or *generator*) by converting one form of energy (mechanical in this particular situation) into electrical energy to perform a measurement (quantitative assessment) operation. A sensor generally aims at measuring the variation of a physical parameter, such as displacement, velocity, acceleration, pressure, electrical resistance, also named *measurand*, by converting that variation into another parameter's variation, which can subsequently be processed more easily; the two parameters are usually connected by a linear relationship which is typical of zero-order systems, see Eq. (1.4) where the input is the quantity to be measured and the output is the parameter that measures (the converted quantity). The two parameters are related by the static sensitivity K, as illustrated in Figure 1.23. Being methods of converting one form of energy into another one, actuation and sensing are known collectively as *transduction* (although in many instances transduction substitutes for sensing). More on transduction is discussed in Chapter 10 and in texts specialized in measurement and instrumentation.

1.7 LINEAR AND NONLINEAR DYNAMIC SYSTEMS

Linearity or nonlinearity of a dynamic system is associated with the differential equation that defines the behavior of that specific system. A SISO system for instance is linear when

1. The coefficients a_i of Eq. (1.2) do not depend on the unknown function (response) $y(t)$.
2. The unknown function and its derivatives in the left-hand side of Eq. (1.2) are first-degree polynomial functions.

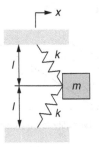

FIGURE 1.24

Mass with Two Springs in a Deformed Position as a Nonlinear Mechanical System.

Systems where the coefficients a_i are not constant (are time variable, for instance) still preserve their linear character. In a SISO mechanical system, nonlinearity can be produced by several factors connected to either mass, stiffness, or damping. Consider the mass of Figure 1.24, which is attached by two identical springs of stiffness k and of undeformed length l.

When the body moved a distance x to the right from the equilibrium position, the elongation of each of the two identical springs is equal to $\sqrt{l^2 + x^2} - l$. By applying Newton's second law of motion and projecting the two identical spring forces on the horizontal motion direction, the following equation results:

$$m\ddot{x}(t) + 2k_{eq}x(t) = 0 \tag{1.27}$$

with the equivalent stiffness being

$$k_{eq} = k\left(1 - \frac{l}{\sqrt{l^2 + x(t)^2}}\right) \tag{1.28}$$

which indicates the stiffness is nonlinear and therefore the whole mechanical system is nonlinear.

The stiffness increase is produced by a "hardening" effect, whereby the stiffness and its slope increase with x. Another variant, where the slope decreases with x, is generated by a "softening" effect. Both behaviors are nonlinear and sketched in Figure 1.25 alongside the characteristic of a linear spring.

Two other types of nonlinearities, saturation and hysteresis, mostly related to material behavior in mechanical and electrical components, are briefly mentioned next. The phenomenon of *saturation* is predominantly encountered in electrical components. Figure 1.26 shows the voltage-current characteristic curve of an inductor; it can be seen that the linear relationship between the voltage and the current changes past a central zone, as the voltage decreases (saturates) when the current increases past

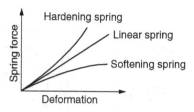

FIGURE 1.25

Linear, Hardening, and Softening Spring Characteristics.

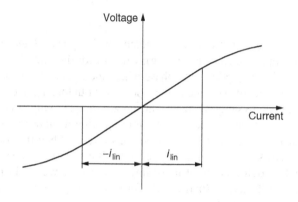

FIGURE 1.26

Saturation-Type Nonlinearity.

the linearity limits. As known from electromagnetism, the voltage across an inductor is defined as

$$v_L(t) = L \frac{di(t)}{dt} \tag{1.29}$$

Obviously, when the current is confined within the $-i_{\text{lin}}$ and $+i_{\text{lin}}$ bounds, the slope di/dt is constant, and the voltage-current relationship is linear. This actually is the range utilized in the majority of electrical system calculations.

Hysteresis nonlinearities are encountered in deformable solids, magnetic materials, and electrical materials. Hysteresis consists mainly in a path-dependence of the load-response characteristic curve depending on whether loading or unloading is applied. The curve of Figure 1.27(a) shows no dependence on the path as the loading and unloading curves are identical. However, for the material of Figure 1.27(b), there is a path dependence as there are different curves for the loading and unloading; therefore, hysteresis has to be accounted for.

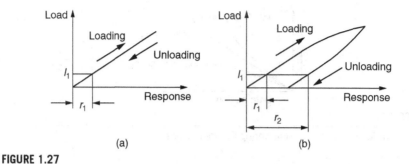

FIGURE 1.27

Hysteresis-Type Nonlinearity: (a) Material with No Hysteresis; (b) Material with Hysteresis.

At a load of l_1, a response r_1 is registered for this phase (*loading* meaning the increase in the applied load or excitation), whereas for the same load, the response is r_2 ($r_2 < r_1$) during *unloading* (the phase where the load and excitation decrease), Figure 1.27(b). Evidently, r_1 is the only response at both loading and unloading for a material showing no hysteresis effects. One good example to illustrate the hysteresis phenomenon is to gradually increase the force pulling a rubber wire and measuring the deformations that correspond to various forces. When the force is reduced gradually, it can be seen that, for a force level that is equal to the one used at loading, the corresponding deformation at unloading is larger. However, in many applications, the effect of hysteresis (particularly for metals) is generally small and can be neglected, as is the case in this text.

Mechanical Systems I

2

OBJECTIVES

In this chapter you will learn about

- Inertia, stiffness, damping, and forcing as lumped-parameter mechanical elements for translation and rotation.
- Application of Newton's second law of motion to formulate the mathematical models of basic, single degree-of-freedom, dynamic mechanical systems.
- Analysis of natural, free-damped, and forced vibrations of single degree-of-freedom mechanical systems.
- Use of MATLAB® as a tool for symbolic calculation and plotting of basic mechanical system response.
- Utilization of Simulink® as a computing environment for graphical modeling and solving differential equations representing dynamic models of basic mechanical systems.

INTRODUCTION

The main objective of this chapter is to derive time-domain mathematical models describing the dynamics of mechanical systems that consist of the basic elements of inertia, damping, stiffness, and forcing (or actuation). The lumped-parameter modeling approach is utilized, according to which these basic element properties are considered pointlike. Studied here are single degree-of-freedom (single-DOF) mechanical systems that depend on one variable and whose mathematical model consists of one differential equation.

Translatory and rotary mechanical motions are modeled by using Newton's second law of motion. The resulting mathematical models are used to analyze the free-undamped (natural), free-damped, and forced responses of single-DOF mechanical systems. Analytical methods as well as MATLAB® and Simulink® are used to solve several examples that complement the theory, including mechanical lever and gear system applications.

DOI: 10.1016/B978-0-240-81128-4.00002-7

2.1 BASIC MECHANICAL ELEMENTS: INERTIA, STIFFNESS, DAMPING, AND FORCING

This section introduces the mechanical elements of inertia, stiffness, damping, and forcing for both translatory and rotary motion.

2.1.1 Inertia Elements

For *rigid bodies*, inertia properties can be considered pointlike; therefore, inertia features corresponding to either translatory or rotary motion are naturally *lumped*. Inertia is represented by *mass* (usually denoted by *m*) in translatory motion and *mechanical* (or *mass*) *moment of inertia* (generally symbolized by *J*) in rotary motion, as sketched in Figure 2.1.

The mechanical moment of inertia of a point mass *m* rotating about a fixed point at a distance *l* as in the simple pendulum of Figure 2.1(b), and the mechanical moment of inertia of a body of mass *m* rotating about an axis, Figure 2.1(c), are calculated as

$$J = ml^2; J = \int_m r^2 dm \qquad (2.1)$$

where *r* is the distance from the rotation axis to an element of mass *dm*.

The *parallel-axis theorem*, which is illustrated in Figure 2.2, gives the mechanical moment of inertia of a rotating body about an axis that does not coincide with the body's centroidal axis in terms of the distance *d* between axes and the body mass *m* as

$$J_\Delta = J + md^2 \qquad (2.2)$$

where *J* is the mechanical moment of inertia of the body about its centroidal axis.

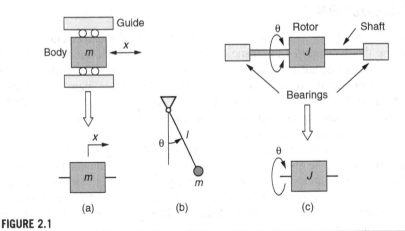

FIGURE 2.1

Lumped-Parameter Mechanical Inertia and Related Symbols: (a) Translatory; (b) and (c) Rotary.

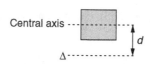

FIGURE 2.2

Schematic for the Parallel-Axis Theorem.

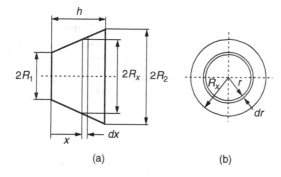

FIGURE 2.3

Frustum of a Right Circular Conical Solid: (a) Side View; (b) Isolated Cross-Section.

Example 2.1

Calculate the mass moment of inertia about the centroidal (symmetry) axis of the right circular cone frustum shown in Figure 2.3(a) in side view and defined by R_1, R_2, and h. Use the obtained result to also calculate the mass moment of inertia of a cylinder, both about its centroidal axis and about a parallel axis that is offset at a distance $d = 2R_2$ from the centroidal axis.

Solution

For a homogeneous cone frustum of mass density ρ, the mechanical moment of inertia is expressed as

$$J = \rho \int_V r^2 dV = \rho \int_0^h \left(\int_A r^2 dA \right) dx \tag{2.3}$$

As shown in Figure 2.3(b), the area of an elementary circular strip of width dr and inner radius r is

$$dA = 2\pi r dr \tag{2.4}$$

By substituting Eq. (2.4) into Eq. (2.3), the mass moment of inertia of the cone frustum becomes

$$J = 2\pi\rho \int_0^h \left(\int_0^{R_x} r^3 dr \right) dx = \frac{\pi\rho}{2} \int_0^h R_x^4 dx \qquad (2.5)$$

The variable external radius, as sketched in Figure 2.3(a), can be calculated as

$$R_x = R_1 + \frac{R_2 - R_1}{h} x \qquad (2.6)$$

Symbolic mathematical calculations can be performed by using MATLAB® Symbolic Math Toolbox™. Integrals are evaluated by means of the int(expr, x, a, b) MATLAB® command, where expr is the integrand (the expression to be integrated), x is the integration variable and a, b are the integration limits. The following MATLAB® sequence is used to calculate the mass moment of inertia of Eq. (2.5):

```
>> syms r1 r2 r h x rho
>> rx = r1+(r2-r1)/h*x;
>> da = 2*pi*r;
>> J = simplify(rho*int(int(r^2*da, r, 0, rx), x, 0, h))
```

which returns

$$J = \frac{\pi\rho h}{10} \left(R_1^4 + R_1^3 R_2 + R_1^2 R_2^2 + R_1 R_2^3 + R_2^4 \right) \qquad (2.7)$$

When $R_1 = R_2 = R$, the cone frustum becomes a cylinder and Eq. (2.7) simplifies to

$$J = \frac{\pi\rho h}{10} (5R^4) = \frac{1}{2}\pi\rho h R^4 = \frac{1}{2}(\pi R^2 h\rho) R^2 = \frac{1}{2}mR^2 \qquad (2.8)$$

with m being the mass of the cylinder. The cylinder's mass moment of inertia about an axis situated at $d = 2R_2$ from its centroidal axis is found from Eq. (2.8) by means of the parallel-axis theorem, Eq. (2.2), as

$$J = \frac{1}{2}mR^2 + m(2R)^2 = \frac{9}{2}mR^2 \qquad (2.9)$$

For translation, the mass is included in the *inertia force*, which is proportional to acceleration (the second time derivative of displacement) and is calculated as

$$f_i(t) = m\frac{d^2 x(t)}{dt^2} = m\ddot{x} \qquad (2.10)$$

where f_i has the same direction as x in Figure 2.1(a). Similarly, an *inertia moment* (or *inertia torque*) is defined in rotary motion as a function of the mechanical moment of inertia and the angular acceleration (the second time derivative of rotation angles):

$$m_i(t) = J\frac{d^2\theta(t)}{dt^2} = J\ddot{\theta} \qquad (2.11)$$

where m_i has the same direction as θ in Figure 2.1(b).

Bodies in motion possess *kinetic energy*, denoted by T, which is expressed for translation and rotation as a function of the corresponding velocity:

$$T = \frac{1}{2}m\left[\frac{dx(t)}{dt}\right]^2 = \frac{1}{2}m\dot{x}^2 \qquad (2.12)$$

$$T = \frac{1}{2}J\left[\frac{d\theta(t)}{dt}\right]^2 = \frac{1}{2}J\dot{\theta}^2 \qquad (2.13)$$

2.1.2 Spring Elements

Springs serving as elastic supports for translatory and rotary motion are studied in this section in relation to their lumped *stiffness* (or *spring constant*), denoted by k. Springs are mechanical elements that generate elastic forces in translatory motion and elastic torques in rotary motion that oppose the spring deformation; these elastic reactions are proportional to the spring deformation (linear or angular displacement). Figure 2.4 sketches a helical spring and gives the stiffness equations corresponding

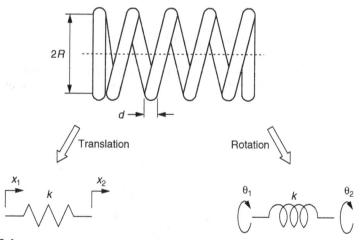

FIGURE 2.4

Helical Spring and Symbols for Translatory or Rotary Motion: Translatory Stiffness, $k_t = Gd^4/(64nR^3)$; Rotary Stiffness, $k_r = Ed^4/(64nR)$.

to either axial (translatory) motion or to torsion-generated (rotary) motion. The parameters defining the helical spring are the radius R, the wire diameter d, the number of active turns n, and the material shear and Young's modulii, G and E (more details on these quantities are given in Appendix D).

For a spring whose end points undergo the displacements x_1 and x_2 (as shown in Figure 2.4), the elastic force developed in the spring is proportional to the spring deformation, which is the difference between the two end point displacements, and can be expressed as

$$f_e(t) = k\Delta x(t) = k[x_1(t) - x_2(t)] \tag{2.14}$$

Similarly, an elastic torque is generated by a spring in rotation whose end points undergo the rotations θ_1 and θ_2, the elastic torque being expressed as

$$m_e(t) = k\Delta\theta(t) = k[\theta_1(t) - \theta_2(t)] \tag{2.15}$$

These equations assume the springs are linear; therefore, the stiffness is constant. Stiffness equations are given in the companion website Chapter 2 for other translatory and rotary springs. A *spiral spring,* for instance, such as the one sketched in Figure 2.5 and whose total length is l, is used in rotary motion applications.

For a translatory spring, the elastic energy stored corresponding to a deformation Δx is

$$U_e = \frac{1}{2}k[\Delta x(t)]^2 \tag{2.16}$$

Similarly, for a rotary spring, the elastic energy relative to an angular deformation $\Delta\theta$ is

$$U_e = \frac{1}{2}k[\Delta\theta(t)]^2 \tag{2.17}$$

Springs can be combined in series or in parallel, as sketched in Figure 2.6, where it has been assumed that the serial and parallel chains are clamped at one end.

Anchored end

FIGURE 2.5

Spiral Torsion Spring for Rotary Motion; Spring Stiffness Is $k = \pi E d^4/(64l)$.

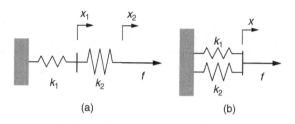

FIGURE 2.6

Translatory Spring Combinations: (a) Serial; (b) Parallel.

For serial springs, the force is the same in each component and is equal to f in Figure 2.6(a), whereas the total deformation is the sum of individual deformations. Conversely, for parallel spring combinations, the displacements are identical for all springs, whereas the sum of individual spring forces equals the externally applied force f at equilibrium. The equivalent series stiffness k_s and the parallel stiffness k_p corresponding to n spring elements are derived in the companion website Chapter 2; their equations are

$$\begin{cases} \dfrac{1}{k_s} = \dfrac{1}{k_1} + \dfrac{1}{k_2} + \cdots + \dfrac{1}{k_n} \\ k_p = k_1 + k_2 + \cdots + k_n \end{cases} \qquad (2.18)$$

Example 2.2

Four identical translatory helical springs are combined in two arrangements such that, in each of the two combinations, there are both series and parallel connections. When the same force is applied separately to each spring arrangement at the free end (the other one being fixed), it is determined that the ratio of the free-end displacements is 25/4. Identify the two spring combinations and calculate the equivalent stiffness for each when $d = 1$ mm, $R = 6$ mm, $n = 10$, and $G = 160$ GPa.

Solution

The largest displacement is obtained when all four springs are coupled in series, because the stiffness is minimal, see first Eq. (2.18). Conversely, the smallest displacement corresponds to a full parallel spring connection when the stiffness is maximal, as indicated by the second Eq. (2.18). However, these connections are not allowed in this example. Two combinations are sketched in Figure 2.7, which are candidates satisfying this example's requirement to mix series and parallel pairs.

The equivalent stiffnesses of the spring connections shown in Figure 2.7 are

$$\begin{cases} k_1 = \dfrac{k}{2} + k + k = \dfrac{5}{2}k \\ \dfrac{1}{k_2} = \dfrac{1}{k} + \dfrac{1}{k} + \dfrac{1}{2k} = \dfrac{5}{2k} \end{cases} \qquad (2.19)$$

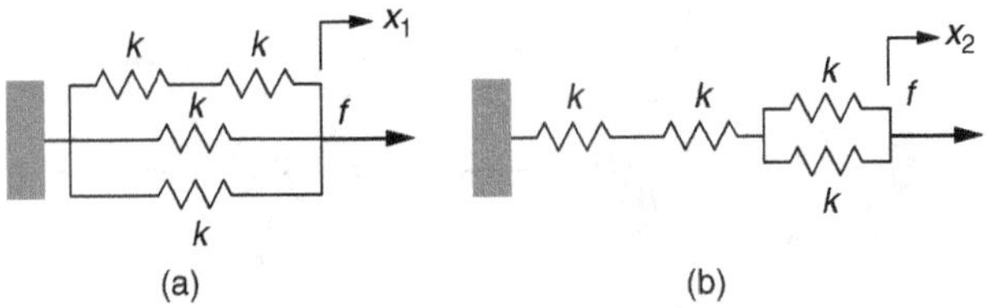

FIGURE 2.7

Translatory Spring Combinations: (a) Three Parallel Branches; (b) Three Series Branches.

Because $f = k_1x_1 = k_2x_2$, Eqs. (2.19) yield

$$\frac{k_1}{k_2} = \frac{x_2}{x_1} = \frac{25}{4} \tag{2.20}$$

which is indeed the displacement ratio given in this example; as a consequence, the spring arrangements are the ones shown in Figure 2.7. By using the equation of a translatory helical spring, as given in the caption of Figure 2.4, and the specified numerical values, the following results are obtained: $k = 1157.4$ N/m, $k_1 = 5k/2 = 2893.5$ N/m, and $k_2 = 2k/5 = 462.96$ N/m. ▪

2.1.3 Damping Elements

Damping is associated with energy losses, and for mechanical systems, the damping mechanisms are mainly viscous, frictional, or internal (hysteretic). We briefly study the viscous and friction losses next.

Viscous Damping

In *viscous damping*, damping forces (or torques) that are proportional *to the relative velocity* are set whenever there is relative motion between a structural member and the surrounding fluid (liquid or gas). For a translatory relative motion, such as the one sketched in Figure 2.8(a), the damping force opposes the direction of motion and is expressed as

$$f_d(t) = cv(t) = c\frac{dx(t)}{dt} = c\dot{x} \tag{2.21}$$

whereas for a rotating body, such as the one sketched in Figure 2.8(b), the damping torque opposes the relative rotation direction and is equal to

$$m_d(t) = c\omega(t) = c\frac{d\theta(t)}{dt} = c\dot{\theta} \tag{2.22}$$

The *viscous damping coefficient* c can be determined as a function of geometrical and material parameters for either translatory or rotary dampers. In both cases, *Newton's law of viscous flow* is applied, which establishes that *shear stresses* τ (more details

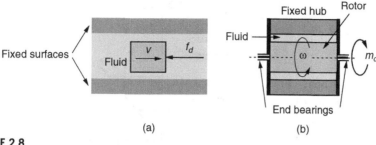

FIGURE 2.8

Damping through Structure-Fluid Relative Motion: (a) Translatory; (b) Rotary.

on stresses are given in Appendix D) are set up between adjacent layers of fluid, according to the equation

$$\tau = \mu \frac{dv(z)}{dz} \qquad (2.23)$$

where μ is the *coefficient of dynamic viscosity* and $dv(z)/dz$ is the gradient of the relative velocity between the moving surface and the fixed one.

Example 2.3

Derive the coefficient of viscous damping c corresponding to a plate that moves parallel to another fixed plate with fluid in between, as sketched in Figure 2.9. Known are the two plates superposition area $A = 1000$ mm², the plates gap $g = 0.1$ mm, and the coefficient of dynamic viscosity $\mu = 0.001$ N-s/m². Assume the fluid velocity varies linearly from zero at the interface with the fixed plate to the velocity of the mobile plate at the interface with it (Couette flow model).

Solution

Figure 2.9 illustrates the velocity gradient under the assumptions of *Couette's flow model*. For this type of flow, the velocity $v(z)$ at a distance z from the fixed surface is determined as a function of the maximum velocity v and the gap g as

$$v(z) = \frac{z}{g}v \qquad (2.24)$$

As a consequence, Newton's law for viscous flow, Eq. (2.23), becomes

$$\tau = \mu \frac{v}{g} \qquad (2.25)$$

By considering now that the contact area between the moving plate and the fluid is A, the *damping (dragging) force* that acts on the moving plate is shear stress times area:

$$f_d = \tau A = \frac{\mu A}{g}v \qquad (2.26)$$

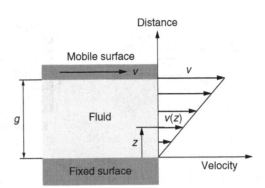

FIGURE 2.9

Linear Profile of Velocity for Couette Flow between Parallel Plates.

Comparison of Eqs. (2.26) and (2.21) indicates that the damping coefficient is

$$c = \frac{\mu A}{g} \tag{2.27}$$

For the numerical values of this problem, a damping coefficient $c = 0.01$ N-s/m is obtained.

The damping coefficient c_t of a piston of diameter D_i and length l, which translates in a cylinder of diameter D_o filled with fluid of density ρ and coefficient of dynamic viscosity μ, as well as the damping coefficient c_r corresponding to the rotary motion of the piston inside the cylinder, are derived in the companion website Chapter 2 as

$$\begin{cases} c_t = \dfrac{2\pi\mu D_i l}{D_o - D_i} + R\dfrac{\pi^2 D_i^4}{16}\rho \\ c_r = \dfrac{\pi\mu D_i^3 l}{2(D_o - D_i)} \end{cases} \tag{2.28}$$

R being a fluid resistance (more on this topic will follow in Chapter 5).

Figure 2.10 illustrates the symbols used for viscous damping elements in translation and rotation.

The energy dissipated through viscous damping is equal to the work done by the damping force in translation and the damping torque in rotation:

$$U_d = c\int \dot{x}\,dx = c\int \dot{x}^2\,dt \tag{2.29}$$

$$U_d = c\int \dot{\theta}\,d\theta = c\int \dot{\theta}^2\,dt \tag{2.30}$$

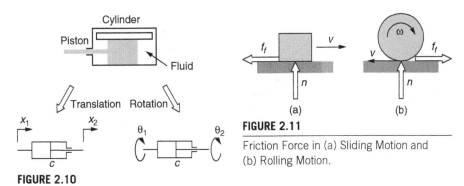

FIGURE 2.10

Viscous Damping Representation for Translatory and Rotary Motions.

FIGURE 2.11

Friction Force in (a) Sliding Motion and (b) Rolling Motion.

Dampers can be combined in series (cascade) or in parallel, similarly to springs. The corresponding series and parallel equivalent damping coefficients are derived in the website companion Chapter 2, and their equations are

$$\frac{1}{c_s} = \frac{1}{c_1} + \frac{1}{c_2} + \cdots + \frac{1}{c_n}; \quad c_p = c_1 + c_2 + \cdots + c_n \tag{2.31}$$

Dry-Friction (Coulomb) Damping

Dry friction (or *Coulomb*) *damping* occurs at the interface between two bodies in relative motion with contact. As indicated in Figure 2.11, the *friction force* opposes the relative velocity direction and depends on the normal force that acts on the body.

The friction force f_f is proportional to the normal force n, the proportionality constant being the kinematic friction coefficient μ_k

$$f_f = \mu_k n \tag{2.32}$$

and is constant when n is also constant. The energy dissipated through dry-friction damping is simply the work done by the force f_f over a distance x:

$$U_d = W_d = f_f x = \mu_k n x \tag{2.33}$$

2.1.4 Actuation (Forcing)

Mechanical systems that are formed of inertia, spring, and damping elements are set into motion by external factors that are the *actuation* or *forcing* elements.

For lumped-parameter modeling, the actuation is represented by forces and torques, as illustrated in Figure 2.12. In translation, the force f (the cause of motion) generates a displacement x (the effect), both in the same direction; similarly, in rotation, the moment (torque) m causes the disc to rotate by an angle θ about a fixed

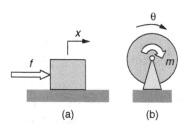

FIGURE 2.12

Actuation (Forcing) Produced by: (a) Force in Translatory Motion; (b) Torque in Rotary Motion.

pivot point, the torque and angle having the same direction. A large variety of actuation sources for translation or rotation exist, comprising mechanical, electrical, and electromechanical means, to cite just a few.

2.2 BASIC MECHANICAL SYSTEMS

Now that the main mechanical elements have been introduced, basic *mechanical systems* formed of various inertia, spring, damping, and actuation elements are modeled and analyzed for both translatory and rotary motion. The motion of these systems is defined by one time-dependent variable (a translatory or rotary displacement) and therefore are *single degree-of-freedom* systems; a more systematic discussion on the degrees of freedom is given in Chapter 3.

2.2.1 Newton's Second Law of Motion Applied to Mechanical Systems Modeling

Newton's second law of motion states that the acceleration of a particle $(a = \ddot{x} = d^2x(t)/dt^2)$ is proportional to the force applied to it and is in the direction of this force. The proportionality constant is the mass m for a constant-mass particle or a rigid body undergoing translatory motion whose inertia can be reduced to a point. In cases where several external forces act on the particle in the motion direction, the acceleration is proportional to the algebraic sum of external forces f_j

$$m\ddot{x} = \sum_{j=1}^{n} f_j \tag{2.34}$$

An equation similar to Eq. (2.34) can be derived from Newton's second law of motion (which is formulated for translation), expressing the dynamic equilibrium of a body of constant mass moment of inertia J rotating about a fixed axis:

$$J\ddot{\theta} = \sum_{j=1}^{n} m_{tj} \tag{2.35}$$

where m_{tj} are the torques acting on the rotating body and the rotary acceleration is $\ddot{\theta} = d^2\theta(t)/dt^2$ (the symbol m_t has been used for moments to avoid confusion with the mass symbol m). Equation (2.35) is also valid for a particle rotating about an external axis.

Equation (2.34) for translation and Eq. (2.35) enable deriving *mathematical models* formed of the differential equation(s) connecting the input and the output through system component parameters. Instrumental to correctly applying these equations is the *free-body diagram*, which isolates a mechanical element from its system using external forcing as well as reaction forces and moments from the elements that have been separated from the studied element. The equations based on Newton's second law of motion and the free-body diagrams are utilized for the remainder of this chapter, as well as in Chapter 3, to determine the free and forced responses of mechanical systems. Other methods, such as the energy method (which is utilized in subsequent chapters) or d'Alembert principle (equation) can also be applied to derive mathematical models for mechanical systems.

2.2.2 **Free Response**

In the *free response*, no forcing is applied to a mechanical system, which consequently undergoes free vibrations when excited externally by an initial displacement or initial velocity. The *natural* (undamped) *response* and the *damped response* are studied in this section.

Natural Response of Conservative Systems

As illustrated next, the natural response of single DOF mechanical systems is concerned with determining the *natural frequencies* (which are the vibration frequencies of free undamped mechanical systems).

Free undamped vibrations and the related response are characteristic of the simplest mechanical systems undergoing vibrations and are formed of a rigid body with inertia and a spring, as shown in Figure 2.13 for a translatory-motion example. According to Newton's second law of motion and based on the free-body diagram of Figure 2.13, the equation of motion is

$$m\ddot{x} = -kx \tag{2.36}$$

which can be written as

$$\ddot{x} + \frac{k}{m}x = 0 \tag{2.37}$$

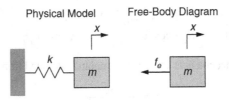

FIGURE 2.13

Mass-Spring System with Free-Body Diagram.

The mathematical model of the free undamped response for a single-DOF system is represented by either Eq. (2.36) or Eq. (2.37).

The solution to a differential equation with no forcing (a *homogeneous equation*) consists only of the *natural part*, and the theory of ordinary differential equations shows (see Appendix A) that this solution is of the type

$$x = X\sin(\omega t) \tag{2.38}$$

Equation (2.38) produces a relationship between acceleration and displacement of the same type as that of Eq. (2.37):

$$\ddot{x} = -\omega^2 x \tag{2.39}$$

Comparison of Eqs. (2.37) and (2.39) yields

$$\omega = \sqrt{\frac{k}{m}} = \omega_n \tag{2.40}$$

The parameter ω_n, which is based on the physical characteristics of the single DOF mass-spring system, is named *natural frequency* and plays an important role in defining the free undamped response of mechanical systems. Equations (2.38) and (2.40) illustrate that the body motion corresponding to the natural response, also known as *modal motion*, is harmonic, has a frequency of oscillation equal to the natural frequency ω_n, and is generated by nonzero initial conditions (displacement or velocity).

The notion of natural frequency can be extended to any second-order homogeneous differential equation of the type

$$a_2 \frac{d^2 x(t)}{dt^2} + a_0 x(t) = 0 \tag{2.41}$$

whose natural frequency is

$$\omega_n = \sqrt{\frac{a_0}{a_2}} \tag{2.42}$$

The amplitude X of Eq. (2.38) can be determined from a known initial condition (usually the displacement x_0 or velocity v_0):

$$\begin{cases} x_0 = X\sin(\omega_n t)\big|_{t=t_0} = X\sin(\omega_n t_0) \\ v_0 = \omega_n X\cos(\omega_n t)\big|_{t=t_0} = \omega_n X\cos(\omega_n t_0) \end{cases} \tag{2.43}$$

When the initial displacement is known, the amplitude is found from the first Eq. (2.43) as

$$X = \frac{x_0}{\sin(\omega_n t_0)} \tag{2.44}$$

When the initial velocity is provided, the amplitude is calculated from the second Eq. (2.43) as

$$X = \frac{v_0}{\omega_n \cos(\omega_n t_0)} \qquad (2.45)$$

Lever systems and gear shaft transmissions are analyzed next in relation to their natural response.

Lever Systems

Mechanical levers are employed to change motion characteristics, such as displacements and forces or moments. For a rotation angle θ of the rod, the free end point C displaces to the position C' in Figure 2.14(a) by following an arc of a circle centered at O. For small angles (normally smaller than 5°), it can be approximated that the length of the arc CC' is equal to the vertical distance z_c (or the length CC''), which is obtained by extending the displaced rod OC' to the intersection with the vertical line taken from C; this means that

$$z_C = CC'' \approx CC' = l_c\theta \qquad (2.46)$$

This approximation, which is known as the *small-angle* (or *small-rotation*) *approximation*, enables transforming the actual rotation motion of the rod into a translation of the end point.

Consider the rigid rod of Figure 2.14(b), which is pinned at O and acted upon by the forces f_A and f_B. Under the assumption of small angular motions, the lever position after the loads f_A and f_B have been applied quasi-statically is determined by points A' and B', whose displacement from the original state is related as

$$\frac{z_B}{z_A} = \frac{l_B}{l_A} = a \qquad (2.47)$$

where a is the *displacement amplification* (also known as *mechanical advantage*). Equation (2.47) indicates that the displacement at B is a times larger than the one at A. Since the forces f_A and f_B act antagonistically, the work done by the two forces is equal:

$$f_A z_A = f_B z_B \qquad (2.48)$$

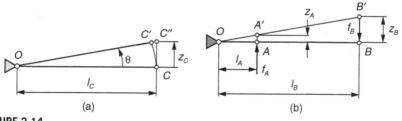

FIGURE 2.14

Single-Stage Lever System: (a) Small-Rotation Approximation; (b) with Two Opposing Forces.

From Eq. (2.47) it follows that the force f_B is a times smaller than the force f_A:

$$f_B = \frac{z_A}{z_B} f_A = \frac{1}{a} f_A \qquad (2.49)$$

such that the lever mechanism of Figure 2.14(b) operates as a *force-reduction device*.

The actual lumped inertia and stiffness properties can be moved (transferred) from their original positions to the mobile end of a lever, for instance. The companion website Chapter 2 gives the derivation for moving masses and stiffnesses using energy principles.

The mass m_B that is dynamically equivalent to the original mass m_A of Figure 2.15 is

$$m_B = \left(\frac{l_A}{l_B}\right)^2 m_A \qquad (2.50)$$

The original stiffness k_A is transferred into the stiffness k_B, see Figure 2.16:

$$k_B = \left(\frac{l_A}{l_B}\right)^2 k_A \qquad (2.51)$$

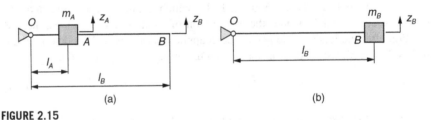

(a)   (b)

FIGURE 2.15

Equivalent Lever-Mass Systems: (a) Original; (b) Transformed.

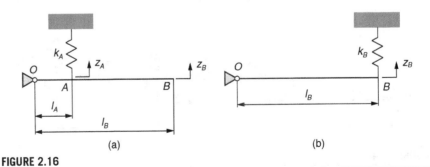

(a)   (b)

FIGURE 2.16

Equivalent Lever-Spring Systems: (a) Original; (b) Transformed.

Example 2.4

Calculate the natural frequency of the lever system of Figure 2.17(a) by transferring the lumped-parameter stiffness and inertia from their locations to the free end point of the rigid lever, as shown in Figure 2.17(b), by assuming that

a. The lever is massless.

b. The lever has a mass $m_l = m_B/4$. Plot the rotation angle θ of the rod in terms of time for $\theta(0) = 4°$ at $t(0) = t_0 = 0.1$ s.

Known are $l_A = 0.5$ m, $l_B = 1$ m, $l_C = 1.5$ m, $k_A = 1000$ N/m, and $m_B = 1$ kg. Assume small lever rotations.

Solution

a. According to Eqs. (2.50) and (2.51), the equivalent mass and stiffness at points C are

$$m_C = \left(\frac{l_B}{l_C}\right)^2 m_B; \quad k_C = \left(\frac{l_A}{l_C}\right)^2 k_A \tag{2.52}$$

The natural frequency of the equivalent mass-spring system is therefore

$$\omega_n = \sqrt{\frac{k_C}{m_C}} = \left(\frac{l_A}{l_B}\right)\sqrt{\frac{k_A}{m_B}} \tag{2.53}$$

with a numerical value of 15.81 rad/s.

b. The free-body diagram of the rod is shown in Figure 2.18, where f_e is the elastic force produced by the equivalent spring k_A.

The following equation describes the rotation of the mechanical system around the pivot point O:

$$J_0\ddot{\theta} = -f_e l_C \tag{2.54}$$

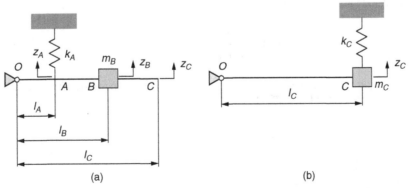

(a)   (b)

FIGURE 2.17

Lever-Mass-Spring System: (a) Actual; (b) Transformed.

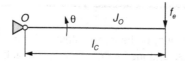

FIGURE 2.18

Free-Body Diagram of a Rotating Rod.

The total mechanical moment of inertia with respect to point O contains contributions from the rod and the lumped-mass m_C, while the elastic force uses the small-angle approximation, which results in

$$\begin{cases} J_O = \dfrac{m_l l_C^2}{3} + m_C l_C^2 = \dfrac{m_B l_C^2}{12} + m_C l_C^2 = \left(\dfrac{1}{12}l_C^2 + l_B^2\right)m_B \\ f_e = k_C z_C = k_C l_C \theta = k_A \dfrac{l_A^2}{l_C^2}l_C\theta = k_A \dfrac{l_A^2}{l_C}\theta \end{cases} \tag{2.55}$$

By substituting Eq. (2.55) in Eq. (2.54), the latter becomes

$$\left(1 + 12\dfrac{l_B^2}{l_C^2}\right)m_B\ddot{\theta} + 12\dfrac{l_A^2}{l_C^2}k_A\theta = 0 \tag{2.56}$$

Equation (2.56) indicates that the natural frequency of the analyzed system is

$$\omega_n^* = \dfrac{2\sqrt{3}\,(l_A/l_C)}{\sqrt{1 + 12(l_B/l_C)^2}}\sqrt{\dfrac{k_A}{m_B}} \tag{2.57}$$

and its numerical value is 14.51 rad/s, which is obviously smaller than ω_n of point (a), where less mass was accounted for. As shown in Eq. (2.44), the rotation angle is the solution to Eq. (2.56) when the initial rotation is nonzero:

$$\theta(t) = \dfrac{\theta(0)}{\sin\left(\omega_n^* t_0\right)}\sin\left(\omega_n^* t\right) \tag{2.58}$$

and Figure 2.19 displays θ as a function of time. The following MATLAB® code enables drawing the plot of Figure 2.19:

```
>> omn = 14.51;
>> t0=0.1;
>> theta0=4;
>> t = 0:0.001:2;
>> theta = theta0/sin(omn*t0)*sin(omn*t);
```

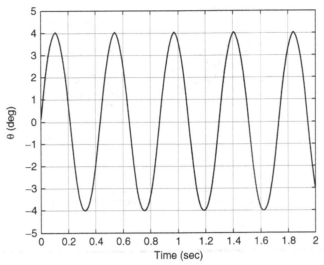

FIGURE 2.19

Free Undamped Response of Mechanical Rod System with Nonzero Initial Rotation Angle.

```
>> plot(t,theta)
>> xlabel('Time (sec)')
>> ylabel('\theta (deg)')
>> grid on
```

Geared Shafts Transmissions

Rotary motion is transmitted through *toothed gears* from one shaft to another in many mechanical engineering applications. Part of a geared wheel is shown in Figure 2.20(a). The median circle is the *pitch circle* (of radius R) and the distance measured on this circle between two consecutive teeth is known as the *circular pitch*, denoted by p. Two gears that *mesh* (or *engage*) are schematically shown in Figure 2.20(b) with their pitch circles being tangent.

The following relationships apply between the parameters of the two gears:

$$\frac{N_1}{N_2} = \frac{R_1}{R_2} = \frac{\theta_2}{\theta_1} = \frac{\dot{\theta}_2}{\dot{\theta}_1} = \frac{\ddot{\theta}_2}{\ddot{\theta}_1} = \frac{m_{t1}}{m_{t2}} \tag{2.59}$$

where N is the number of teeth; R, the radius of the pitch circle; m_t, the torsion moment; and θ, the rotation angle. The sequence of Eq. (2.59) is instrumental in transferring inertia, stiffness, and damping properties from one shaft to another as shown in the following example.

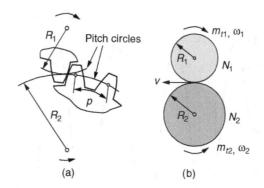

FIGURE 2.20

Toothed Gears: (a) Main Parameters of a Toothed Gear; (b) Schematic of Meshing Gears.

Example 2.5

Two elastic shafts of negligible inertia are connected through two meshing gears, as shown by the spring-inertia lumped-parameter model of Figure 2.21(a). Known are $N_1 = 32$, $N_2 = 26$, $J_1 = 0.001$ kg-m^2, $J_2 = 0.0008$ kg-m^2, $k_1 = 80$ N-m, and $k_2 = 200$ N-m. Find the natural frequency of this system. Graphically analyze the change in the natural frequency when k_1 increases gradually up to 120 N-m and k_2 decreases to 140 N-m. Assume small rotations to preserve shaft integrity.

Solution

The potential elastic energy of this system is stored by the two deforming springs and is

$$U_e = \frac{1}{2}k_1\theta_1^2 + \frac{1}{2}k_2\theta_2^2 \qquad (2.60)$$

From Eq. (2.59) it follows that

$$\theta_2 = \frac{N_1}{N_2}\theta_1 \qquad (2.61)$$

which, substituted in Eq. (2.47), results in

$$U_e = \frac{1}{2}\left[k_1 + \left(\frac{N_1}{N_2}\right)^2 k_2\right]\theta_1^2 \qquad (2.62)$$

Equation (2.62) suggests that, from the viewpoint of stiffness, the actual system is equivalent to another system that is reduced (or transferred) to the first shaft and whose equivalent stiffness is

$$k_{e,1} = k_1 + \left(\frac{N_1}{N_2}\right)^2 k_2 \qquad (2.63)$$

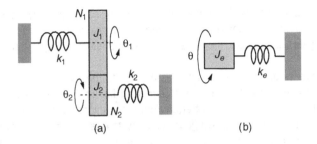

FIGURE 2.21

Two Shaft-Gear System: (a) Actual Lumped-Parameter Model; (b) Equivalent Lumped-Parameter Model.

When θ_1 is expressed in terms of θ_2 from Eq. (2.59) and then substituted into Eq. (2.62), a similar result is obtained and the equivalent stiffness transferred to the second shaft is

$$k_{e,2} = k_2 + \left(\frac{N_2}{N_1}\right)^2 k_1 \qquad (2.64)$$

The kinetic energy of the shaft-gear system is

$$T = \frac{1}{2}J_1\omega_1^2 + \frac{1}{2}J_2\omega_2^2 = \frac{1}{2}J_1\dot{\theta}_1^2 + \frac{1}{2}J_2\dot{\theta}_2^2 \qquad (2.65)$$

The angular velocity ω_2 is expressed as a function of ω_1 and the number of teeth N_1 and N_2 from Eq. (2.59) and substituted in Eq. (2.65), resulting in

$$T = \frac{1}{2}\left[J_1 + \left(\frac{N_1}{N_2}\right)^2 J_2\right]\dot{\theta}_1^2 \qquad (2.66)$$

which indicates that the actual system's inertia is equivalent to the following inertia when reduction to the first shaft is utilized:

$$J_{e,1} = J_1 + \left(\frac{N_1}{N_2}\right)^2 J_2 \qquad (2.67)$$

A similar result is obtained when the actual inertia is reduced to the second shaft by expressing the angular velocity ω_1 as a function of ω_2 and the number of teeth N_1 and N_2 from Eq. (2.59):

$$J_{e,2} = J_2 + \left(\frac{N_2}{N_1}\right)^2 J_1 \qquad (2.68)$$

Equations (2.63), (2.64), (2.67), and (2.68) indicate a simple rule of transferring inertia or stiffness from one shaft to another: Either of these properties can be transferred on a connecting shaft through multiplication of the original amount by the square of the ratio of the destination gear teeth number to the source gear teeth number. The transferred parameter algebraically adds to the corresponding parameter that resides on the destination shaft.

When stiffness and inertia reduction is applied to the same shaft, say shaft 1, the actual system is equivalent to the one sketched in Figure 2.21(b), which consists of a spring of stiffness $k_{e,1}$ and a gear of equivalent mechanical moment of inertia $J_{e,1}$. This also proves that the two-shaft, two-gear actual system is a single DOF system, and its natural frequency corresponding to torsional vibration is

$$\omega_n = \sqrt{\frac{k_{e,1}}{J_{e,1}}} = \sqrt{\frac{k_1 + \left(\frac{N_1}{N_2}\right)^2 k_2}{J_1 + \left(\frac{N_1}{N_2}\right)^2 J_2}} = \sqrt{\frac{k_2 + \left(\frac{N_2}{N_1}\right)^2 k_1}{J_2 + \left(\frac{N_2}{N_1}\right)^2 J_1}} = \sqrt{\frac{k_{e,2}}{J_{e,2}}} \qquad (2.69)$$

A numerical value of 416.1 is obtained for ω_n. The variation of ω_n of Eq. (2.69) in terms of k_1 and k_2 is plotted in Figure 2.22. As expected, an increase in k_1 results in an increase of ω_n, whereas a decrease in k_2 produces smaller natural frequencies. The MATLAB®

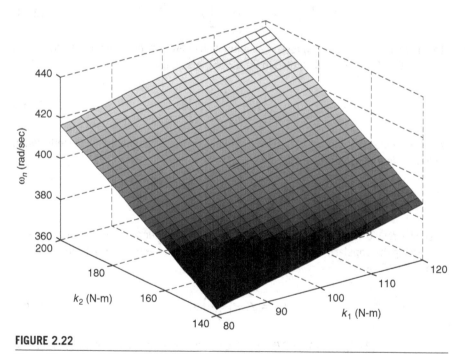

FIGURE 2.22

Three-Dimensional Plot of Natural Frequency as a Function of the Two Shafts' stiffnesses.

commands `meshgrid` and `surf` (or `mesh`), and the following code have been used to produce Figure 2.22:

```
>> j1=0.001;
>> j2=0.0008;
>> n1=32;
>> n2=26;
>> [k1,k2] = meshgrid(80:2:120,140:2:200);
>> omn = sqrt((k1+n1^2/n2^2*k2)/(j1+n1^2/n2^2*j2));
>> surf(k1,k2,omn)
>> colormap(gray)
>> xlabel('k_1 (N-m)')
>> ylabel('k_2 (N-m)')
>> zlabel('\omega_n (rad/s)')
```

Free Damped Response

When viscous damping is added to a mass-spring setting, a mechanical system such as the one sketched in Figure 2.23(a) is obtained.

By applying Newton's second law of motion to the free-body diagram of the mechanical system of Figure 2.23(b), the following equation is obtained:

$$m\ddot{x} = -c\dot{x} - kx \tag{2.70}$$

After dividing the left- and right-hand sides of Eq. (2.70) by m and utilizing the definitions of the natural frequency ω_n and *damping ratio* ξ ($c/m = 2\xi\omega_n$, as introduced in Chapter 1), Eq. (2.70) changes to

$$\ddot{x} + 2\xi\omega_n\dot{x} + \omega_n^2 x = 0 \tag{2.71}$$

Three cases of damping are possible, depending on how the damping coefficient relates to 1. The cases of *overdamping* ($\xi > 1$) and *critical damping* ($\xi = 1$) are analyzed in the companion website Chapter 2. *Underdamping*, which is frequently encountered in engineering applications, refers to situations where $0 < \xi < 1$ and in this case, the solution to Eq. (2.71) is:

$$x(t) = Xe^{-\xi\omega_n t}\sin\left(\sqrt{1-\xi^2}\,\omega_n t + \varphi\right) \tag{2.72}$$

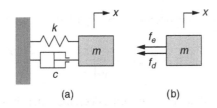

(a)   (b)

FIGURE 2.23

Free Mechanical System: (a) Lumped-Parameter Model; (b) Free-Body Diagram.

Details on deriving Eq. (2.72) are given in the companion website Chapter 2. Using known initial conditions for this solution, such as initial displacement x_0 and velocity v_0, the *amplitude X* and *phase angle* φ can be determined as

$$
\begin{cases}
X = \dfrac{1}{\omega_n} \sqrt{\dfrac{v_0^2 + 2\xi v_0 \omega_n x_0 + (\omega_n x_0)^2}{1 - \xi^2}} \\
\varphi = \tan^{-1}\left(\dfrac{\omega_n x_0 \sqrt{1 - \xi^2}}{v_0 + \xi \omega_n x_0}\right)
\end{cases}
\tag{2.73}
$$

The quantity ω_d, defined as

$$
\omega_d = \sqrt{1 - \xi^2}\,\omega_n
\tag{2.74}
$$

is the *damped frequency* of the mass-dashpot system, and it can be seen that $\omega_d < \omega_n$ since $\xi < 1$.

Figure 2.24 shows the plot of the free response given in Eq. (2.72) for some arbitrary values of $\xi = 0.2$, $\omega_n = 100$ rad/s, $x_0 = 0.05$ m, and $v_0 = 0$. The plot displays an amplitude-decaying sine curve having a (damped) period T_d equal to

$$
T_d = \frac{2\pi}{\omega_d} = \frac{2\pi}{\sqrt{1 - \xi^2}\,\omega_n}
\tag{2.75}
$$

which is equal to 0.064 s for the particular numerical values of the plot.

An amount that characterizes the underdamping is the *logarithmic decrement*, which is defined as the natural logarithm of two successive decaying amplitudes:

$$
\delta = \ln\frac{X_{k-1}}{X_k}
\tag{2.76}
$$

The two successive amplitudes of Eq. (2.76) can be written by means of Eq. (2.72) as

$$
\begin{cases}
X_{k-1} = Xe^{-\xi\omega_n t_{k-1}} \\
X_k = Xe^{-\xi\omega_n(t_{k-1} + T_d)}
\end{cases}
\tag{2.77}
$$

and therefore the logarithmic decrement of Eq. (2.76) becomes

$$
\delta = \xi\omega_n T_d = \frac{2\pi\xi}{\sqrt{1 - \xi^2}}
\tag{2.78}
$$

Equation (2.78) shows that the logarithmic decrement depends only on the damping ratio.

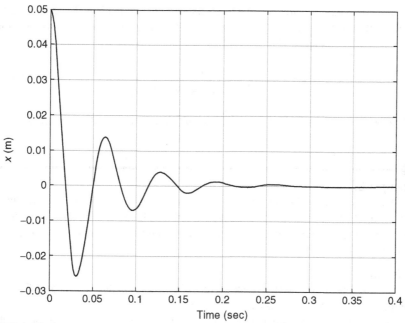

FIGURE 2.24

Plot of Free Underdamped Response.

Example 2.6

A single DOF mass-damper-spring system undergoes free damped vibrations, which are relatively fast. A sensing system is utilized to measure the vibration amplitudes, which can capture successive data only every five vibration cycles, whereby a constant decay of 50% is noted between two consecutive measurements. Determine the damping ratio that corresponds to this system's vibrations.

Solution

The ratio of two consecutively measured amplitudes can be written as

$$\frac{X_k}{X_{k+5}} = \left(\frac{X_k}{X_{k+1}}\right)\left(\frac{X_{k+1}}{X_{k+2}}\right)\left(\frac{X_{k+2}}{X_{k+3}}\right)\left(\frac{X_{k+3}}{X_{k+4}}\right)\left(\frac{X_{k+4}}{X_{k+5}}\right) \tag{2.79}$$

Applying the natural logarithm to Eq. (2.79) results in

$$\ln\left(\frac{X_k}{X_{k+5}}\right) = \ln\left(\frac{X_k}{X_{k+1}}\right) + \ln\left(\frac{X_{k+1}}{X_{k+2}}\right) + \ln\left(\frac{X_{k+2}}{X_{k+3}}\right)$$

$$+ \ln\left(\frac{X_{k+3}}{X_{k+4}}\right) + \ln\left(\frac{X_{k+4}}{X_{k+5}}\right) = 5\delta \tag{2.80}$$

On the other hand, according to the given data, $X_k/X_{k+5} = 1/0.5 = 2$ and $\ln 2 = 0.693$. Equation (2.80) results in $\delta = 0.139$ and Eq. (2.78) yields the damping ratio–logarithmic decrement relationship:

$$\xi = \frac{\delta}{\sqrt{4\pi^2 + \delta^2}} \tag{2.81}$$

Using the value of $\delta = 0.139$ in Eq. (2.81) results in a damping ratio of $\xi = 0.022$. ■

Example 2.7

For the geared shaft system of Figure 2.25(a), with $N_1 = 48$ and $N_2 = 36$, determine the total (equivalent) viscous damping coefficient reduced to the second shaft when the two shafts are supported on bearings with viscous damping of coefficients $c_1 = 20$ N-m-s and $c_2 = 40$ N-m-s.

Solution

The damped mechanical system is sketched in Figure 2.25(a) and its single-DOF lumped-parameter counterpart is shown in Figure 2.25(b).

Equation (2.59) is the relationship between the angular velocities and the numbers of teeth for two meshing gears, whereas Eq. (2.30) expresses the energy lost through viscous damping. The total viscous-damping-type energy being lost by the mechanical system of Figure 2.25(a) is therefore

$$U_d = c_1 \int \dot{\theta}_1^2 dt + c_2 \int \dot{\theta}_2^2 dt = c_1 \left(\frac{N_2}{N_1}\right)^2 \int \dot{\theta}_2^2 dt + c_2 \int \dot{\theta}_2^2 dt \tag{2.82}$$

The total equivalent viscous-damping energy reduced to the second shaft is

$$U_{d,e} = c_e \int \dot{\theta}_2^2 dt \tag{2.83}$$

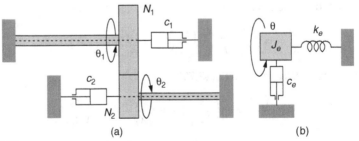

(a)   (b)

FIGURE 2.25

Two-Shaft Gear System with Damping: (a) Actual Representation; (b) Transformed Lumped-Parameter Model.

By equating Eqs. (2.82) and (2.83), the following equivalent damping coefficient is obtained:

$$c_e = c_1 \left(\frac{N_2}{N_1} \right)^2 + c_2 \tag{2.84}$$

with a numerical value of 51.25 N-m-s, and Figure 2.25(b) illustrates the lumped-parameter damped mechanical model. Equation (2.84) shows that damping can be transferred from one shaft to another similarly to the rule that applied for inertia or stiffness relocation. The transferred damping coefficient is equal to the original damping coefficient multiplied by the square of the ratio of the destination gear teeth number to the source gear teeth number. The transferred damped coefficient algebraically adds to the one existing on the destination shaft.

2.2.3 Forced Response with Simulink®

This section studies the forced response of first- and second-order single DOF mechanical systems. Solutions and time-domain solution plots of the resulting mathematical models are obtained using the program Simulink®, which is a graphical user interface (GUI) module built atop of MATLAB® that employs diagrams with various signal-connected operators (or blocks).

First-Order Mechanical Systems

As introduced in Chapter 1, *first-order systems* are described by the differential equation:

$$\tau \dot{x}(t) + x(t) = Kf(t) \tag{2.85}$$

where τ is the *time constant* and K is the *static sensitivity* (or *gain*). The input (or forcing) is $f(t)$ and the output is $x(t)$ in Eq. (2.85).

Example 2.8

Derive the mathematical model of the lumped-parameter mechanical system sketched in Figure 2.26(a), and use Simulink® to draw the simulation diagram corresponding to the mathematical model and to plot the system response for $f = 40$ N. Consider zero initial conditions and three spring-damper combinations: $k_1 = 100$ N/m, $c_1 = 20$ N-s/m; $k_2 = 200$ N/m, $c_2 = c_1 = 20$ N-s/m; and $k_3 = k_1 = 100$ N/m, $c_3 = 10$ N-s/m.

Solution

Based on the free-body diagram of Figure 2.26(b), the following equation can be written by applying Newton's second law of motion:

$$0 = f - f_d - f_e \tag{2.86}$$

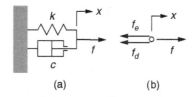

(a)   (b)

FIGURE 2.26

Spring-Damper Mechanical System under the Action of a Force: (a) Lumped-Parameter Model; (b) Free-Body Diagram.

where the damping force and the elastic force are

$$f_d = c\dot{x}; \ f_e = kx \tag{2.87}$$

Combining Eqs. (2.86) and (2.87) yields the following first-order differential equation:

$$c\dot{x} + kx = f \tag{2.88}$$

which can be written in the generic form of a first-order system as given in Eq. (2.85) with the time constant and static sensitivity being

$$\tau = \frac{c}{k}; \ K = \frac{1}{k} \tag{2.89}$$

The units of the time constant are seconds, as expected, because (Ns/m)/(N/m) = s, whereas the static sensitivity of the first-order mechanical system is measured in $N^{-1}m$. Equation (2.88) is reformulated as

$$\dot{x} = -\frac{k}{c}x + \frac{1}{c}f \tag{2.90}$$

which is particularly useful for obtaining the Simulink® solution.

To access Simulink®, type the word simulink at the MATLAB® prompt, then click on File, New, and Model to open a new *model window*. By using blocks dragged into the model window from the Simulink Library Browser, the simulation diagram of Figure 2.27 can be constructed, which solves the differential Eq. (2.90) and plots the solution $x(t)$.

Before explaining the construction of the diagram, please note that the diagram integrates the differential Eq. (2.90) to get and plot the response $x(t)$. At the *summing point* (the circle with two inputs and one output), the signals 0.05 f (this is f/c_1 of Eq. (2.90)) and −5 (which is $-k_1/c_1$ of Eq. (2.90)) do arrive, and the result of this algebraic summation is the derivative of x, in accordance with the same Eq. (2.90). This signal is subsequently passed through the *integrator* (it is shown in Chapter 6 that integration in the time domain corresponds to division by the complex variable s in the Laplace domain, which explains the symbol used for the integration operator, but Simulink® actually applies integration in the time domain with this integration operator) to obtain the output x. The effect of a Gain block is to multiply the input to obtain a specific output.

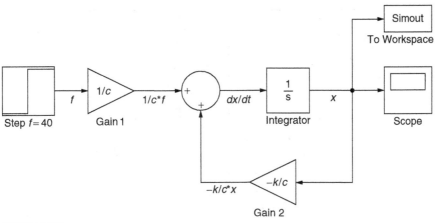

FIGURE 2.27

Simulink® Diagram for Solving and Plotting the Time-Response Corresponding to a
First-Order Differential Equation Defining a Mechanical System.

The diagram contains five blocks that are dragged from different sections of the
`Library Browser` and subsequently defined and connected as shown in Figure 2.27.
The `Step` block is dragged from the `Sources` section of the library. Double clicking on
the block symbol opens a window where a `Step time` of 0 and a `Final value` of 40
need to be specified. A nonzero value of the `Step Time` would consider that the step
input is applied using a delay equal to that time value. The gain blocks come from the
`Commonly Used Blocks` section of the library. A value of 0.05 is specified for the top
`Gain` block and a value of −5 for the bottom one; the bottom `Gain` block needs to be
first flipped by clicking `Format, Flip Block` in the model window. The `Sum` block, as
well as the `Integrator` and the `Scope` blocks, are dragged from the same `Commonly
Used Blocks` library. No parameter specifications for those blocks need be made. The
blocks are then connected by arrows: left click on the source block and then on the
destination block while pressing down the Control (Ctrl) key. To generate the branch connection to the bottom `Gain` block, left click on a point on the existing top arrow and, while
holding down Ctrl, move the mouse (while pressing the left mouse button) to the right side
of the bottom `Gain` block, then release the left mouse button. By double clicking on each
arrow, text can be input, as seen in Figure 2.27. Once the diagram is completed, select
`Simulation, Configuration Parameters, Stop Time`, and select a value of
1.5 s. Click `Simulation` again, then `Start`. To visualize the results of the simulation,
double click the `Scope` block in the model window, and the plot of Figure 2.28(a) is
obtained (you need to click the binoculars icon on the plot window to get the conveniently
scaled plot). The other simulations are shown in Figure 2.28(b) and 2.28(c). As Figure
2.28(b) shows, increasing the spring constant from 100 N-s/m to 200 N-s/m reduces
both the sensitivity (together with the output level) and the time constant, such that the
steady-state response levels at 0.2 m instead of 0.4 m and the time constant reduces by

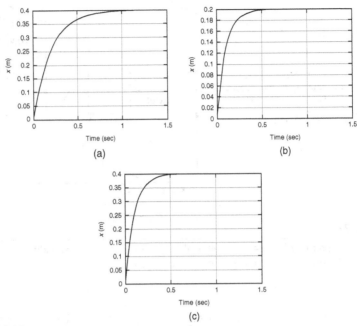

FIGURE 2.28

Plots of Time Response for First-Order Mechanical System: (a) $k_1 = 100$ N/m, $c_1 = 20$ N-s/m; (b) $k_2 = 200$ N/m, $c_2 = 20$ N-s/m; (c) $k_3 = 100$ N/m, $c_3 = 10$ N-s/m.

half. Figure 2.28(c) corresponds to a damping coefficient, which is half its original value, and displays a response with the same steady-state value of 0.4 s as the original one but with a time constant that is half the original time constant. The plots of Figure 2.28 are typical first-order system responses for step (constant) input. Instead of using the Scope to directly plot the time response, the time history is exported to MATLAB® Workspace by means of the To Workspace block dragged from the Source library. Its parameters needing configuration are Variable name: simout (this is the default, but a different name can be used), Save format: Array. Also under Configuration Parameters, Data Import/Export, Save to workspace: check the time box and use a variable to name time (such as "tout"), then check output box and use simout to name the output. After running the simulation, saved on the workspace will be the time vector and the displacement vector. A simple command plot (tout, simout), plus the axes labeling commands, generate the plot of Figure 2.28. ∎

Second-Order Mechanical Systems

Second-order systems, as introduced in Chapter 1, are defined by the following differential equation:

$$\ddot{x}(t) + 2\xi\omega_n\dot{x}(t) + \omega_n^2 x(t) = \omega_n^2 Kf(t) \tag{2.91}$$

where ξ is the *damping ratio*, ω_n is the *natural frequency*, and K is the *static sensitivity*. The input (or forcing) is $f(t)$ and the output is $x(t)$ in Eq. (2.91).

Example 2.9

Derive the mathematical model of the mechanical system shown in Figure 2.29, which contains a rotary two-disc pulley, a horizontally translating body with a spring and a damper attached to it, and two rigid massless rods (the vertical one has a spring connected to it). The pulley is formed of two concentric disks that form a solid piece. Assume that the pulley and the translatory body undergo small motions.

Solution

Figure 2.30 presents the free body diagrams of the rotary cylinder and translatory mass.

Newton's second law of motion is applied to the rotary disc and the translatory mass by taking into account the coordinates and forces shown in Figure 2.30, which results in

$$\begin{cases} J\ddot{\theta} = f_t R_1 - f_{e1} R_2 \\ m_3 \ddot{x} = f - f_t - f_d - f_{e2} \end{cases} \qquad (2.92)$$

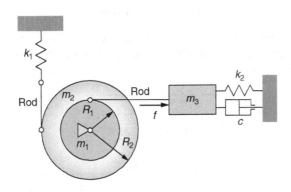

FIGURE 2.29

Mechanical System with Disc Pulley, Translatory Mass, Linear-Motion Springs, and Damper.

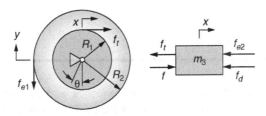

FIGURE 2.30

Free-Body Diagrams of the Rotary Disc and Translatory Mass.

where f_t is the tension in the horizontal rod, f_{e1}, f_{e2} are spring elastic forces, and f_d is the damping force. For small rotations θ of the disc, the tangential displacements x and y can be approximated as

$$x \approx R_1\theta; \quad y \approx R_2\theta \qquad (2.93)$$

By using these approximations, the elastic and damping forces of Eqs. (2.92) are formulated as

$$\begin{cases} f_{e1} = k_1 y = k_1 R_2\theta \\ f_{e2} = k_2 x = k_2 R_1\theta \\ f_d = c\dot{x} = cR_1\dot{\theta} \end{cases} \qquad (2.94)$$

The tension f_t is substituted from the second Eq. (2.92) into the first Eq. (2.92) and Eqs. (2.94) are also used, which yields the differential equation

$$\left[\left(\frac{1}{2}m_1 + m_3\right)R_1^2 + \frac{1}{2}m_2 R_2^2\right]\ddot{\theta} + cR_1^2\dot{\theta} + (k_1 R_2^2 + k_2 R_1^2)\theta = fR_1 \qquad (2.95)$$

Equation (2.95) can be written in short form as

$$a_2\ddot{\theta} + a_1\dot{\theta} + a_0\theta = m_a \qquad (2.96)$$

where $m_a = fR_1$ is the actuation moment and plays the role of $f(t)$ in the generic Eq. (2.91). The natural frequency, damping ratio, and static sensitivity are obtained by comparing Eqs. (2.91), (2.95), and (2.96):

$$\omega_n = \sqrt{\frac{k_1 R_2^2 + k_2 R_1^2}{\left(\frac{1}{2}m_1 + m_3\right)R_1^2 + \frac{1}{2}m_2 R_2^2}};$$

$$\xi = \frac{cR_1^2}{2\sqrt{(k_1 R_2^2 + k_2 R_1^2)\left[\left(\frac{1}{2}m_1 + m_3\right)R_1^2 + \frac{1}{2}m_2 R_2^2\right]}};$$

$$K = \frac{1}{k_1 R_2^2 + k_2 R_1^2} \qquad (2.97)$$

Example 2.10

Determine the static sensitivity, the natural frequency, and the damping ratio values corresponding to the mechanical system of Example 2.9 and sketched in Figure 2.29 for $m_1 = 0.5$ kg, $m_2 = m_3 = 1$ kg, $k_1 = 120$ N/m, $k_2 = 100$ N/m, $R_1 = 0.02$ m, $R_2 = 0.04$ m, $c = 25$ N-s/m, and $f = 0.25$ N. Use Simulink® to plot the disc rotation angle $\theta(t)$ for $\theta(0) = -5°$.

Solution

For the given numerical parameters, the coefficients of Eq. (2.96) assume the values $a_0 = 0.232$, $a_1 = 0.01$, and $a_2 = 0.0013$. Equations (2.97) yield the following values for the quantities of interest: $\omega_n = 13.36$ rad/s, $\xi = 0.29$, and $K = 4.31$.

To use Simulink®, Eq. (2.96) is reformulated as

$$\ddot{\theta} = \frac{fR_1}{a_2} - \frac{a_1}{a_2}\dot{\theta} - \frac{a_0}{a_2}\theta \tag{2.98}$$

With the numerical parameters of the example, Eq. (2.98) becomes

$$\ddot{\theta} = 3.8462 - 7.69\dot{\theta} - 178.46\theta \tag{2.99}$$

Figure 2.31 displays the Simulink® block diagram corresponding to the integration of Eq. (2.99); this time two integrations are necessary to ultimately obtain and plot $\theta(t)$. To

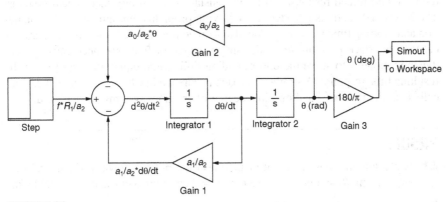

FIGURE 2.31

Simulink® Diagram for Solving and Plotting the Time-Response Corresponding to a Second-Order Differential Equation Defining a Mechanical System.

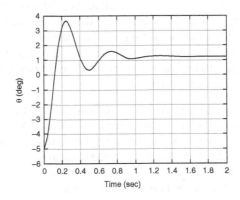

FIGURE 2.32

Plot of Time-Response for Second-Order Mechanical System.

obtain the three input ports of the summing point of Figure 2.31, the following sequence $-|+|-$ needs to be input in the `Function Block Parameters: Sum`, where the vertical lines indicate port separators. The nonzero angle initial condition has to be specified in the `Integrator1` operator under `Initial condition`. The angle θ is converted to degrees by means of the $180/\pi$ `Gain` block and its variation with time as displayed in Figure 2.32. The time response is typical for a second-order system that is subject to a step (constant) input. ■

SUMMARY

This chapter introduces the mechanical elements (inertia, stiffness, damping, and forcing) that are the main components in the dynamic modeling of mechanical systems. By using the lumped-parameter approach, you have learned to model the dynamics of basic, single degree-of-freedom mechanical systems. Newton's second law of motion is applied to derive the mathematical models for the natural, free-damped, and forced responses of basic translatory and rotary mechanical systems. The MATLAB® code is introduced for solving problems through symbolic calculation and plotting time responses. You also learned how to use Simulink® as a graphical user interface program to model single degree-of-freedom mechanical system dynamic models, to solve the corresponding differential equations, and to plot the resulting time response. A similar approach is pursued in Chapter 3, where mechanically compliant devices and multiple degree-of-freedom systems are analyzed.

PROBLEMS

2.1 Calculate the mass moment of inertia of the pulley shown in Figure 2.33 for $R_1 = 2$ cm, $R_2 = 1$ cm, $h_1 = 0.8$ cm, $h_2 = 2$ cm, $h_3 = 3$ cm, and $\rho = 7800$ kg/m³.

2.2 The plate of Figure 2.34 has a width $w = 40$ mm and a thickness $h = 2$ mm. The mass moment of inertia of the plate with respect to the central axis x is 20 times smaller than it needs to be. The necessary increase in the mass moment of inertia can be realized by either enlarging the width (no more than 50% width increase is allowed) or calculating the moment with respect to a parallel axis Δ situated at a distance d from x (d can be no larger than w) or by a combination of these two methods. Select a method that will result in the necessary mechanical moment of inertia and determine the corresponding parameter changes to realize this condition.

2.3 A rigid bar has two identical helical torsion springs at its ends. The system is subjected to a torque that causes the bar to rotate about its longitudinal axis. To save space, the regular helical springs are replaced by planar spiral springs. Consider that both springs are made of the same material, their total wire length is the same, and the circular cross-sections are also identical. Study the relative change in stiffness when Poisson's ratio ranges between 0.25 and 0.5. Hint: Use connection between the E and G modulii (see Appendix D).

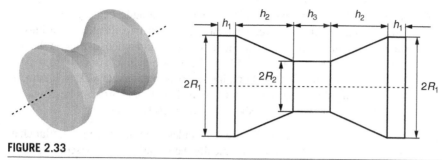

FIGURE 2.33

Three-Dimensional View and Front View with Dimensions of a Pulley.

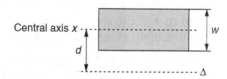

FIGURE 2.34

Plate with Modified Axial Mass Moment of Inertia.

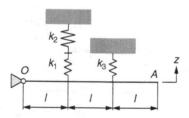

FIGURE 2.35

Springs Acting on a Small-Rotation Pinned Rod.

2.4 For small rotations of the rod sketched in Figure 2.35, determine the stiffness of a spring to be placed at point A and deforming about the z direction that is equivalent to the actual springs acting on the rod. Known are $k_1 = 80$ N/m, $k_2 = 130$ N/m, and $k_3 = 60$ N/m.

2.5 A piston of length $l = 30$ mm and diameter D_i, which is not known precisely, can move in a very long cylinder of inner diameter $D_o = 14$ mm. When the piston cylinder translates into the cylinder with a velocity of 0.5 m/s, the damping force is 0.05 N, and when the piston rotates with $n = 2000$ rot/min, the damping torque is 9.2×10^{-4} N-m. Evaluate the dynamic viscosity of the liquid in the cylinder, as well as the piston diameter.

2.6 A long shaft is supported by three identical hydraulic bearings and meshes through gears with a shorter shaft supported on two identical bearings, as shown in Figure 2.36. Calculate the total damping moment acting on the long shaft, knowing $N_1 = 64$, $N_2 = 42$, the bearing dynamic viscosity $\mu = 0.002$ N-s/m^2, shaft diameters $D_i = 1$ cm, interior bearing diameters $D_o = 1.1$ cm, the length of long shaft bearings $l_1 = 3$ cm, the length of short shaft bearings $l_2 = 2$ cm, and angular velocity of short shaft $n_2 = 200$ rot/min.

2.7 The parallelipipedic plate of Figure 2.37 slides in a fluid within a similar channel of very large width. Study graphically the variation of the viscous damping coefficient with the transverse position of the plate in the channel and determine the specific positions that yield extreme values of the damping coefficient. Known are the dynamic viscosity $\mu = 0.002$ N-s/m^2, the plate length $l = 0.05$ m, the width $w = 0.008$ m, the thickness $h = 0.001$ m, and the channel thickness $g_0 = 0.01$ m. It is also known that the minimum gap between the body and the internal wall of the channel is $(g_0 - h)/20$.

2.8 The pendulum of Figure 2.38 is formed of a rigid rod and a point body of mass $M = 0.5$ kg. Assume small displacements and calculate the natural frequency of this mechanical system when known are the rod's length $l = 0.03$ m, its cross-sectional diameter $d = 0.005$ m, as well as the rod's material mass density $\rho = 7800$ kg/m^3.

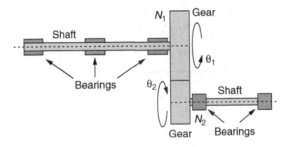

FIGURE 2.36

Meshing Gears with Shafts and Hydraulic Bearings.

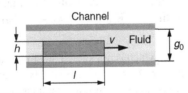

FIGURE 2.37

Body Sliding in a Channel with Viscous Damping.

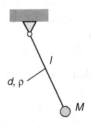

FIGURE 2.38

Pendulum with Rigid Rod and Point Mass.

2.9 The mass m is relocated to a different position on the rod to increase the natural frequency of the two-lever mechanical system sketched in Figure 2.39 by 20%. Determine the new position of the mass m. Consider small displacements and massless rods.

2.10 Consider the mechanical system sketched in Figure 2.40, consisting of a horizontal massless rod that connects at one end (through another massless rod) to a wheel of radius R, mass m, and mass moment of inertia J, and at the other end to a vertical spring of stiffness k_2. The wheel can roll without slippage on a vertical wall under small angular motions of the rod. A pointlike body of mass m is also placed on the horizontal rod. At the rod's pivot point is a torsional (spiral) spring of stiffness k_1. Ignoring gravitational effects and considering small motions, derive the mathematical model for this system.

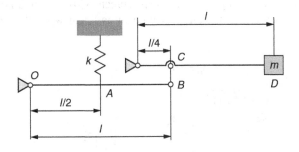

FIGURE 2.39

Two-Lever Mechanical System.

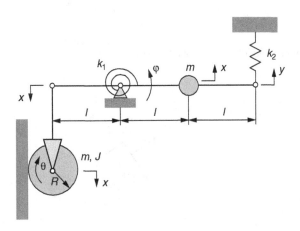

FIGURE 2.40

Mechanical System with Planar Motion.

2.11 Calculate the natural frequency of the mechanical system shown in Figure 2.40 for $m = 1$ kg, $l = 0.5$ m, $k_1 = 50$ N-m, and $k_2 = 110$ N/m. What change in the mass m reduces the natural frequency by 10%?

2.12 Find the natural frequency of the gear shaft mechanical system of Figure 2.41 by reducing the relevant parameters to the middle rigid shaft. Known are $N_1 = 48$, $N_2 = 24$, $N_3 = 32$, $N_4 = 26$, $J_1 = 0.008$ kg-m^2, $J_2 = 0.001$ kg-m^2, $J_3 = 0.006$ kg-m^2, $J_4 = 0.002$ kg-m^2, $k_1 = 100$ N-m, and $k_2 = 140$ N-m. What change in k_1 increases the natural frequency by 2%?

2.13 Use Newton's second law of motion to derive the mathematical model of the mass-spring system shown in Figure 2.42 by considering gravity and ignoring the friction between the body and the incline.

2.14 A mechanical system consists of a translating mass m and a linear spring k whereby the mass is subjected to dry-friction (Coulomb) damping of coefficient μ. Considering an initial displacement $x(0) = x_0$ and no initial velocity, demonstrate that the amplitude decay is linear when the system vibrates freely.

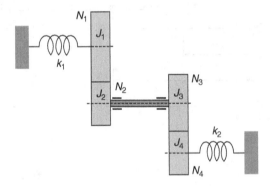

FIGURE 2.41

Lumped-Parameter Model of Three-Shaft Geared System.

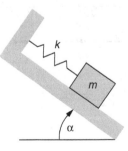

FIGURE 2.42

Mass-Spring Mechanical System on an Incline.

2.15 A mass $m = 1$ kg slides with friction (the kinematic coefficient of friction is $\mu_k = 0.5$ and the static coefficient of friction is $\mu_s = 0.6$) on a horizontal surface. The mass is connected by means of a spring of constant $k = 1200$ N/m to a fixed vertical wall, and is displaced initially by a distance $x_0 = 0.02$ m. Find the total distance traveled by the mass until it stops.

2.16 The rotary mechanical system shown in Figure 2.43 is composed of a cylinder, a torsional damper, and a torsional spring. Calculate the damped frequency of the system, as well as the number of vibration cycles of the cylinder necessary to reduce the amplitude of the free vibrations by a factor of n. Known are $J = 0.01$ kg-m^2, $c_t = 1.5$ N-m-s, $k_t = 600$ N-m, and $n = 100$.

2.17 Consider a translatory mechanical system that is composed of rigid body of mass m and a helical spring with $n = 10$ turns, $G = 1.6$ GPa, $d = 1$ mm, and $R = 5$ mm. The spring is anchored at one end and connects to the body at the other end. When placed in vacuum, the natural period of the system is 0.7 s, and when placed in a fluid of unknown viscous damping coefficient, the period of the free vibrations becomes 0.75 s. Calculate the viscous damping coefficient of the fluid.

2.18 Derive a mathematical model for the translatory mechanical system of Figure 2.44. Use Simulink® to plot the time variation of coordinate x for $k_1 = 90$ N/m, $k_2 = 120$ N/m, $k_3 = 100$ N/m, $c_1 = 12$ N-s/m, and $c_2 = 5$ N-s/m when the system is subjected to

(a) The initial condition $x(0) = 0.02$ m.

(b) The force $f = 50$ N (acting at coordinate x) and zero initial conditions.

2.19 Derive the mathematical model for the small-motions mechanical system of Figure 2.45. Determine the equivalent natural frequency, the damping ratio,

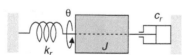

FIGURE 2.43

Mechanical System Undergoing Damped Rotary Vibrations.

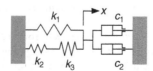

FIGURE 2.44

Spring-Damper Mechanical System.

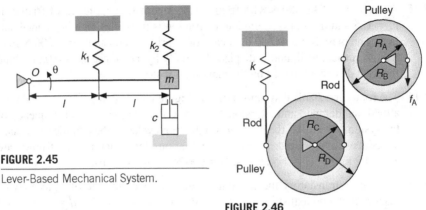

FIGURE 2.45

Lever-Based Mechanical System.

FIGURE 2.46

Rotary Mechanical System with Disc-Pulleys, Rods, and Spring.

and the static sensitivity, knowing the mass of the rod $m_r = 0.6$ kg, $m = 0.4$ kg, $k_1 = 110$ N/m, $k_2 = 80$ N/m, and $c = 16$ N-s/m. Use Simulink® to model and plot the time response of the system when an initial rotation angle $\theta(0) = 3°$ is applied to the rod.

2.20 For the small-motion pulley-rods system of Figure 2.46,

(a) Derive the mathematical model. Use Simulink® to find the system's time response when known are the mass moments of inertia of the upper pulley, $J_{AB} = 0.0003$ kg-m² and of the lower pulley, $J_{CD} = 0.001$ kg-m², the radii $R_A = 0.05$ m, $R_B = 0.07$ m, $R_C = 0.06$ m, $R_D = 0.09$ m, the stiffness $k = 160$ N/m, and the force f_A, which varies linearly from 0 to 80 N in 1 s and then drops to 0 and remains constant. Hint: Use the Wrap to Zero function from the Discontinuities library.

(b) Find the system's natural frequency when the force f_A is removed.

Suggested Reading

A. M. Wahl, *Mechanical Springs*, 2nd Ed. McGraw-Hill, New York, 1963.

F. Beer, E. Russell Johnston, W. Clausen, E. Eisenberg, and P. Cornwell, *Vector Mechanics for Engineers: Dynamics*, 9th Ed. McGraw-Hill, New York, 2009.

W. T. Thomson, *Theory of Vibration with Applications*, 3rd Ed. Prentice-Hall, Englewood Cliffs, NJ, 1988.

J. P. Den Hartog, *Mechanical Vibrations*. Dover, New York, 1985.

D. J. Inman, *Engineering Vibration*, 3rd Ed. Prentice-Hall, Englewood Cliffs, NJ, 2007.

A. D. Nashif, D. I. G. Jones, and J. P. Anderson, *Vibration Damping*. John Wiley & Sons, New York, 1985.

Mechanical Systems II

OBJECTIVES

Developing the concepts of basic mechanical systems modeling introduced in Chapter 2, this chapter focuses on

- Equivalent inertia and stiffness of compliant (flexible) mechanical elements, such as beams and bars.

- Modeling simple compliant (flexible) mechanisms and micro or nano (very small-scale) devices as lumped-parameter mechanical systems.

- Calculation of the natural frequencies of single degree-of-freedom (DOF) compliant mechanisms.

- Application of Newton's second law of motion and the energy method to formulate the mathematical models of conservative multiple-DOF mechanical systems.

- Application of Newton's second law of motion to derive mathematical models of forced multiple-DOF mechanical systems.

- Analysis of free undamped (natural) and forced vibrations of multiple-DOF mechanical systems.

- Introduction to the notions and calculation of natural frequencies (eigenvalues) and modal motions (eigenmodes) of multiple-DOF mechanical systems by means of MATLAB®.

- Using Simulink® for graphical modeling and solution of differential equations systems corresponding to multiple-DOF mechanical systems.

INTRODUCTION

Compliant (or flexible) mechanical systems, which are increasingly utilized in precision devices and small-scale systems such as micro- and nano-electromechanical systems (MEMS and NEMS), can also be modeled as lumped-parameter systems through an energy equivalence process, as illustrated in this chapter. Specifically, lumped inertia and stiffness properties can be used to approximate the actual corresponding distributed parameters of beams in bending and bars in torsion. The natural

DOI: 10.1016/B978-0-240-81128-4.00003-9

response of a few single-DOF compliant mechanical microsystems is studied, including mass detection by the frequency shift method.

Mathematical models are derived for conservative multiple-DOF mechanical systems by means of Newton's second law of motion and the energy method, while Newton's second law of motion is utilized to generate the forced response mathematical models of mechanical systems. MATLAB® and the eigenvalue matrix method are applied to determine the natural response, whereas Simulink® is used to graphically model and solve the forced response of multiple-DOF mechanical systems.

3.1 LUMPED INERTIA AND STIFFNESS OF COMPLIANT ELEMENTS

Compliant mechanical systems, as shown in the introductory Chapter 1, comprise or are entirely formed of flexible parts that elastically deform during motion and vibrations. Compliant mechanisms are frequently utilized in precision mechanical devices, as well as in a wide variety of micro- and nano-electromechanical systems, where they operate as mechanical, electrical, electromagnetic, piezoelectric, thermal, optical actuators, and sensors. Compliant elements have a continuous structure, as their inertia and stiffness properties are inherently distributed. Through an energy method, it is possible to obtain approximate pointlike inertia and stiffness properties that are equivalent to the original, distributed-parameter properties. Consider, for instance, the fixed-free elastic member of Figure 3.1, which can bend out-of-the-plane (and behave as a *cantilever*) or can torque about the x axis (and be a *bar in torsion*).

With regards to the out-of-plane bending, the distributed-parameter cantilever can be converted in a lumped-parameter system (as shown in Figure 3.2), which is formed of an equivalent mass m_e and a spring of equivalent stiffness k_e, both placed at the free end. The main condition is that that the z-axis deflection of the cantilever's free end (denoted by u in Figure 3.1) is identical to the motion of the lumped mass, denoted by u in Figure 3.2.

When the same elastic structure of Figure 3.1 is subjected to torsion, it undergoes angular deformations (rotations) about the x axis. The distributed-parameter bar in torsion can be transformed into a lumped-parameter rotary mechanical system

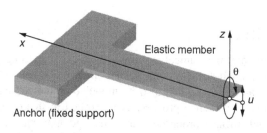

FIGURE 3.1

Fixed-Free Elastic Member with Translatory and Rotary Motions at the Free End.

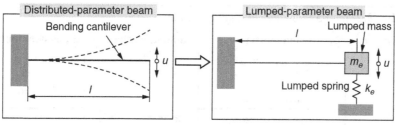

FIGURE 3.2

Distributed-Parameter Bending Cantilever and Its Equivalent Lumped-Parameter Translatory Dynamic System.

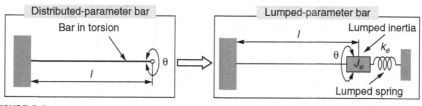

FIGURE 3.3

Distributed-Parameter Fixed-Free Bar in Torsion and Its Equivalent Lumped-Parameter Rotary Dynamic System.

comprising a body of equivalent mass moment of inertia J_e and a torsional spring of equivalent stiffness k_e, both placed at the bar's free end, as shown in Figure 3.3. The condition, again, is that the actual and equivalent systems have the same rotation θ at the free end.

Values of the equivalent inertia and stiffness properties (which are determined from the actual system's parameters) are provided next for several boundary conditions in addition to the fixed-free one. Details on the derivation of these lumped parameters are given in the companion website Chapter 3.

 ### 3.1.1 Inertia Elements

For elastic members it is possible to transform the distributed inertia into an approximate equivalent lumped-parameter *inertia parameter* by equating the kinetic energies of the actual and equivalent systems.

Table 3.1 illustrates the result of substituting the distributed inertia of *beams in bending* and *bars in torsion* by equivalent pointlike (lumped-parameter) inertia properties, which are located at convenient points on the original flexible members. The corresponding inertia fractions (mass m_e for beams and mass moment of inertia J_e for bars) are formulated in the last column of Table 3.1 and are derived in the companion website Chapter 3. They depend on the member total mass m (for bending beams) and mechanical moment of inertia J about the longitudinal axis (for torsion bars).

Table 3.1 Inertia Properties of Beams and Bars

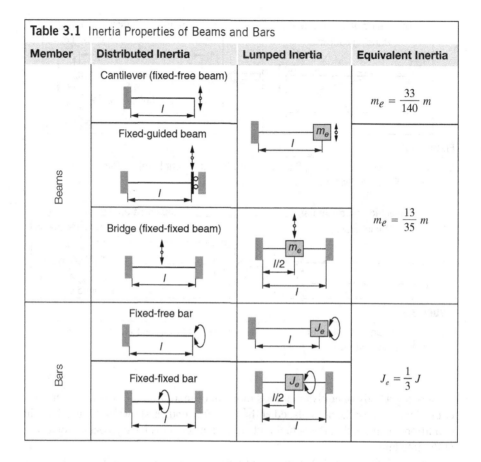

Member	Distributed Inertia	Lumped Inertia	Equivalent Inertia
Beams	Cantilever (fixed-free beam)		$m_e = \dfrac{33}{140}\,m$
	Fixed-guided beam		$m_e = \dfrac{13}{35}\,m$
	Bridge (fixed-fixed beam)		
Bars	Fixed-free bar		
	Fixed-fixed bar		$J_e = \dfrac{1}{3}\,J$

Example 3.1

The microsensor of Figure 3.4, which is formed of two flexible end segments and a central rigid plate, is used to sense vibrations that occur rotationally about the x and y directions. Calculate the rotation inertia about each of the two axes taking into account inertia contributions from the flexible parts; compare these values to the ones that do not consider the flexible members inertia contributions. The circular cross-section bars have a diameter $d = 20\ \mu m$; the other parameters are $l = 310\ \mu m$, $a = 150\ \mu m$, $b = 320\ \mu m$, $h = 50\ \mu m$ (plate thickness), and $\rho = 5600\ kg/m^3$ (mass density).

Solution

The plate rotation about the x-axis involves torsion of the two end bars; therefore, the equivalent mass moment of inertia is composed of the plate's own mass moment of inertia and inertia contributions from the two bars undergoing torsion, as given in Table 3.1:

$$J_{e,x} = J_{p,x} + 2 \times \frac{J_{b,x}}{3} = \frac{\rho abh(a^2 + h^2)}{12} + 2 \times \frac{1}{3} \times \frac{\rho l \pi d^4}{32}$$

$$= \frac{\rho}{12}\left[abh(a^2 + h^2) + \frac{\pi l d^4}{4} \right] \tag{3.1}$$

where the subscripts p and b denote plate and bar, respectively; the plate's and cylinder's mass moments of inertia are given in Appendix D. When the bar's inertia is not accounted for, the x-axis rotation inertia comes only from the plate and is

$$J^*_{e,x} = J_{p,x} = \frac{\rho abh(a^2 + h^2)}{12} \tag{3.2}$$

With the numerical values of the example, the mass moments of inertia given in Eqs. (3.1) and (3.2) are $J_{e,x} = 2.802 \times 10^{-17}$ kg-m² and $J^*_{e,x} = 2.8 \times 10^{-17}$ kg-m²; it can be seen that the bars inertia contribution is very small.

The total inertia in the system's rotation about the y-axis results from the two end beams bending about this axis, as well as from the plate's own mass moment of inertia about that axis. This case is depicted in Figure 3.5, which shows a side view of the plate and the two equivalent point masses (m_e) corresponding to the distributed inertia of the two beams.

The equivalent mass moment of inertia corresponding to the rotation about the y-axis is

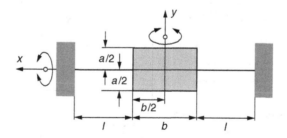

FIGURE 3.4

Mechanical Microsensor with Flexible Segments and Central Rigid Plate.

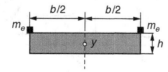

FIGURE 3.5

Inertia Contributions by the End Beams and the Central Plate for the y-Axis Rotation.

$$J_{e,y} = J_{p,y} + 2m_{e,y}\left(\frac{b}{2}\right)^2 = \frac{\rho abh\,(b^2 + h^2)}{12} + 2 \times \frac{13}{35} \times \rho l \times \frac{\pi d^2}{4} \times \frac{b^2}{4}$$

$$= \frac{\rho b}{4}\left[\frac{ah\,(b^2 + h^2)}{3} + \frac{13\pi lbd^2}{70}\right] \tag{3.3}$$

where the plate's mass moment of inertia has been calculated again as indicated in Appendix D and the equivalent mass corresponding to the bending of a clamped-guided beam, $m_{e,y}$, is given in Table 3.1. The y-axis rotation inertia without the beams considered is simply

$$J_{e,y}^* = J_{p,y} = \frac{\rho abh\,(b^2 + h^2)}{12} \tag{3.4}$$

By using the numerical values of this problem, the mass moment of inertia of Eqs. (3.3) and (3.4) are $J_{e,y} = 1.28 \times 10^{-16}$ kg-m^2 and $J_{e,y}^* = 1.17 \times 10^{-16}$ kg-m^2; this time, as can be seen from these figures, the inertia contribution by the beams in the total inertia is more significant. ■

Table 3.2 Stiffness Properties of Beams and Bars

Member	Distributed Stiffness	Lumped Stiffness	Equivalent Stiffness
Beams	Cantilever (fixed-free beam)		$k_e = \dfrac{3EI}{l^3}$
	Fixed-guided beam		$k_e = \dfrac{12EI}{l^3}$
	Bridge (fixed-fixed beam)		$k_e = \dfrac{192EI}{l^3}$
Bars	Fixed-free bar		$k_e = \dfrac{GI_t}{l}$
	Fixed-fixed bar		$k_e = \dfrac{2GI_t}{l}$

3.1.2 Spring Elements

For bending beams and torsion bars, equivalent lumped-parameter stiffness properties are available, as summarized next in Table 3.2. The particular equations of these equivalent stiffnesses formulated in the last column of the table are derived in the companion website Chapter 3 using an equivalence process involving the elastic potential (strain) energy. In Table 3.2, E is the *modulus of elasticity* (or *Young's modulus*), G is the *shear modulus*—these are material properties, I is the *area moment of inertia*, and I_t is the *torsion area moment of inertia*; details on these parameters are provided in Appendix D.

Example 3.2

Determine the lumped-parameter stiffness of the MEMS spring shown in Figure 3.6(a) about its midpoint and the x direction. The spring is formed of two serpentine units, one is shown in Figure 3.6(b), each consisting of three flexible beams of circular cross-section with diameter $d = 2$ μm. Known are also $l = 100$ μm and Young's modulus $E = 160$ GPa.

Solution

The three beams making up the serpentine spring unit of Figure 3.6(b) act as springs in series, and because the two serpentine units have the same displacement at their connection point, they behave as springs in parallel. As a consequence, the lumped-parameter spring model is that of Figure 3.6(c). The stiffness of one serpentine unit is

$$\frac{1}{k_{e,s}} = \frac{2}{k_s} + \frac{1}{k_l} \tag{3.5}$$

where k_s and k_l are the stiffnesses of the short and long beams, respectively, calculated according to the fixed-guided beam condition of Table 3.2:

$$k_s = 12\frac{EI_y}{l^3}; \ k_l = 12\frac{EI_y}{(2l)^3} = \frac{3EI_y}{2l^3} \tag{3.6}$$

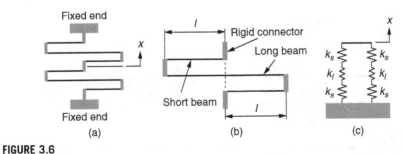

Fixed end

Rigid connector

Long beam

Short beam

Fixed end

(a)   (b)   (c)

FIGURE 3.6

Microspring for Translatory Motion: (a) Spring Configuration; (b) Serpentine Unit; (c) Equivalent Lumped-Parameter Model.

By combining Eqs. (3.5) and (3.6), the stiffness of one serpentine spring is

$$k_{e,s} = \frac{6EI_y}{5l^3} \tag{3.7}$$

The parallel-combination total (equivalent) stiffness of the microspring of Figure 3.6(c) is obtained as

$$k_e = k_{e,p} = 2k_{e,s} = \frac{12EI_y}{5l^3} = \frac{3\pi Ed^4}{80l^3} \tag{3.8}$$

In Eq. (3.8), the area moment of inertia of the circular cross-section has been calculated as $I_y = \pi d^4/64$ (see Appendix D). With the given numerical values of this example, the stiffness is $k_e = 0.3$ N/m.

Example 3.3
The torsional microspring of Figure 3.7(a) can be used in redirecting incoming rays in MEMS applications; it is composed of four flexible bars assembled symmetrically with respect to the midpoint O. The short bars have a diameter $d_1 = 2$ μm and a length $l_1 = 100$ μm whereas the long ones have a diameter $d_2 = 1$ μm and a length $l_2 = 140$ μm. The shear modulus is $G = 110$ GPa. Calculate the torsion moment m_x that has to be applied at O to produce a maximum rotation angle $\theta_{x,max} = 5°$ at the same point.

Solution
The two bars on the left of point O are serially connected, and the same applies for the two mirrored bars on the right of O; the stiffness of each of these two segments with respect to O is

$$k_s = \frac{k_1 k_2}{k_1 + k_2} \tag{3.9}$$

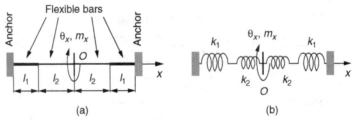

(a)   (b)

FIGURE 3.7

Torsional Microspring for Rotary Motion: (a) Schematic Configuration; (b) Equivalent Lumped-Parameter Model.

Because these two segments are jointed at O, they undergo the same rotation and therefore are coupled in parallel. The lumped-parameter model illustrating these connections is shown in Figure 3.7(b), and the equivalent stiffness of the microspring with respect to the midpoint is

$$k_e = 2k_s = 2\frac{k_1 k_2}{k_1 + k_2} \tag{3.10}$$

Table 3.2 gives the torsional stiffness of the long and short flexible bars as

$$k_1 = \frac{GI_{t1}}{l_1}; \; k_2 = \frac{GI_{t2}}{l_2} \tag{3.11}$$

where the torsion-related area moments of inertia of the two bars are (see Appendix D)

$$I_{t1} = \frac{\pi d_1^4}{32}; \; I_{t2} = \frac{\pi d_2^4}{32} \tag{3.12}$$

By substituting Eqs. (3.12) and (3.11) in Eq. (3.10), the spring equivalent stiffness becomes

$$k_e = \frac{2GI_{t1}I_{t2}}{I_{t1}l_2 + I_{t2}l_1} = \frac{\pi G d_1^4 d_2^4}{16(l_1 d_2^4 + l_2 d_1^4)} \tag{3.13}$$

The maximum torque to be applied at the midpoint O is

$$m_{x,\max} = k_e \theta_{x,\max} = \frac{\pi G d_1^4 d_2^4 \theta_{x,\max}}{16(l_1 d_2^4 + l_2 d_1^4)} \tag{3.14}$$

With the numerical values of this example, the equivalent stiffness is $k_e = 1.48 \times 10^{-10}$ N-m and the maximum moment is $m_{x,\max} = 1.29 \times 10^{-11}$ N-m (or 12.9 µN-µm). ■

3.2 NATURAL RESPONSE OF COMPLIANT SINGLE DEGREE-OF-FREEDOM MECHANICAL SYSTEMS

The natural response of compliant mechanical systems that can be modeled as single-DOF lumped-parameter systems is studied here based on a couple of examples. One particular example studies the detection of minute amounts of mass that deposits on micro and nano flexible members and alter their natural frequencies.

■

Example 3.4
The MEMS accelerometer of Figure 3.8(a), to sense mechanical motion, is formed of two pairs of beams and a shuttle mass. Determine the corresponding lumped-parameter physical model corresponding to the x direction taking into account the inertia contributions

from the long beams in addition to the shuttle mass. All beams have square cross-sections with a side length $a = 3$ µm. Known also are $m = 8 \times 10^{-11}$ kg, $l_1 = 220$ µm, $l_2 = 80$ µm, Young's modulus $E = 160$ GPa, and mass density $\rho = 5200$ kg/m³.

Solution

One long beam and the corresponding short beam on each part of the symmetry (dashed) line in Figure 3.8(a) are coupled in series. The resulting pairs are connected in parallel, as illustrated in Figure 3.8(b). As a consequence, the equivalent stiffness of the microdevice of Figure 3.8(a) is

$$k_e = 2\frac{k_1 k_2}{k_1 + k_2} \tag{3.15}$$

The beams are fixed-guided, and therefore, as seen in Table 3.2, the stiffnesses k_1 and k_2 are

$$k_1 = \frac{12EI}{l_1^3}; \; k_2 = \frac{12EI}{l_2^3} \tag{3.16}$$

where I is the cross-sectional area moment of inertia, $I = a^4/12$. By substituting the stiffnesses of Eqs. (3.16) together with the particular moment of inertia, the equivalent stiffness becomes

$$k_e = \frac{24EI}{l_1^3 + l_2^3} = \frac{2Ea^4}{l_1^3 + l_2^3} \tag{3.17}$$

The equivalent mass m_e is

$$m_e = m + 2m_{b,e} \tag{3.18}$$

where $m_{b,e}$ is the bending-equivalent mass of a long beam, which is provided in Table 3.1 as

$$m_{b,e} = \frac{13}{35}m_b = \frac{13}{35}\rho a^2 l_1 \tag{3.19}$$

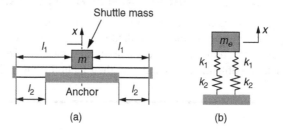

(a)   (b)

FIGURE 3.8

Spring-Mass Microaccelerometer for Translatory Motion: (a) Schematic Representation; (b) Lumped-Parameter Model.

and the equivalent mass of Eq. (3.18) becomes

$$m_e = m + \frac{26}{35}\rho a^2 l_1 \tag{3.20}$$

The natural frequency is therefore expressed by means of Eqs. (3.17) and (3.20) as

$$\omega_n = \sqrt{\frac{k_e}{m_e}} = a^2 \sqrt{\frac{2E}{(l_1^3 + l_2^3)\left(m + \frac{26}{35}\rho a^2 l_1\right)}} \tag{3.21}$$

The natural frequency that does not include inertia contributions from the beams is

$$\omega_n^* = \sqrt{\frac{k_e}{m}} = a^2 \sqrt{\frac{2E}{m(l_1^3 + l_2^3)}} \tag{3.22}$$

By using the numerical values of this example, the natural frequencies of Eqs. (3.21) and (3.22) are $\omega_n = 162{,}780$ rad/s, $\omega_n^* = 170{,}390$ rad/s. The relative error that occurs by not considering the inertia contribution from the long beam is therefore $(\omega_n^* - \omega_n)/\omega_n^* = 0.045$ or 4.5%. It is of interest to note that the bending-equivalent mass is $2m_{b,e} = 7.65 \times 10^{-12}$ kg, which is one order of magnitude smaller than the shuttle mass. ■

Mass Detection by the Natural Frequency Shift Method in MEMS

An interesting application in micro- and particularly nano-systems regards the possibility of detecting very small amounts of substance that deposits on a cantilever or bridge and changes its natural frequencies in bending or torsion.

Comparison between the original and altered natural frequencies enables detecting the quantity of deposited mass. More details on how this method can physically be implemented using the frequency response are offered in Chapters 9 and 10. Suppose, for instance, that a particle lands on a cantilever, such as the one sketched in Figure 3.9(a). The corresponding lumped-parameter model of this system with

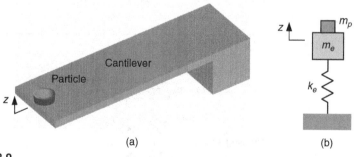

(a)   (b)

FIGURE 3.9

Cantilever with Added Particle: (a) Three-Dimensional View; (b) Equivalent Lumped-Parameter Model.

respect to out-of-plane bending consists of a spring and mass contributions from the original cantilever, plus the added mass, as modeled in Figure 3.9(b).

Example 3.5

A particle of unknown mass attaches to the paddle portion of a nanobridge as shown in Figure 3.10. The out-of-plane bending and torsional natural and modified frequencies are monitored experimentally. Determine the mass of the attached particle, as well as its position parameter b. Consider that the paddle is rigid and parallelipipedic having a thickness h; the nanowires are circular of diameter d. Known are $l_1 = 100$ μm, $h = d = 20$ μm, $l = 200$ μm, $w = 100$ μm, $E = 180$ GPa, $G = 140$ GPa, and $\rho = 5000$ kg/m³. Through mass addition, the altered bending frequency is 4.1139×10^6 rad/s and the altered torsional frequency is 5.0222×10^6 rad/s. Ignore inertia contributions from the end flexible segments.

Solution

Figure 3.11 shows the equivalent lumped-parameter model of the bridge-particle system undergoing out-of-plane bending vibrations. The equivalent mass consists of the plate mass m and the particle mass m_p. The two springs represent the stiffness contributions of the two nanowires. The original bending natural frequency is

$$\omega_{n,b} = \sqrt{\frac{2k_b}{m}} \qquad (3.23)$$

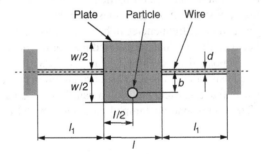

FIGURE 3.10

Top View of Paddle Nanobridge with Attached Particle.

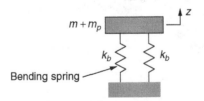

FIGURE 3.11

Side View of Equivalent Lumped-Parameter Model in Bending.

where k_b is given in Table 3.2 for a fixed-guided beam with $I_y = \pi d^4/64$, and the mass is $m = \rho l w h$. Numerically, the natural bending frequency is $\omega_{n,b} = 4.1188 \times 10^6$ rad/s. The modified bending frequency is

$$\omega_{n,b}^* = \sqrt{\frac{2k_b}{m + m_p}} \qquad (3.24)$$

Combining Eqs. (3.23) and (3.24) yields the mass of the added particle as

$$m_p = \frac{\left(\omega_{n,b}\right)^2 - \left(\omega_{n,b}^*\right)^2}{\left(\omega_{n,b}^*\right)^2} m \qquad (3.25)$$

with a numerical value of $m_p = 4.78 \times 10^{-12}$ kg.

Torsion, whose vibrations are also monitored experimentally, can be modeled by the equivalent lumped-parameter model of Figure 3.12. By using a reasoning similar to the one applied in bending, the mechanical moment of inertia expressing the torsion contribution of the added particle, J_p, is related to the plate's mechanical moment of inertia J as

$$J_p = \frac{\left(\omega_{n,t}\right)^2 - \left(\omega_{n,t}^*\right)^2}{\left(\omega_{n,t}^*\right)^2} J = \frac{\left(\omega_{n,t}\right)^2 - \left(\omega_{n,t}^*\right)^2}{\left(\omega_{n,t}^*\right)^2} \times \frac{m}{12} \times (w^2 + h^2) \qquad (3.26)$$

where

$$\omega_{n,t} = \sqrt{\frac{2k_t}{J}} \qquad (3.27)$$

The spring constant is given in Table 3.2 with $I_t = \pi d^4/32$. The numerical value of the natural frequency in torsion is $\omega_{n,t} = 5.0373 \times 10^6$ rad/s. The mechanical moment of inertia of the particle with respect to the rotation axis is

$$J_p = m_p b^2 \qquad (3.28)$$

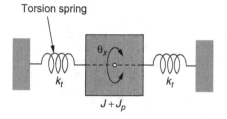

Torsion spring

θ_x

k_t k_t

$J + J_p$

FIGURE 3.12

Equivalent Lumped-Parameter Model in Torsion.

Combining Eqs. (3.26) and (3.28) results in

$$b = \frac{1}{2}\sqrt{\frac{\rho whl\,(w^2 + h^2)}{3m_p} \times \left[\frac{(\omega_{n,t})^2}{(\omega_{n,t}^*)^2} - 1\right]} \qquad (3.29)$$

The numerical value of the offset is therefore $b = 46.75\ \mu m$, which indicates the particle deposited almost at the edge of the middle plate (whose half width is of 50 μm). ■

3.3 MULTIPLE DEGREE-OF-FREEDOM MECHANICAL SYSTEMS

Systems defined by more than one variable are known as *multiple degree-of-freedom systems*. The notion of degrees of freedom is discussed followed by derivation of mathematical models for conservative mechanical systems by means of the energy method and for nonconservative ones using Newton's second law of motion. The natural and forced responses of multiple-DOF mechanical systems (including compliant ones) are further investigated employing MATLAB® and Simulink®.

3.3.1 Configuration, Degrees of Freedom

The notion of degrees of freedom is extremely important in correctly modeling dynamic systems. In general, the DOFs represent the *minimum number of independent parameters that fully define the configuration of a system*. Consider for instance a simple system made up of a mass with a spring and a damper in parallel, such as the one of Figure 3.13. The configuration of this system is fully defined by the parameter x, which locates the mass m at all times. As a consequence, this system has one DOF (it is called a *single-DOF system*).

A particle in three-dimensional space is defined by three parameters—its coordinates x, y, and z—therefore, this system has three DOFs. Similarly, a system of two particles possesses $2 \times 3 = 6$ DOFs. However, if the two particles are connected by means of a segment whose length is l, the following constraint does apply:

$$l = \sqrt{(x_1 - x_2)^2 + (y_1 - y_2)^2 + (z_1 - z_2)^2} \qquad (3.30)$$

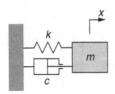

FIGURE 3.13

Single-DOF Mechanical System.

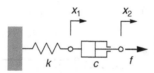

FIGURE 3.14

Two-DOF Mechanical System with Series Spring-Damper Connection.

so that, for this system, the number of independent parameters fully defining its configuration is $6 - 1 = 5$, and therefore the system is a five-DOF one. This remark can be generalized by concluding that *the number of DOFs of a system is equal to the number of apparent (candidate) DOF parameters minus the number of constraints relating those apparent DOF parameters.*

Figure 3.14 shows a system composed of a spring and a damper connected in series; the system is a two-DOF one, as it needs two parameters to define its configuration at any moment in time (the displacements x_1 and x_2 that are measured from the fixed positions shown in the figure).

Example 3.6

Establish the DOFs of the mechanical system sketched in Figure 3.15, which is formed of a two-cylinder pulley rolling without slippage on a horizontal surface, a lever-rod acting on the inner pulley cylinder, another rod and mass rigidly acting on the outer pulley cylinder, and a horizontal spring attached to the pulley's center. Assume small displacements.

Solution

The apparent coordinates that describe the motion of the mechanical system of Figure 3.15 are the rotation angles φ (vertical lever) and θ (pulley) and the linear motions x_1 (pulley center), x_2 (lever-rod connection point), and y (mass), which indicates there are five apparent DOFs. The following constraints can be written between these coordinates under the assumption of small motions and based on the fact that various points on the wheel undergo pure rotations with respect to the instant center of rotation (the contact point between the pulley and the horizontal surface):

$$\begin{cases} x_2 = l\varphi \\ x_2 = (R_1 + R_2)\theta \\ x_1 = R_1\theta \\ y = 2R_1\theta \end{cases} \qquad (3.31)$$

Because the difference between the apparent number of DOFs (five) and the number of constraints (four) is 1, this system is a single-DOF system. ■

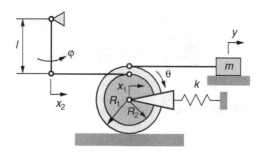

FIGURE 3.15

Pulley-Lever-Rod-Body Mechanical System.

3.3.2 **Conservative Mechanical Systems**

This section introduces the energy method to derive mathematical models for multiple-DOF conservative (free undamped) mechanical systems. The resulting mathematical models are further utilized to calculate the *natural frequencies* and qualify the corresponding *modal motions* either analytically or using MATLAB®.

Mathematical Modeling by the Energy Method

Mathematical models for conservative mechanical systems can also be derived by means of *energy methods* based on *kinetic* and *potential energies* (as shown here in a direct approach and for conservative systems but also in the companion website Chapter 3, where *Lagrange's equations method* is introduced). Energy methods are particularly advantageous to apply to mechanical systems formed of several bodies where all kinetic energy contributions are collected in one total kinetic energy and all potential energy fractions are summed in one potential energy; the result is one equation stating that the total energy is constant. On the other hand, application of Newton's second law of motion in such cases appears to be somewhat more difficult, as it implies formulating a number of equations equal to the number of DOFs and also using the reactions resulting from adjacent DOFs—this method is applied exclusively to study the forced response of multiple-DOF mechanical systems.

In the case of bodies changing their vertical position and when the gravitational effect is taken into account, the *gravitational potential energy* needs to be considered. A body of mass m situated above a datum line (an arbitrary reference line) at a distance h, see Figure 3.16, has a gravitational potential energy of

$$U_g = mgh \tag{3.32}$$

It is known from dynamics that, for systems acted upon only by elastic-type and gravitational forces, the total energy conserves. For such systems, which are called *conservative*, the total energy is

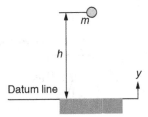

FIGURE 3.16

Body in a Gravitational Field.

$$E = T + U_e + U_g \tag{3.33}$$

where T stands for kinetic energy, U_e is the elastic (spring-type) potential energy, and U_g is the gravitational potential energy. A direct consequence of Eq. (3.33) is that the time derivative of the total energy for a conservative system is zero:

$$\frac{dE}{dt} = \frac{d}{dt}(T + U_e + U_g) = 0 \tag{3.34}$$

Equation (3.34), as shown next, provides the mathematical model of multiple-DOF mechanical systems.

Example 3.7

Derive the mathematical model for the cart-pendulum shown in Figure 3.17 using the energy method under the assumption of small motions. Known are the masses of the cart, m_1, and of the bob, m_2, the spring stiffness, k, and the pendulum length, l. The system is in a vertical plane and the cart moves horizontally.

Solution

The system has two DOFs: the linear motion coordinate x and the angular coordinate θ. The total (absolute) velocity v of the bob has two components: the tangential (horizontal) velocity of the cart v_t and the one resulting from the pendulum rotation about the cart, denoted v_r and perpendicular to the pendulum rod, as illustrated in Figure 3.17.

The kinetic energy of the mechanical system is

$$T = \frac{1}{2}m_1\dot{x}^2 + \frac{1}{2}m_2 v^2 \tag{3.35}$$

According to the cosine theorem, the total bob velocity is

$$v^2 = v_t^2 + v_r^2 - 2v_t v_r \cos(\pi - \theta) = v_t^2 + v_r^2 + 2v_t v_r \cos\theta \tag{3.36}$$

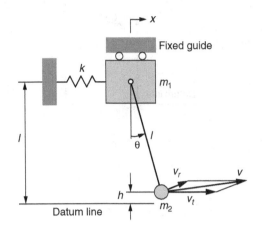

FIGURE 3.17

Cart-Pendulum Mechanical System.

It is also known that

$$\begin{cases} v_t = \dot{x} \\ v_r = l\dot{\theta} \end{cases} \tag{3.37}$$

Equations (3.36) and (3.37) are now combined and the result is substituted into the kinetic energy of Eq. (3.35). Considering small angular displacements (which lead to the approximation $\cos\theta \cong 1 - \theta^2/2$) result in the following expression of the kinetic energy (where the term in θ^2, which is very small, has been neglected):

$$T = \frac{1}{2}\left[(m_1 + m_2)\dot{x}^2 + 2m_2 l\dot{x}\dot{\theta} + m_2 l^2\dot{\theta}^2\right] \tag{3.38}$$

The total potential energy is produced by elastic and gravitational contributions:

$$U = U_e + U_g = \frac{1}{2}kx^2 + m_2 gl(1 - \cos\theta) \approx \frac{1}{2}kx^2 + \frac{1}{2}m_2 gl\theta^2 \tag{3.39}$$

The total energy E of the mechanical system is determined by summing the kinetic energy component of Eq. (3.38) and the potential energy of Eq. (3.39). The mechanical system of Figure 3.17 is conservative; therefore, the time derivative of the total energy is zero, which yields

$$\frac{dE}{dt} = \frac{d}{dt}(T + U)$$

$$= \left[(m_1 + m_2)\ddot{x} + m_2 l\ddot{\theta} + kx\right]\dot{x} + \left[m_2 l\ddot{x} + m_2 l^2\ddot{\theta} + m_2 gl\theta\right]\dot{\theta} \tag{3.40}$$

Because the velocities $\dot{x}$, $\dot{\theta}$ cannot be zero at all times, Eq. (3.40) is valid when

$$\begin{cases} (m_1 + m_2)\ddot{x} + m_2 l\ddot{\theta} + kx = 0 \\ \ddot{x} + l\ddot{\theta} + g\theta = 0 \end{cases} \tag{3.41}$$

The two Eqs. (3.41) form the mathematical model of the cart-pendulum mechanical system. ■

Natural Response and the Modal Problem

As introduced in Chapter 2, the natural response of single-DOF mechanical systems is concerned with determining the *natural frequencies* (which are the frequencies of free undamped mechanical vibrations). For *multiple-DOF mechanical systems*, the number of *natural frequencies* is equal to the number of DOFs, except for the repeated natural frequencies. At each natural frequency, the ensemble of bodies vibrates at that particular frequency and the resulting system motion is a combination of individual motions performed by system components. The overall system motion corresponding to a natural frequency is called *modal motion* or *mode shape*. Unless initial conditions are specified, the absolute motion amplitudes of each component are not known; only their relative ratios can be determined precisely. Discussed next are the analytical and the MATLAB® modal approaches.

Analytical Approach

The following example takes you through the main steps involved in formulating and solving a *modal problem* (which consists in calculating the natural frequencies and characterizing the corresponding modal motions).

■───

Example 3.8

The mechanical system of Figure 3.18 consists of two rigid disks and two torsional springs (which are modeling two elastic shafts). Find its natural frequencies and determine the corresponding modes for $k_1 = 884$ N-m, $k_2 = k_1/2$, $J_1 = 0.002$ kg-m², $J_2 = J_1/2$.

Solution

The system is a two-DOF one because the two angular parameters θ_1 and θ_2, which indicate the absolute angular positions of the disks, fully define the configuration of the system.

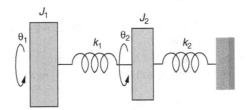

FIGURE 3.18

Rotary Mechanical System with Two Disks and Two Torsional Springs.

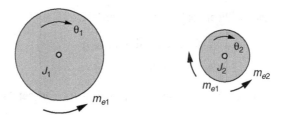

FIGURE 3.19

Free-Body Diagrams of the Two-Disk Mechanical System of Figure 3.18.

The energy method is applied to derive the mathematical model of this mechanical system. The sum of kinetic and potential energy of the mechanical system is

$$T + U = \frac{1}{2}J_1\dot{\theta}_1^2 + \frac{1}{2}J_2\dot{\theta}_2^2 + \frac{1}{2}k_1(\theta_1 - \theta_2)^2 + \frac{1}{2}k_2\theta_2^2 \tag{3.42}$$

The mechanical system is conservative and therefore the time derivative of the total energy of Eq. (3.42) is zero:

$$\dot{\theta}_1[J_1\ddot{\theta}_1 + k_1(\theta_1 - \theta_2)] + \dot{\theta}_2[J_2\ddot{\theta}_2 - k_1(\theta_1 - \theta_2) + k_2\theta_2] = 0 \tag{3.43}$$

Because the angular velocities of the two discs cannot be zero at all times, Eq. (3.43) is valid only when the two brackets are zero, and this results in

$$\begin{cases} J_1\ddot{\theta}_1 + k_1\theta_1 - k_1\theta_2 = 0 \\ J_2\ddot{\theta}_2 - k_1\theta_1 + (k_1 + k_2)\theta_2 = 0 \end{cases} \tag{3.44}$$

which form the mathematical model of the two-DOF mechanical system of Figure 3.18.

Newton's second law can also be applied and written for the two disks undergoing rotary motion, based on the free-body diagrams of Figure 3.19.

The two dynamic equations are

$$\begin{cases} J_1\ddot{\theta}_1 = -m_{e1} \\ J_2\ddot{\theta}_2 = m_{e1} - m_{e2} \end{cases} \tag{3.45}$$

The elastic torques are expressed as

$$\begin{cases} m_{e1} = k_1(\theta_1 - \theta_2) \\ m_{e2} = k_2\theta_2 \end{cases} \tag{3.46}$$

Substituting Eqs. (3.46) into Eqs. (3.45) results in Eqs. (3.44), which have already been obtained using the energy method.

As shown in the section dedicated to the free undamped vibrations of single-DOF systems in Chapter 2, the solution θ_1 and θ_2 of Eqs. (3.44) needs to be harmonic:

$$\begin{cases} \theta_1 = \Theta_1\sin(\omega t) \\ \theta_2 = \Theta_2\sin(\omega t) \end{cases} \tag{3.47}$$

Equations (3.44) and (3.47) are combined, which results in

$$\begin{cases} \left[(-\omega^2 J_1 + k_1)\Theta_1 - k_1\Theta_2\right]\sin(\omega t) = 0 \\ \left[-k_1\Theta_1 + (-\omega^2 J_2 + k_1 + k_2)\Theta_2\right]\sin(\omega t) = 0 \end{cases} \tag{3.48}$$

Equations (3.48) could be satisfied at all times only when

$$\begin{cases} (-\omega^2 J_1 + k_1)\Theta_1 - k_1\Theta_2 = 0 \\ -k_1\Theta_1 + (-\omega^2 J_2 + k_1 + k_2)\Theta_2 = 0 \end{cases} \tag{3.49}$$

The unknowns in the algebraic Eqs. (3.49) are the amplitudes Θ_1 and Θ_2, and they are nonzero when

$$\begin{vmatrix} -\omega^2 J_1 + k_1 & -k_1 \\ -k_1 & -\omega^2 J_2 + k_1 + k_2 \end{vmatrix} = 0 \tag{3.50}$$

Equation (3.50), known as *characteristic equation* (or *frequency equation*), can be solved for the *natural frequencies*. Expanding the determinant of Eq. (3.50) results in the algebraic equation

$$J_1 J_2 \omega^4 - [J_1(k_1 + k_2) + J_2 k_1]\omega^2 + k_1 k_2 = 0 \tag{3.51}$$

For the particular parameters of the example, the solution to Eq. (3.51) is

$$\begin{cases} \omega_{n1} = \sqrt{(2 - \sqrt{3})\dfrac{k_2}{J_2}} = 344.14 \text{ rad/s} \\ \omega_{n2} = \sqrt{(2 + \sqrt{3})\dfrac{k_2}{J_2}} = 1284.4 \text{ rad/s} \end{cases} \tag{3.52}$$

It can be seen that $\omega_{n2}/\omega_{n1} = 3.732$; therefore the second natural frequency is almost four times larger than the first one.

As mentioned at the beginning of this section, for each of the two natural frequencies of Eq. (3.52), the two disks undergo rotary vibrations at that frequency. The relative directions of motion and amplitudes can be determined by analyzing the ratio Θ_1/Θ_2 from either of the Eqs. (3.49):

$$\begin{cases} r_1 = \dfrac{\Theta_1}{\Theta_2}\bigg|_{\omega=\omega_{n1}} = \dfrac{\Theta_{11}}{\Theta_{21}} = \dfrac{k_1}{-\omega_{n1}^2 J_1 + k_1} = \dfrac{-\omega_{n1}^2 J_2 + 3k_2}{2k_2} = \dfrac{1}{\sqrt{3}-1} \\ r_2 = \dfrac{\Theta_1}{\Theta_2}\bigg|_{\omega=\omega_{n2}} = \dfrac{\Theta_{12}}{\Theta_{22}} = \dfrac{k_1}{-\omega_{n2}^2 J_1 + k_1} = \dfrac{-\omega_{n2}^2 J_2 + 3k_2}{2k_2} = -\dfrac{1}{\sqrt{3}+1} \end{cases} \tag{3.53}$$

where the natural frequencies of Eqs. (3.52) have been used. The fact that, once the natural frequencies are determined, Eqs. (3.49) yield only the amplitude ratio and not the amplitudes themselves, is a direct consequence of the fact that the amplitudes are dependent, so one can be chosen arbitrarily.

The ratios of Eqs. (3.53) allow qualifying the modal motions of the mechanical system for each natural frequency. The ratio r_1 of the first Eq. (3.53) is a positive number, which suggests that both disks rotate in the same direction (either the positive direction identified in the diagrams of Figure 3.19 or the opposite direction). Because $r_1 > 1$, the amplitude Θ_1 is always larger than amplitude Θ_2 during the system's rotary vibrations at the first natural frequency ω_1. The ratio r_2 indicates the two disks rotate in opposite directions and the amplitude Θ_1 is smaller than the amplitude Θ_2. Since one amplitude can be chosen arbitrarily, the value of 1 rad can be assigned to Θ_2 for each of the two modes, which results in the amplitude of Θ_1 being

$$\begin{cases} \Theta_1 \big|_{\omega=\omega_{n1}} = \dfrac{1}{\sqrt{3}-1} \approx 1.366 \\ \Theta_1 \big|_{\omega=\omega_{n2}} = -\dfrac{1}{\sqrt{3}+1} \approx -0.366 \end{cases} \tag{3.54}$$

As a consequence, the following two vectors (also named *eigenvectors*) can be formulated: $\{V\}_1 = \{1.366 \quad 1\}^t$ for the first natural frequency and $\{V\}_2 = \{-0.366 \quad 1\}^t$ for the second natural frequency. The two disks' vibrations during the first modal motion can therefore be expressed as

$$\begin{cases} \theta_{11} = \dfrac{1}{\sqrt{3}-1} \sin(\omega_{n1} t) \\ \theta_{21} = \sin(\omega_{n1} t) \end{cases} \tag{3.55}$$

whereas during the second modal motion, the two disks' vibrations are described by

$$\begin{cases} \theta_{12} = -\dfrac{1}{\sqrt{3}+1} \sin(\omega_{n2} t) \\ \theta_{22} = \sin(\omega_{n2} t) \end{cases} \tag{3.56}$$

The two modal motions of Eqs. (3.55) and (3.56) are plotted in Figure 3.20, which uses the following MATLAB® code:

```
>> om1=344;
>> om2=1284;
>> t=0:0.00001:0.06;
>> theta11=180/pi/(sqrt(3)-1)*sin(om1*t);
>> theta21=180/pi*sin(om1*t);
>> subplot(1,2,1)
```

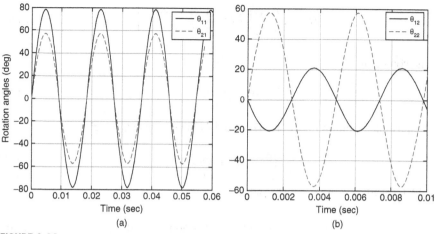

FIGURE 3.20

Modal Vibrations of Two Disks at the (a) First Natural Frequency; (b) Second Natural Frequency.

```
>> plot(t,theta11,t,theta21,'--')
>> xlabel('Time(sec)')
>> ylabel('Rotation angles (deg)')
>> legend('\theta_1_1','\theta_2_1')
>> t=0:0.00001:0.012;
>> theta12=-180/pi/(sqrt(3)+1)*sin(om2*t);
>> theta22=180/pi*sin(om2*t);
>> subplot(1,2,2)
>> plot(t,theta12,t,theta22,'--')
>> xlabel('Time(sec)')
>> legend('\theta_1_2','\theta_2_2')
```

MATLAB® Approach—the Dynamic Matrix and the Eigenvalue Problem

The natural response of multiple-DOF systems can also be formulated as an *eigenvalue problem* (see Appendix B for more details). This topic is studied more thoroughly in disciplines such as vibrations, but provided here is an introduction aiming to facilitate understanding a modeling procedure that utilizes a vector-matrix formulation and also is implemented in MATLAB®. Let us reformulate the example of the two-disk system of Figure 3.18, by pointing out that the dynamic equations of motion, Eqs. (3.44), can be written in vector-matrix form as

$$[M]\{\ddot{\theta}\} + [K]\{\theta\} = \{0\} \tag{3.57}$$

where $[M]$ and $[K]$ are the *mass (inertia)* and *stiffness matrices*, collected from Eqs. (3.44) as

$$[M] = \begin{bmatrix} J_1 & 0 \\ 0 & J_2 \end{bmatrix}; \ [K] = \begin{bmatrix} k_1 & -k_1 \\ -k_1 & k_1 + k_2 \end{bmatrix} \tag{3.58}$$

The coordinate vector is

$$\{\theta\} = \{\theta_1 \ \theta_2\}' \tag{3.59}$$

A sinusoidal solution for Eq. (3.57) of the following form is sought:

$$\{\theta\} = \{\Theta\} \sin(\omega t) \tag{3.60}$$

which substituted in Eq. (3.57) yields

$$(-\omega^2 [M]\{\Theta\} + [K]\{\Theta\}) \sin(\omega t) = \{0\} \tag{3.61}$$

As per a previous discussion, $\sin(\omega t)$ cannot be zero at all times; therefore,

$$-\omega^2 [M]\{\Theta\} + [K]\{\Theta\} = \{0\} \tag{3.62}$$

Left multiplication in Eq. (3.62) by $[M]^{-1}$ and rearranging of the resulting equation produces

$$[M]^{-1} [K]\{\Theta\} = \omega^2 \{\Theta\} \tag{3.63}$$

Equation (3.63) is a typical *eigenvalue formulation* of the type

$$[D]\{\Theta\} = \lambda \{\Theta\} \tag{3.64}$$

The matrix $[D] = [M]^{-1}[K]$ is the *dynamic matrix* and $\lambda = \omega^2$ is the *eigenvalue*. Equation (3.64) further leads to

$$([D] - \lambda [I])\{\Theta\} = \{0\} \tag{3.65}$$

where [I] is the identity matrix. Equation (3.65) can have a nontrivial solution only when

$$\det ([D] - \lambda [I]) = 0 \tag{3.66}$$

and this is another form of the *characteristic equation* given for this particular case in Eq. (3.50). It can be checked that Eq. (3.66) results in Eq. (3.50) when $\lambda = \omega^2$.

MATLAB® has the command eig(D), which returns the eigenvalues of a nonsingular square matrix. The MATLAB® command [V, D] = eig(D) returns the *modal matrix V*, where the columns are the *eigenvectors* (each containing the relative vibration amplitudes), and the diagonal matrix *D* with the eigenvalues on the main diagonal. Let us check that, indeed, the natural frequencies determined analytically in the previous example can also be found by using MATLAB®.

Example 3.9

Calculate the dynamic matrix corresponding to Example 3.8, then determine their eigenvalues (with the related natural frequencies) and eigenvectors using MATLAB®.

Solution

Equations (3.58) provide the inertia and stiffness matrices of Example 3.8. The following MATLAB® code can be used to determine the natural frequencies and the eigenvectors:

```
>> j1 = 0.002;
>> j2 = 0.001;
>> k1 = 884;
>> k2 = 442
>> in = [j1,0;0,j2];
>> stiff = [k1,-k1;-k1,k1+k2];
>> d = inv(in)*stiff;
>> [V,D] = eig(d)
V =
  -0.8069 0.3437
  -0.5907 -0.9391
D =
  1.0e+006 *
    0.1184           0
         0     1.6496
```

It can easily be checked that the eigenvalues returned by MATLAB® are the squares of the natural frequencies obtained in Example 3.8, namely, $\omega_{n1} = 344$ rad/s and $\omega_{n2} = 1284$ rad/s. The columns of V (the eigenvectors) appear different from the ones obtained in Example 3.8, although the ratio of each vector's components are those of Eqs. (3.53). It is known however that the eigenvectors are not unique, because one of the eigenvector's components is chosen arbitrarily. MATLAB® defines the eigenvectors by using a norm of 1; for instance, $\sqrt{(-0.8069)^2 + (-0.5907)^2} = 1$ and $\sqrt{(0.3427)^2 + (-0.9391)^2} = 1$ for this example. Such eigenvectors are known as *unit-norm* (or *normalized*) *eigenvectors.* ■

3.3.3 Forced Response with Simulink®

Simulink® was introduced in Chapter 2 and is used here as a graphical tool of modeling and solving systems of ordinary differential equations that correspond to the forced response of multiple-DOF mechanical systems. The mathematical models are derived by using Newton's second law of motion.

Example 3.10

The mechanical filter microsystem of Figure 3.21(a) is driven by a force $f = 2e^{-0.005t} \mu N$. Find a lumped-parameter physical model of the original system, and plot the system response consisting of the time-domain displacements of the two shuttle masses for $m_1 = 3.6 \times 10^{-10}$ kg, $m_2 = 2.5 \times 10^{-10}$ kg, $l_1 = 160$ μm, $l_2 = 100$ μm, $l = 80$ μm (see Figure 3.6(b) for serpentine spring dimensions), $d_1 = d_2 = d = 2$ μm (circular cross-section diameter of all flexible members), and $E = 165$ GPa. Consider zero initial conditions.

Solution

The beam springs act as two parallel springs on each of the shuttle masses of Figure 3.21(a), and the serpentine spring realizes the elastic coupling between m_1 and m_2. The lumped-parameter physical model equivalent to the actual MEMS device is shown in Figure 3.21(b). This system has two DOFs, the displacements x_1 and x_2.

Based on the free-body diagrams of Figure 3.22 and applying Newton's second law of motion, the mathematical model corresponding to the lumped-parameter model of Figure 3.21(b) is

$$\begin{cases} m_1 \ddot{x}_1 = f - 2f_{e1} - f_e \\ m_2 \ddot{x}_2 = f_e - 2f_{e2} \end{cases} \tag{3.67}$$

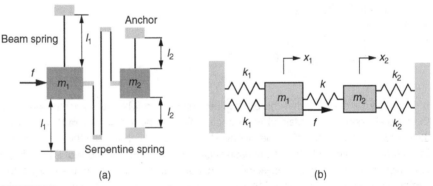

(a)   (b)

FIGURE 3.21

Linear-Motion Mechanical MEMS Device: (a) Schematic of MEMS; (b) Lumped-Parameter Model.

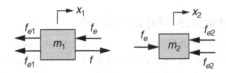

FIGURE 3.22

Shuttle Masses Free-Body Diagrams.

By taking into account that

$$f_{e1} = k_1 x_1; \quad f_e = k(x_1 - x_2); \quad f_{e2} = k_2 x_2 \tag{3.68}$$

Eqs. (3.67) are rewritten in the form

$$\begin{cases} \ddot{x}_1 = -\dfrac{2k_1 + k}{m_1}x_1 + \dfrac{k}{m_1}x_2 + \dfrac{1}{m_1}f \\ \ddot{x}_2 = \dfrac{k}{m_2}x_1 - \dfrac{2k_2 + k}{m_2}x_2 \end{cases} \tag{3.69}$$

The stiffnesses k_1 and k_2 are given in Table 3.2 for fixed-guided beams, whereas the stiffness of the serpentine spring is expressed in Eq. (3.7); and based on Figure 3.6, they are

$$k_1 = \frac{12EI}{l_1^3}; \quad k_2 = \frac{12EI}{l_2^3}; \quad k = \frac{6EI}{5l^3} \tag{3.70}$$

where l is the length of the short leg of the serpentine spring and I is the circular cross-sectional area moment of inertia, $I = \pi d^4/64$. Equations (3.69) can be written as

$$\begin{cases} \ddot{x}_1 = -a_{11}x_1 + a_{12}x_2 + \dfrac{1}{m_1}f \\ \ddot{x}_2 = a_{21}x_1 - a_{22}x_2 \end{cases} \tag{3.71}$$

With the numerical values of this example, the parameters of Eqs. (3.70) and (3.71) are $k_1 = 0.38$ N/m, $k_2 = 1.55$ N/m, $k = 0.304$ N/m, $a_{11} = 2.9529 \times 10^9$, $a_{12} = 8.4369 \times 10^8$, $a_{21} = 1.2149 \times 10^9$, $a_{22} = 1.3656 \times 10^{10}$, $1/m_1 = 2.7778 \times 10^9$ kg^{-1}. The block diagram of Figure 3.23 integrates Eqs. (3.71).

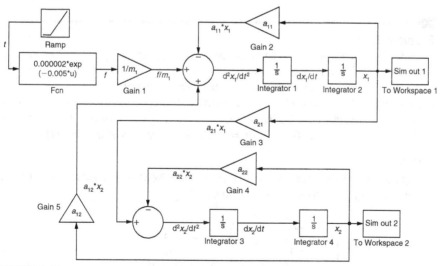

FIGURE 3.23

Simulink® Block Diagram for Solving the Differential Eqs. (3.71).

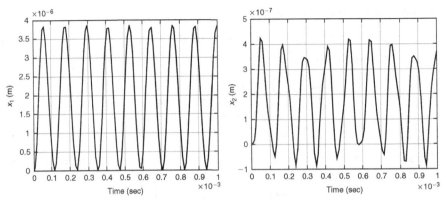

FIGURE 3.24

Time Variation of the Shuttle Masses Displacements (in Meters) for the Mechanical System of Figure 3.23.

The input force on the mass m_1 is produced by combining the Ramp block from the Source library and the Math Function from the Math Operations library. While the unit ramp provides the time input (which is interpreted as a variable u), the following Expression needs to be specified for the Math Function: 0.000002*exp(-0.005*u), which constitutes $f(t)$ of the example. The remaining blocks have been introduced to solve differential equations of single-DOF mechanical systems in Chapter 2. As can be seen in Figure 3.23, the stiffness coupling (which is realized through k) is reflected in the cross-connections between x_1 and x_2. Figure 3.24 contains the plots displaying x_1 and x_2 as functions of time after they have been exported to MATLAB®. ▇

Example 3.11

a. Derive the mathematical model of the gear-train mechanical system of Figure 3.25, where an actuation torque m_a drives the input shaft and a load torque m_l acts on the output shaft. Assume the rotary inertia of the gears is negligible.

b. Use Simulink® to generate the corresponding simulation diagram and plot $\omega_2(t)$, $\omega(t)$, $\omega_3(t)$ for $N_1 = 44$, $N_2 = 38$, $N_3 = 38$, $N_4 = 24$, $J = 0.002$ kg-m², $c_1 = c = c_3 = 240$ N-m-s, $k_1 = k_2 = 3500$ N-m, $m_a = 10000$ N-m, and $m_l = 200 \sin(10t)$ N-m.

Solution

a. The equations of motion are formulated by means of Newton's second law of motion for the three disks placed on the intermediate shaft:

$$\begin{cases} 0 = \dfrac{N_2}{N_1}m_a - \left(\dfrac{N_2}{N_1}\right)^2 c_1\dot{\theta}_2 - k_1(\theta_2 - \theta) \\[2mm] J\ddot{\theta} = -c\dot{\theta} - k_1(\theta - \theta_2) - k_2(\theta - \theta_3) \\[2mm] 0 = -\dfrac{N_3}{N_4}m_l - \left(\dfrac{N_3}{N_4}\right)^2 c_3\dot{\theta}_3 - k_2(\theta_3 - \theta) \end{cases} \qquad (3.72)$$

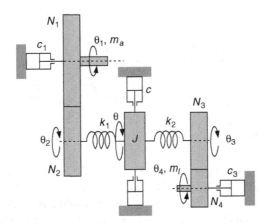

FIGURE 3.25

Gear-Shaft Rotary Mechanical System.

In the first and third Eqs. (3.72), the driving and load torques, m_a and m_l, have been transferred from their original shafts to the middle shaft on the corresponding gears; the same procedure has been applied for the damping coefficients c_1 and c_3. Equations (3.72) can also be written in vector-matrix form as

$$
\begin{bmatrix} 0 & 0 & 0 \\ 0 & J & 0 \\ 0 & 0 & 0 \end{bmatrix} \begin{Bmatrix} \ddot{\theta}_2 \\ \ddot{\theta} \\ \ddot{\theta}_3 \end{Bmatrix} + \begin{bmatrix} \left(\dfrac{N_2}{N_1}\right)^2 c_1 & 0 & 0 \\ 0 & c & 0 \\ 0 & 0 & \left(\dfrac{N_3}{N_4}\right)^2 c_3 \end{bmatrix} \begin{Bmatrix} \dot{\theta}_2 \\ \dot{\theta} \\ \dot{\theta}_3 \end{Bmatrix}
$$

$$
+ \begin{bmatrix} k_1 & -k_1 & 0 \\ -k_1 & k_1 + k_2 & -k_2 \\ 0 & -k_2 & k_2 \end{bmatrix} \begin{Bmatrix} \theta_2 \\ \theta \\ \theta_3 \end{Bmatrix} = \begin{Bmatrix} \dfrac{N_2}{N_1} m_a \\ 0 \\ -\dfrac{N_3}{N_4} m_l \end{Bmatrix} \tag{3.73}
$$

The first matrix on the left-hand side of Eq. (3.73) is the *inertia matrix*, the next one is the *damping matrix*, and the last one is the *stiffness matrix*. The vector on the right-hand side is the *load vector*. Similar formulations are used in subsequent chapters for other modeling procedures, such as transfer function (Chapter 7) and state space modeling (Chapter 8).

b. Equations (3.72) can also be reformulated in a way that would render use of Simulink® applicable:

$$
\begin{cases} \dot{\theta}_2 = a_{11}\theta - a_{11}\theta_2 + a_{12}m_a \\ \ddot{\theta} = -b_{11}\dot{\theta} - b_{12}\theta + b_{13}\theta_2 + b_{14}\theta_3 \\ \dot{\theta}_3 = c_{11}\theta - c_{11}\theta_3 - c_{12}m_l \end{cases} \tag{3.74}
$$

with

$$\begin{cases} a_{11} = \dfrac{k_1}{c_1}\left(\dfrac{N_1}{N_2}\right)^2; \quad a_{12} = \dfrac{1}{c_1} \times \dfrac{N_1}{N_2} \\[4mm] b_{11} = \dfrac{c}{J}; \quad b_{12} = \dfrac{k_1 + k_2}{J}; \quad b_{13} = \dfrac{k_1}{J}; \quad b_{14} = \dfrac{k_2}{J} \\[4mm] c_{11} = \dfrac{k_2}{c_3}\left(\dfrac{N_4}{N_3}\right)^2; \quad c_{12} = \dfrac{1}{c_3} \times \dfrac{N_4}{N_3} \end{cases} \qquad (3.75)$$

The numerical values of the coefficients given in Eq. (3.75) are $a_{11} = 19.55$, $a_{12} = 0.0048$, $b_{11} = 120{,}000$, $b_{12} = 3{,}500{,}000$, $b_{13} = b_{14} = 1{,}750{,}000$, $c_{11} = 5.82$, $c_{12} = 0.0026$. The Simulink® diagram corresponding to the differential Eqs. (3.75) is shown in Figure 3.26. Each of the three differential equations is represented by a horizontal chain plus the connections between chains because of the coupling. The input to the first chain (differential equation) is a $\texttt{Step}$ block, which can be dragged from the $\texttt{Source}$ library and simply configured by specifying the final value of 250. The signals' names have not been specified in the block diagram to keep the figure legible.

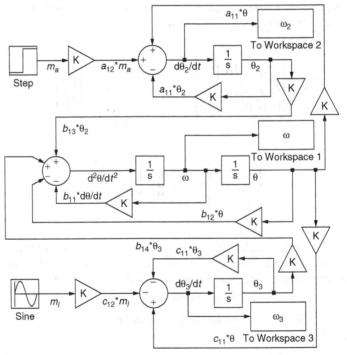

FIGURE 3.26

Simulink® Diagram for Solving the Differential Eqs. (3.74) and Plotting the Time-Response of a Gear-Shaft Rotary Mechanical System.

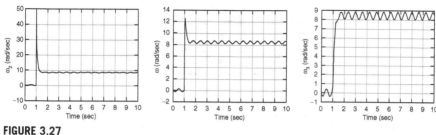

FIGURE 3.27

Time Variation of the Angular Velocities for the Mechanical System of Figure 3.25.

The three outputs are exported and plotted into MATLAB® as shown in the plots of Figure 3.27.

SUMMARY

This chapter utilizes the notions introduced in Chapter 2 for basic mechanical elements and systems to study compliant and multiple-DOF mechanical systems. Equivalent inertia and stiffness parameters are derived for distributed-parameter bending beams and bars in torsion, which subsequently are incorporated into dynamic models of mechanical systems. The natural and forced responses of multiple-DOF mechanical systems are studied by applying MATLAB® and Simulink® in order to determine the corresponding time-domain solutions. Chapters 4 and 5 follow a similar approach and presentation to obtain the mathematical models and responses of electrical, fluid, and thermal systems.

PROBLEMS

3.1 A cantilever has an equivalent mass with respect to its free end that is equal to the equivalent mass of a bridge with respect to its midpoint in terms of out-of-plane bending. Is it possible that the two members also have identical equivalent mechanical moments of inertia about their respective points with regard to torsion if they are made of the same material and have rectangular cross-sections and identical thicknesses?

3.2 Design the four beams of the microaccelerometer of Figure 3.28 such that their inertia contribution to the total (equivalent) inertia with respect to the motion about the y direction does not exceed 10%. The plate's thickness is h, and the beams have identical circular cross-section of diameter $d = h/5$.

3.3 A constant circular cross-section bridge of diameter d, length l, and mass density $\rho = 6300$ kg/m^3 is replaced by a similar bridge system formed of a rigid cylinder of diameter d_c, length l_c, mass density $\rho_c = 7800$ kg/m^3, and two

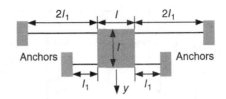

FIGURE 3.28

Four-Beam Accelerometer.

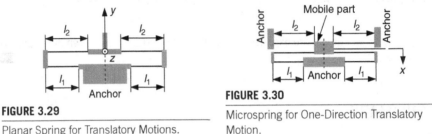

FIGURE 3.29

Planar Spring for Translatory Motions.

FIGURE 3.30

Microspring for One-Direction Translatory Motion.

identical flexible segments having the cross-section and material properties identical to the original bridge and are attached to the cylinder at one end and clamped at the opposite end; the total length of the altered bridge remains l. Determine the torsional change in both stiffness and inertia with respect to the bridge's midpoint that occurred through the cylinder addition by graphically analyzing the stiffness and inertia ratios for $0.2l < l_c < 0.5l$ and $0.2d_c < d < 0.5d_c$. Calculate the respective changes for the particular case $l = 4l_c$ and $d_c = 5d$. Consider the inertia contribution from the flexible segments.

3.4 The planar MEMS spring of Figure 3.29 needs to have stiffness in the y direction 4 times larger than the stiffness in the out-of-plane z direction. The thicker segments of the spring are rigid, and the thinner ones are flexible segments of identical rectangular cross-section with a thickness $h = 200$ nm (h is measured perpendicular to the figure plane). Use lumped-parameter modeling to design the flexible segments.

3.5 The microspring of Figure 3.30 is formed of three pairs of beams, two side rigid connectors, and a rigid moving part. Determine the corresponding lumped-parameter physical model and calculate the spring stiffness in terms of the moving part and about the y direction. The short beams have a diameter $d_1 = 2$ μm and the long ones have a diameter $d_2 = 3$ μm. Known also are $l_1 = 100$ μm, $l_2 = 140$ μm, and Young's modulus $E = 155$ GPa.

3.6 The microsensor of Figure 3.31(a) is changed into the one of Figure 3.31(b). Consider that the two vertical pushrods in Figure 3.31(b) are attached to the corresponding bridges and the plate. Compare the stiffnesses of the two designs in the motion direction using lumped-parameter modeling.

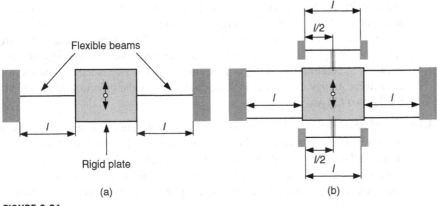

FIGURE 3.31

Micromechanism with Plate and (a) Two Flexure Hinges; (b) Six Flexure Hinges.

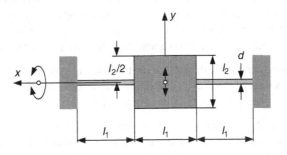

FIGURE 3.32

Microbridge with Central Plate and Two Side Flexible Supports.

3.7   A particle can attach either at the free end of a microcantilever or at the midpoint of and axially on a microbridge of identical dimensions and material properties as the microcantilever. Which of these two detectors has better detection precision in using the out-of-plane bending natural frequency shift method? Consider the design with $\rho = 6000$ kg/m^3, $w = 20$ μm, $h = 2$ μm (w and h are dimensions of the rectangular cross-section, with h being the out-of-plane dimension), and $l = 120$ μm.

3.8   A particle of unknown mass attaches to the midpoint of a rectangular cross-section microbridge at an unknown off-axis distance. Determine the particle's mass and the off-axis position knowing that the out-of-plane bending natural frequency shift is 25 Hz and the torsion natural frequency shift is 15 Hz. Known also are the geometrical and material parameters of the microbridge: $l = 800$ μm, $w = 200$ μm, $h = 8$ μm, $E = 190$ GPa, $G = 130$ GPa, and $\rho = 6000$ kg/m^3.

3.9   The microbridge system of Figure 3.32 can rotate about the x axis and translate about the y axis. It is composed of a rigid central plate and two identical flexible hinges. Consider that the hinge cross-section is circular with a diameter $d = 2$ μm. The plate and hinges are constructed from the same material with $E = 160$ GPa, and $\rho = 5600$ kg/m³. Known also are $l_1 = 250$ μm, $l_2 = 220$ μm, and $h = 20$ μm (the plate thickness). Find the ratio between the natural frequency corresponding to the y-axis translation and the natural frequency of the x-axis rotation. Consider the effect of added inertia by the two hinges.

3.10 The plate of Figure 3.32 is now supported by four hinges identical to the ones of Problem 3.9, as shown in the top view sketch of Figure 3.33. Analyze and describe the plate translation (out of the plane) along the z axis as well as the plate (in-plane) rotation about the same axis; calculate the two natural frequencies corresponding to these motions by using all the numerical parameters given in Problem 3.9.

3.11 Calculate the bending-related natural frequency of the cantilever beam system of Figure 3.34 when considering the inertia and stiffness of the circular constant cross-section beam in addition to the spring stiffness $k = 10$ N/m. Known are the beam's length $l = 260$ μm, its diameter $d = 20$ μm, as well as Young's modulus $E = 170$ GPa and mass density $\rho = 6200$ kg/m³.

3.12 Derive the mathematical model of the mechanical system shown in Figure 3.35, which consists of a pulley, a mass, three rigid rods, and two springs. Consider small motions; ignore gravity and rods' masses.

3.13 The two-DOF mechanical system shown in Figure 3.36 is the simplified lumped-parameter model of a car's suspension and body. It consists of two point masses m_1 and m_2 connected by a rigid, massless rod of length l, and two

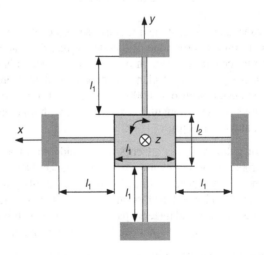

FIGURE 3.33

Microbridge with Central Plate and Four Flexible Supports.

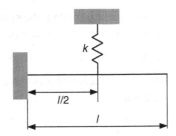

FIGURE 3.34

Cantilever and Spring Mechanical System.

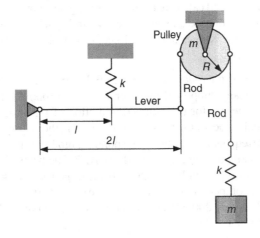

FIGURE 3.35

Pulley-Lever-Rod Mechanical System.

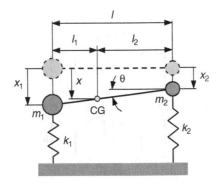

FIGURE 3.36

Two-DOF Lumped-Parameter Model of a Car's Suspension.

linear springs of stiffnesses k_1 and k_2. Obtain the system's mathematical model corresponding to the following coordinate selection:

(a) The coordinates are the mass displacements, x_1 and x_2.

(b) The coordinates are the displacement x of the center of gravity (CG) and the tilt angle θ of the rod.

3.14 Calculate the natural frequencies and determine the associated modes for the mechanical system of Figure 3.35 of Problem 3.12, both analytically and using MATLAB®. Consider the following numerical values: $m = 0.6\,\text{kg}, k = 110\,\text{N-m}, l = 0.15$ m.

3.15 Use analytic calculation to determine the natural frequencies, describe the corresponding modes, and calculate the unit-norm eigenvectors for the cart-pendulum mechanical system of Example 3.7, shown in Figure 3.17. Verify your results using MATLAB®. Known are $m_1 = 1$ kg, $m_2 = 0.2$ kg, $k = 210$ N/m, $l = 0.8$ m, and $g = 9.8$ m/s².

3.16 The mechanical microdevice of Figure 3.37 comprises two identical shuttle masses elastically supported by several beams. Propose a lumped-parameter mass-spring system for the x-direction motion of this system and derive the mathematical model of the natural response using the energy method. Calculate the natural frequencies and the corresponding modes (unit-norm eigenvectors) both analytically and by means of MATLAB®. The beams have identical rectangular cross-section with a width (in-plane dimension) $w = 10$ μm and thickness (dimension perpendicular to the plane) $h = 1$ μm. Known also are $m = 1.2 \times 10^{-10}$ kg, $l_1 = 200$ μm, $l_2 = 120$ μm, Young's modulus $E = 150$ GPa, and mass density $\rho = 5600$ kg/m³.

3.17 (a) Find the mathematical model of the double pendulum of Figure 3.38 assuming small rotations. Consider that the mechanical system moves in a fluid that opposes the bob's motion by a force of viscous damping nature; the coefficient of damping is c.

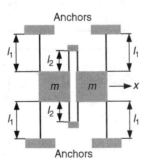

FIGURE 3.37

Spring-Mass Mechanical Microsystem for Translatory Motion.

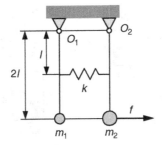

FIGURE 3.38

Double Pendulum with Spring and Viscous Damping.

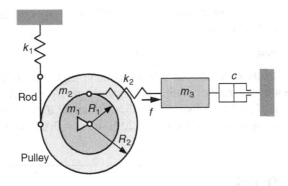

FIGURE 3.39

Mechanical System with Disc-Pulley, Rod, Translating Mass, and Linear-Motion Springs.

(b) Use Simulink® to find the time response of the system for $m_1 = 0.5$ kg, $m_2 = 1.2$ kg, $k = 190$ N/m, $c = 110$ N-s/m, $l = 0.8$ m, $g = 9.8$ m/s², $f = 100 \sin(10t)$ N.

3.18 Use the matrix formulation to derive the mathematical model of the mechanical system shown in Figure 3.39. The pulley is formed of two concentric disks that form a solid piece. Known are $m_2 = 2m_1 = 2m_3 = 1$ kg, $k_1 = 120$ N/m, $k_2 = 100$ N/m, $c = 32$ N-s/m, $R_2 = 2R_1 = 0.03$ m, and $f = 5$ N for the first 3 s and $f = 0$ thereafter. Plot the system's time response by means of Simulink®. Hint: Use the Signal Builder block in the Sources library.

3.19 Consider that a spring of stiffness $k_3 = 150$ N/m is placed instead of the damper and no force is acting on the translating body of the mechanical system shown in Figure 3.39 of Problem 3.18. Calculate the eigenvalues and unit-norm eigenvectors, and describe the modes of the mechanical system. Use the MATLAB® specialized command to check the analytical eigenvalues and eigenvectors.

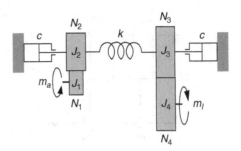

FIGURE 3.40

Gear System with Inertia, Damping, and Stiffness Properties.

3.20 For the geared system shown in Figure 3.40, where m_a is the actuation torque and m_l is the load torque,

    (a) Formulate the mathematical model.

    (b) Use Simulink® to plot the relevant rotation angles, as well as the angular velocities as functions of time when known are: $N_1 = 24$, $N_2 = 30$, $N_3 = 36$, $N_4 = 42$, $J_1 = 4 \times 10^{-5}$ kg-m^2, $J_2 = 5 \times 10^{-5}$ kg-m^2, $J_3 = 7 \times 10^{-5}$ kg-m^2, $J_4 = 9 \times 10^{-5}$ kg-m^2, $c = 2$ N-m-s, $k = 20$ N-m, $m_a = 300$ N-m, $m_l = 40$ N-m.

Suggested Reading

R. D. Blevins, *Formulas for Natural Frequency and Mode Shape*, Van Nostrand Reinhold Company, New York, 1979.

S. H. Crandall, D. C. Karnopp, E. F. K. Kurtz, Jr., and D. C. Pridmore-Brown, *Dynamics of Mechanical and Electromechanical Systems*, McGraw-Hill, New York, 1968.

W. T. Thomson, *Theory of Vibration with Applications*, 3rd Ed/Prentice-Hall, Englewood Cliffs, NJ, 1988.

J. P. Den Hartog, *Mechanical Vibrations*, Dover, New York, 1985.

D. J. Inman, *Engineering Vibration*, 3rd Ed. Prentice-Hall, Englewood Cliffs, NJ, 2007.

N. Lobontiu, *Compliant Mechanisms: Design of Flexure Hinges*, CRC Press, Boca Raton, FL, 2002.

N. Lobontiu and E. Garcia, *Mechanics of Microelectromechanical Systems*, Kluwer, New York, 2004.

H. Klee, *Simulation of Dynamic Systems with MATLAB® and Simulink®*, CRC Press, Boca Raton, FL, 2007.

K. Ogata, *MATLAB® for Control Engineers*, Prentice Hall, Saddle River, NJ, 2008.

Electrical Systems

4

OBJECTIVES

By using an approach similar to that used in Chapters 2 and 3, this chapter addresses the modeling of electrical systems by studying the following topics:

- Electrical elements—voltage and current source, resistance, capacitance, inductance, and operational amplifier.
- Methods of electrostatic actuation and sensing used in micro-electromechanical systems (MEMS).
- Application of Ohm's law, Kirchhoff's laws, the energy method, the mesh analysis method, and the node analysis method to derive mathematical models for dynamic electrical systems.
- Use of MATLAB® for symbolic and numerical calculation of the electrical systems' eigenvalues and eigenvectors, of the free response, and of the forced response.
- Application of Simulink® to modeling and solving differential equations for electrical systems, including systems with nonlinearities.

INTRODUCTION

Electrical systems, also named *circuits* or *networks*, are designed as combinations of mainly three fundamental components—resistor, capacitor, and inductor—which are correspondingly defined by resistance, capacitance, and inductance, generally considered to be lumped parameters. In addition to these primary electrical components, in this chapter, we also discuss the operational amplifier. Producing the electron motion or voltage difference in an electrical circuit are the voltage or current sources, which are the counterparts of forces or moments in mechanical systems. The focus in this chapter is the formulation of mathematical models using methods of electrical circuit analysis. Examples will be analyzed using MATLAB® and Simulink® to determine the natural, free, and forced responses of electrical systems.

DOI: 10.1016/B978-0-240-81128-4.00004-8

4.1 ELECTRICAL ELEMENTS: VOLTAGE AND CURRENT SOURCES, RESISTOR, CAPACITOR, INDUCTOR, OPERATIONAL AMPLIFIER

This section introduces the voltage and current sources, resistor, capacitor, inductor, and operational amplifier as the basic electrical elements relating the voltage and current variables. In addition to these basic electrical components, the companion website Chapter 4 presents the electrical transformer.

4.1.1 Voltage and Current Source Elements

Similar to mechanical systems, where forces for translation and torques for rotation generate the system dynamics (mechanical motion), voltage and current sources are the elements producing the dynamic response of electrical systems. We discuss here *ideal* (also named *independent*) *sources* in the following sense: An *ideal voltage source* delivers a certain voltage independent of the current passing through the source and is set up in the circuit by the other electrical components. Similarly, an *ideal current source* produces a specified current that does not affect the voltage across the source and is not altered by the voltages across other electrical components in the circuit. Figure 4.1 shows a schematic representations of voltage and current sources. Both sources can provide constant signals (in which case they are termed *dc*, *direct current*, sources) or time-varying ones, particularly sinusoidal (known as *ac*, *alternating current* sources).

4.1.2 Resistor Elements

The *resistor*, symbolized as in Figure 4.2, is an electrical element for which the voltage across it, v, is proportional to the current passing through it, i, according to *Ohm's law*:

$$v(t) = Ri(t) \tag{4.1}$$

The proportionality constant is the *resistance*, and for a wire of length l and cross-sectional area A, the electrical resistance is calculated as

$$R = \frac{\rho l}{A} \tag{4.2}$$

where ρ denotes the *electric resistivity*.

Variable (or active) resistances are encountered in *potentiometers*, where a contact wiper can move over a resistor and change the resistor length connected to voltage, resulting in a change in resistance, as seen in Eq. (4.2). A variable resistor (or translation potentiometer), a schematic representation of which is given in Figure 4.2(c), is illustrated with a bit more detail in Figure 4.3.

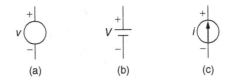

FIGURE 4.1

Voltage and Current Source Representations: (a) Variable Voltage; (b) Constant Voltage (Battery); (c) Current.

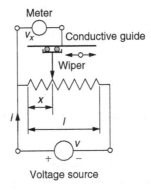

FIGURE 4.2

Resistor Representation: (a) in a Circuit Segment; (b) Symbol for Constant Resistance; (c) Symbol for Variable Resistance.

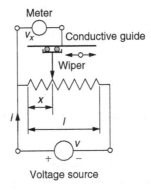

FIGURE 4.3

Translation Potentiometer.

The voltage v is supplied to the resistor of length l, and the translatory-motion wiper, through its mobile contact, separates a voltage v_x. Two voltage-current relationships can be written according to Ohm's law:

$$\begin{cases} v = Ri = \dfrac{\rho l}{A} i \\[2mm] v_x = R_x i = \dfrac{\rho x}{A} i \end{cases} \tag{4.3}$$

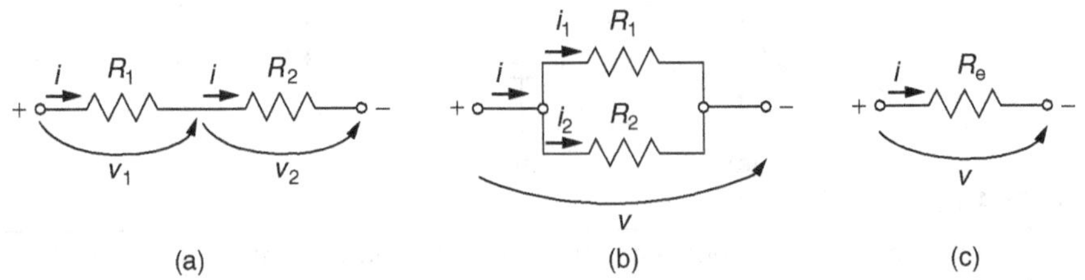

FIGURE 4.4

Resistor Connections: (a) Series; (b) Parallel; (c) Equivalent Resistance.

Combining Eqs. (4.3) yields

$$v_x = \frac{R_x}{R}v = \frac{x}{l}v \qquad (4.4)$$

The resistor dissipates energy in a circuit, similar to damping elements in mechanical systems. The energy dissipated in time t by a resistor transforms into heat, according to the *Joule effect*:

$$E = Ri^2 t = vit = \frac{v^2}{R}t \qquad (4.5)$$

Because energy is power multiplied by time ($E = Pt$), the electric power absorbed by a resistor is

$$P = Ri^2 = vi = \frac{v^2}{R} \qquad (4.6)$$

Equations (4.5) and (4.6) considered the voltage v is constant.

Resistors can be combined in series (as sketched in Figure 4.4(a)), in parallel (as illustrated in Figure 4.4(b)), or in a mixed (series-parallel) manner.

The companion website Chapter 4 contains the derivation of the equivalent series (R_s) and parallel (R_p) resistances, Figure 4.4(c), whose equations are

$$\begin{cases} R_s = R_1 + R_2 + \cdots + R_n \\ \dfrac{1}{R_p} = \dfrac{1}{R_1} + \dfrac{1}{R_2} + \cdots + \dfrac{1}{R_n} \end{cases} \qquad (4.7)$$

Example 4.1

Determine the resistors forming the reversed-Y combination of Figure 4.5(b), which is equivalent to the original Δ (delta) combination of Figure 4.5(a), such that the resistance between any two vertices of one combination is equal to the resistance in the other combination between the same vertices. Numerical application: $R_{AB} = 250\ \Omega$, $R_{BC} = 220\ \Omega$, $R_{CA} = 280\ \Omega$.

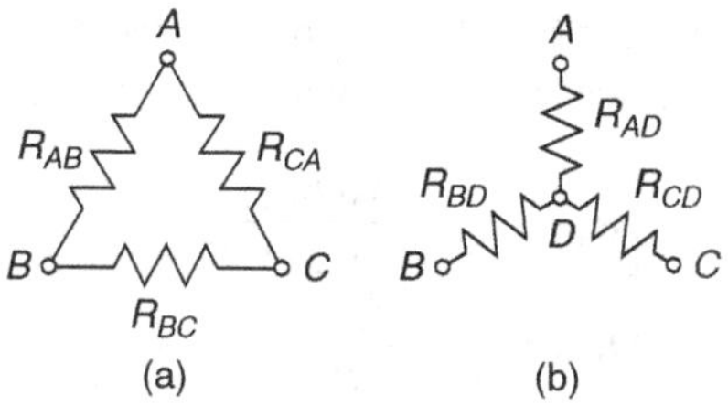

FIGURE 4.5

Three-Resistor Connections: (a) Δ (Delta) Combination; (b) Reversed-Y Combination.

Solution

The Δ arrangement of Figure 4.5(a) is more difficult to manipulate and decompose in series and parallel connections in a complex electrical circuit; therefore, the equivalent (reversed-Y) circuit of Figure 4.5(b) could provide an alternative. Assuming that voltage is applied across the conductor connecting nodes A and B in Figure 4.5(a), the resistor R_{AB} is connected in parallel with the series combination of R_{CA} and R_{BC}; therefore, the equivalent resistance between nodes A and B is calculated as

$$\frac{1}{R_{AB}^{\Delta}} = \frac{1}{R_{AB}} + \frac{1}{R_{BC} + R_{CA}} \tag{4.8}$$

which results in

$$R_{AB}^{\Delta} = \frac{R_{AB}(R_{BC} + R_{CA})}{R_{AB} + R_{BC} + R_{CA}} \tag{4.9}$$

When voltage is applied between nodes A and B of Figure 4.5(b), the only resistances involved are R_{AD} and R_{BD}, which are connected in series; therefore, the equivalent resistance between these two points is

$$R_{AB}^{Y} = R_{AD} + R_{BD} \tag{4.10}$$

Because the problem requires that $R_{AB}^{\Delta} = R_{AB}^{Y}$, it follows from Eqs. (4.9) and (4.10) that

$$R_{AD} + R_{BD} = \frac{R_{AB}(R_{BC} + R_{CA})}{R_{AB} + R_{BC} + R_{CA}} \tag{4.11}$$

By performing a similar analysis for the node pairs B-C and C-A, the following equations are obtained:

$$\begin{cases} R_{BD} + R_{CD} = \dfrac{R_{BC}(R_{CA} + R_{AB})}{R_{AB} + R_{BC} + R_{CA}} \\[4mm] R_{CD} + R_{AD} = \dfrac{R_{CA}(R_{AB} + R_{BC})}{R_{AB} + R_{BC} + R_{CA}} \end{cases} \tag{4.12}$$

Equations (4.11) and (4.12) can be solved for the reversed-Y resistances in terms of the Δ arrangement components using MATLAB® Symbolic Math Toolbox™, for instance, as

$$\begin{cases} R_{AD} = \dfrac{R_{AB}R_{CA}}{R_{AB} + R_{BC} + R_{CA}} \\[2mm] R_{BD} = \dfrac{R_{BC}R_{AB}}{R_{AB} + R_{BC} + R_{CA}} \\[2mm] R_{CD} = \dfrac{R_{CA}R_{BC}}{R_{AB} + R_{BC} + R_{CA}} \end{cases} \tag{4.13}$$

The numerical values of these resistances are R_{AD} = 93.33 Ω, R_{BD} = 73.33 Ω, R_{CD} = 82.13 Ω. ◼

4.1.3 Capacitor Elements

Capacitor elements in electrical systems store electrostatic energy and, therefore, can be considered functionally similar to springs in mechanical systems, which store elastic potential energy. The parameter characterizing a capacitor is the *capacitance*, C, which can be constant (with the symbols shown in Figure 4.6(b)) or variable (as symbolized in Figure 4.6(c)). In the International System (SI) of units, capacitance is measured in farads (F).

The capacitance is related to the current and voltage (these variables are indicated in Figure 4.6(a)) by the equation

$$i(t) = C\frac{dv(t)}{dt} \tag{4.14}$$

Equation (4.14) indicates that, when a capacitor is connected to a constant-voltage source, the current passing through it is zero. Integration of Eq. (4.14) results in

$$v(t) = \frac{1}{C}\int_0^t i(t)\,dt \tag{4.15}$$

Equation (4.15), which shows that the voltage reaches a value v over a period of time t, indicates that the voltage across a capacitor cannot change abruptly (instantaneously). Combining Eq. (4.15) with the current-charge relationship,

$$i(t) = \frac{dq(t)}{dt} \tag{4.16}$$

results in

$$q(t) = Cv(t) \tag{4.17}$$

A common capacitor configuration has two planes, parallel plates, and a dielectric between them. For this configuration, the capacitance is calculated as

$$C = \frac{\varepsilon A}{g} \tag{4.18}$$

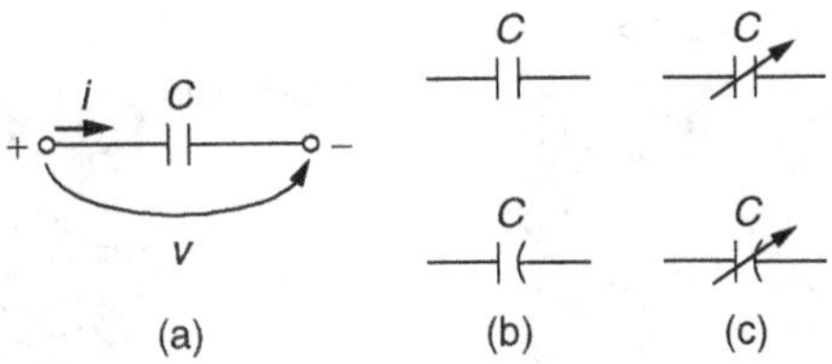

FIGURE 4.6

Capacitor Representation (a) in a Circuit Segment; (b) Symbols for Constant Capacitance; (c) Symbols for Variable Capacitance.

where ε is the *dielectric permittivity*, A is the two plates' superposition area, and g (the *gap*) is the distance between the plates.

The capacitor electric power is

$$P(t) = v(t)i(t) = v(t)C\frac{dv(t)}{dt} \tag{4.19}$$

The energy stored by a capacitor in an infinitesimal time dt is $dE = Pdt$, which, by integration and consideration of Eq. (4.19), yields

$$E(t) = \int_0^t P(t)dt = \int_0^t Cv(t)\frac{dv(t)}{dt}dt = \int Cv(t)dv(t) = \frac{1}{2}Cv(t)^2$$

$$= \frac{1}{2}q(t)v(t) = \frac{1}{2}\frac{q(t)^2}{C} \tag{4.20}$$

The assumption has been made in Eq. (4.20) that the capacitance is constant and $v(0) = 0$. A capacitor stores energy and does not dissipate it as long as it is not connected to other components. However, when a charged capacitor is connected to a resistor, for instance, a current flow is produced through the components until the whole energy stored in the capacitor is dissipated as heat on the resistor.

Actuation and Sensing in Microelectromechanical Systems

Variable capacitors are increasingly utilized for either actuation or sensing purposes in microelectromechanical systems (MEMS). Variable-plate capacitors can be designed using two principles, both based on the fact that one plate can move with respect to the other, which is fixed. According to one design principle (named *parallel-plate* or *transverse-motion*), the mobile plate moves in a direction perpendicular to the two plates such that the common area A remains constant but the gap varies, Figure 4.7(a). The second principle (known as *comb-drive* or *longitudinal-motion*) utilizes the motion of the mobile plate that is parallel to the other plate whereby the gap is preserved but the common area varies, Figure 4.7(b). Each of these design principles is illustrated through a few examples where either *actuation* or *sensing* (together known as *transduction*) is studied.

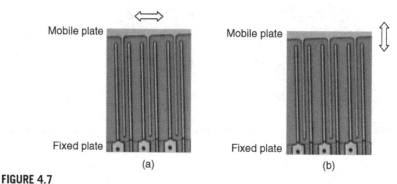

FIGURE 4.7

Capacitive Transduction in MEMS: (a) Parallel Plate; (b) Comb Drive.

Example 4.2

Determine the force f that acts on the mobile plate of the capacitor shown in Figure 4.8. A constant voltage v is applied externally between the two plates and the mobile plate moves by remaining parallel to the fixed one. The initial gap is g_0, the common plate area is A, and the electric permittivity of the dielectric is ε. Plot the f/f_0 ratio in terms of the x/g_0 ratio (f_0 is the initial force and x is the mobile plate displacement).

Solution

The attraction electrostatic force between the two plates generates the motion of the mobile plate toward the fixed one. In MEMS, this principle is used to generate actuation utilizing several parallel-connected pairs of the type shown in Figure 4.8 (a microphotograph of several such pairs is shown in Figure 4.7(a)) to multiply the force generated by a single pair. The capacitance after the mobile plate has displaced a distance x (and the gap is g) is

$$C = \frac{\varepsilon A}{g} = \frac{\varepsilon A}{(g_0 - x)} \qquad (4.21)$$

The electrostatic force can be expressed as the partial derivative of the electrostatic energy in terms of mechanical displacement (see details in the companion website Chapter 4) as

$$f = \frac{\partial E}{\partial x} = \frac{\partial}{\partial x}\left(\frac{Cv^2}{2}\right) = \frac{v^2}{2} \times \frac{\partial C}{\partial x} = \frac{v^2}{2} \times \frac{\partial}{\partial x}\left(\frac{\varepsilon A}{g_0 - x}\right) = \frac{\varepsilon A v^2}{2(g_0 - x)^2} \qquad (4.22)$$

As Eq. (4.22) indicates, the actuation force is proportional to the applied voltage squared and increases nonlinearly when the gap decreases. Initially, when $x = 0$, the electrostatic force is

$$f_0 = \frac{\varepsilon A v^2}{2g_0^2} \qquad (4.23)$$

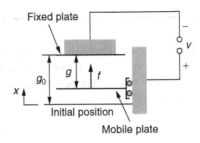

FIGURE 4.8

Variable-Gap, Transverse-Motion (Parallel-Plate) Variable Capacitor.

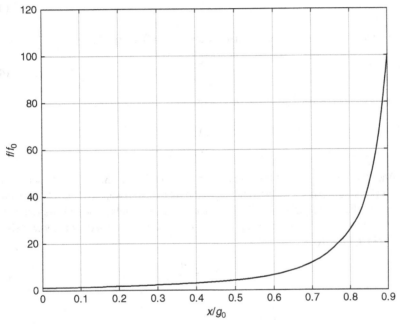

FIGURE 4.9

Plot of Force Ratio in Terms of Nondimensional Gap.

Combination of Eqs. (4.22) and (4.23) results in

$$\frac{f}{f_0} = \frac{1}{\left(1 - x/g_0\right)^2} \tag{4.24}$$

and the corresponding plot is shown in Figure 4.9. For values of x close to g_0, the force f reaches very large values (it is infinite when $x \to g_0$, just before the mobile plate collides with the fixed one). To avoid capacitor plate collision, the mobile plate in MEMS is connected to a spring that limits its motion range. ∎

Example 4.3

Consider that the variable capacitor of Figure 4.8 was designed to sense the displacement x of the mobile plate. Find the relationship between the voltage variation and the mechanical displacement. Assume that a bias dc voltage v_b is applied to the variable capacitor circuit, as indicated in Figure 4.10, as mechanical motion is applied externally to the mobile plate.

Solution

With no displacement of the capacitor mobile plate, the bias voltage fully loads the capacitor, generating the charge

$$q = C_0 v_b \tag{4.25}$$

The original capacitance is determined by taking $x = 0$ in Eq. (4.21):

$$C_0 = \frac{\varepsilon A}{g_0} \tag{4.26}$$

Combination of Eqs. (4.25) and (4.26) results in

$$q = \frac{\varepsilon A v_b}{g_0} \tag{4.27}$$

After the mobile plate of the capacitor moves a distance Δx (Δx is the same as x in the previous example; it is used to highlight the relationship with Δv, the voltage variation) toward the fixed the plate, the voltage changes by a quantity Δv and the capacitance changes from C_0 to C. As a consequence, the total voltage sensed by the low-resistance meter of Figure 4.10 is

$$v_b + \Delta v = \frac{q}{C} \tag{4.28}$$

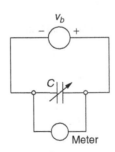

FIGURE 4.10

Electrical Circuit with Variable-Gap Capacitor and Bias Voltage Used as a Displacement Sensor.

By substituting the capacitance C from Eq. (4.21) and the charge q from Eq. (4.27) into Eq. (4.28), the last equation becomes

$$v_b + \Delta v = \frac{(g_0 - \Delta x)v_b}{g_0} \qquad (4.29)$$

Equation (4.29) allows expressing the voltage variation in terms of the mechanical displacement Δx:

$$\Delta v = -\frac{v_b}{g_0}\Delta x \qquad (4.30)$$

which shows that the voltage variation opposes the bias voltage; that is, the total voltage $v_b + \Delta v$ decreases. ∎

Example 4.4

Find the relationship between the electrostatic attraction force and the displacement of the mobile plate in a comb-drive (constant-gap, longitudinal-motion) capacitive actuator, such as the one sketched in Figure 4.11, assuming a constant dc voltage is applied between the two capacitor plates.

Solution

When the two plates are superimposed over a length of x, the capacitance is

$$C = \frac{\varepsilon w x}{g} \qquad (4.31)$$

where g is the constant gap and w is the width (dimension perpendicular to the drawing plane, assuming the plates are of rectangular shape) of the two capacitor plates. The electrostatic force is produced by the electrical field lines, as shown in Figure 4.11, and is

$$f = \frac{\partial E}{\partial x} = \frac{\partial}{\partial x}\left(\frac{Cv^2}{2}\right) = \frac{v^2}{2} \times \frac{\partial C}{\partial x} = \frac{\varepsilon w v^2}{2g} \qquad (4.32)$$

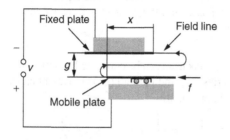

FIGURE 4.11

Fixed-Gap, Longitudinal-Motion (Comb-Drive) Variable Capacitor.

As Eq. (4.32) indicates, the electrostatic force is constant for a comb-drive capacitive actuator and is proportional to the square of the applied voltage. ■

Example 4.5
The MEMS of Figure 4.11 is considered to function as a detector and senses the displacement x of the mobile capacitor plate, which is produced externally. Determine the connection between the mechanical displacement and the voltage variation produced by it. Assume again that a bias dc voltage v_b is applied to the capacitor circuit as shown in Figure 4.10 and that an initial superposition of length x_0 between the two plates occurs before the mechanical motion is to be measured.

Solution
The initial capacitance of the variable capacitor is

$$C_0 = \frac{\varepsilon w x_0}{g} \tag{4.33}$$

The constant charge can be expressed as

$$q = C_0 v_b = \frac{\varepsilon w x_0 v_b}{g} \tag{4.34}$$

The total voltage that results when the mobile plate moved a distance Δx is the sum of v_b and Δv (the voltage variation that results from the mobile plate displacement from its initial position), which is equal to the ratio of the charge q to the modified capacitance C:

$$v_b + \Delta v = \frac{q}{C} = \frac{\varepsilon w x_0 v_b}{g} \times \frac{g}{\varepsilon w (x_0 + \Delta x)} = \frac{x_0}{x_0 + \Delta x} v_b \tag{4.35}$$

Equation (4.35) can be written as

$$\Delta v = -\frac{\Delta x}{x_0 + \Delta x} v_b = -\left(1 - \frac{x_0}{x_0 + \Delta x}\right) v_b \tag{4.36}$$

which shows that the voltage variation opposes the bias voltage (as was the case with the transverse capacitive sensing) and the value of this voltage increases as the plate superposition length increases. ■

Similar to resistors, capacitors can be connected in series or in parallel, as illustrated in Figure 4.12. The series (C_s) and parallel (C_p) equivalent capacitances, derived in the companion website Chapter 4, are

$$\begin{cases} \dfrac{1}{C_s} = \dfrac{1}{C_1} + \dfrac{1}{C_2} + \cdots + \dfrac{1}{C_n} \\ C_p = C_1 + C_2 + \cdots + C_n \end{cases} \tag{4.37}$$

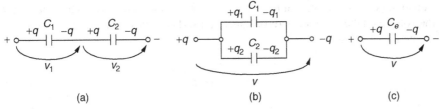

FIGURE 4.12

Capacitor Connections: (a) Series; (b) Parallel; (c) Equivalent Capacitance.

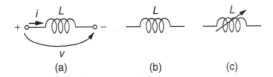

FIGURE 4.13

Inductor Representation: (a) in a Circuit Segment; (b) Symbol for Constant Inductance; (c) Symbol for Variable Inductance.

4.1.4 Inductor Elements

An *inductor*, schematic representations of which are shown in Figure 4.13, consists of a coil with N turns, where a variable current generates a voltage, which in the case of a linear inductor is proportional to the current rate.

The proportionality constant, L, is the *inductance*, which is a measure of this electrical component's capacity to store magnetic energy. For a cylindrical inductor with N turns, coil diameter D, total wire length l, and *magnetic permeability* μ, the inductance is expressed, as

$$L = \frac{\pi \mu N^2 D^2}{4l} \tag{4.38}$$

In SI, the inductance unit is the henry (H). The total magnetic flux, $N\Phi$, is proportional to the generated current:

$$N\Phi(t) = Li(t) \tag{4.39}$$

where L is the inductance. By *Faraday's law*, the flux variation generates a voltage on the inductor, which is

$$v(t) = \frac{d}{dt}\left[N\Phi(t)\right] \tag{4.40}$$

Combining Eqs. (4.39) and (4.40) results in

$$v(t) = L\frac{di(t)}{dt} \tag{4.41}$$

Equation (4.41) indicates that for a constant-current source connected to an inductor, the voltage on the inductor is zero. Equation (4.41) enables expressing the current as a function of voltage:

$$i(t) = \frac{1}{L} \int_0^t v(t)dt \tag{4.42}$$

Equation (4.42) shows that the current (and the magnetic flux, by Eq. (4.39)) cannot change instantaneously, as it needs a time interval to modify its value.

The power related to an inductor is expressed as

$$P(t) = v(t)i(t) = L\frac{di(t)}{dt}i(t) \tag{4.43}$$

The magnetic energy stored by the inductor over a period of time t and assuming $i(0) = 0$ is

$$E(t) = \int_0^t P(t)dt = \int Li(t)di(t) = \frac{1}{2}Li(t)^2 \tag{4.44}$$

Inductors, too, can be connected in series or in parallel, and an equivalent inductance can be computed for either case, as derived in the website companion Chapter 4:

$$\begin{cases} L_s = L_1 + L_2 + \cdots + L_n \\ \dfrac{1}{L_p} = \dfrac{1}{L_1} + \dfrac{1}{L_2} + \cdots + \dfrac{1}{L_n} \end{cases} \tag{4.45}$$

where the subscripts s and p stand for series and parallel, respectively.

4.1.5 Operational Amplifiers

Operational amplifiers (or simply *op amps*) are components that can be connected with other electrical components in circuits to amplify voltage, isolate circuits, count signals, or perform arithmetical and mathematical operations (addition, integration, differentiation, etc.). The symbol of an op amp, which is shown in Figure 4.14, indicates its main feature of having two input ports (a negative one and a positive one) and therefore *differential input voltages*. Voltages are usually measured with respect to the ground (which has zero voltage).

As its name suggests, the op amp amplifies the differential input voltage $v_2 - v_1$ to an output voltage by means of a factor K, known as *gain* or *amplification*:

$$v_o = Kv_i = K(v_2 - v_1) = -K(v_1 - v_2) \tag{4.46}$$

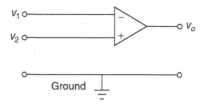

FIGURE 4.14

Schematic Representation of an Operational Amplifier (Op Amp).

The operational amplifier has some basic properties:

- High input impedance (the notion of impedance is studied more thoroughly in Chapter 7, but it mainly refers to the electrical resistance posed to the passing of current), which is ideally equal to infinity.
- Low output impedance (which is ideally zero).
- High gain of 10^5–10^6 (infinity ideally; actually the gain depends on frequency, when harmonic or sinusoidal signals are involved, and it can decrease significantly with the frequency increase).

4.2 ELECTRICAL CIRCUITS AND NETWORKS

Electrical elements are connected in *electrical systems*, known as *circuits* or *networks*, whose dynamic behavior is described by mathematical models expressed as differential equations, similarly to mechanical systems. *Kirchhoff's laws* are presented in this chapter together with procedures derived from Kirchhoff's laws, such as the *mesh analysis method* (the most popular probably) and the *node analysis method*. Applying the *energy method* to conservative electrical systems also is included in this chapter, while the companion website Chapter 4 presents the use of *Lagrange's equations method*.

4.2.1 Kirchhoff's Laws

The two Kirchhoff's laws provide basic procedures for deriving mathematical models for electrical systems (networks). The *first Kirchhoff's law*, also known as the *node* or *current* law (with the acronym KCL from *Kirchhoff's current law*), states that the algebraic sum of currents corresponding to branches that converge into a node is zero (Figure 4.15 illustrates this principle), and this is basically an expression of the charge conservation. For the particular case of Figure 4.15, the first Kirchhoff's law is

$$(i_1 + i_2) - (i_3 + i_4) = 0 \qquad (4.47)$$

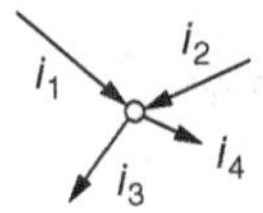

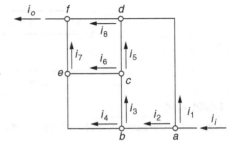

FIGURE 4.15

Illustration of the First Kirchhoff's (Node or Current) Law.

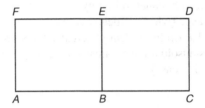

FIGURE 4.16

Electrical Network with Nodes and Currents.

```
F              E              D


A              B              C
```

FIGURE 4.17

Two Meshes and Three Loops in an Electrical Network.

Example 4.6

Using KCL, determine the relationship between the input current i_i and the output current i_o of the electrical network of Figure 4.16.

Solution

Based on KCL, the following current balance equations can be written at nodes a, b, c, d, e, and f of Figure 4.16:

$$i_i = i_1 + i_2; \; i_2 = i_3 + i_4; \; i_3 = i_5 + i_6; \; i_8 = i_1 + i_5; \; i_7 = i_4 + i_6; \; i_o = i_7 + i_8 \quad (4.48)$$

Successive substitutions, going from the last Eq. (4.48) to the second one, result in

$$i_o = (i_4 + i_6) + (i_1 + i_5) = i_4 + i_1 + i_3 = i_1 + i_2 = i_i \quad (4.49)$$

Equation (4.49) shows that, as expected, the current entering the electrical network, i_i, is equal to the current that exits the network, i_o, because the current (and the charge) is ideally a conservative amount.

It should be mentioned here that a *loop* in an electrical network is any closed contour of circuit branches, whereas a *mesh* is a loop that contains no other loop within it.

Figure 4.17 illustrates these definitions by using a simple sketch showing just the circuit branches (lines). According to these definitions, the network is formed of two meshes (*ABEF* and *BCDE*) and three loops (the two meshes and *ACDF*).

The *second Kirchhoff's law*, also known as the *mesh* or *voltage* law (its acronym being KVL from *Kirchhoff's voltage law*), states that, in a mesh, the sum of voltages across individual electrical components is equal to the sum of source voltage.   ■

Example 4.7

Use KVL to derive the mathematical model of the single-mesh electrical circuit of Figure 4.18, which is formed of two voltage sources, a resistor, an inductor, and a capacitor.

Solution

The source v_1 produces a positive voltage in the direction of the arbitrarily chosen direction of the current *i*. The other source, v_2, opposes the current direction; it therefore has a minus sign in the corresponding KVL equation:

$$v_L + v_R + v_C = L\frac{di(t)}{dt} + Ri(t) + \frac{1}{C}\int i(t)\,dt = v_1 - v_2 \tag{4.50}$$

Equation (4.50) is the mathematical model for the electrical system of Figure 4.18. A model that is expressed in terms of charge instead of current is obtained using the charge-current relationship, Eq. (4.16):

$$L\frac{d^2q(t)}{dt^2} + R\frac{dq(t)}{dt} + \frac{1}{C}q(t) = v_1 - v_2 \text{ or } L\ddot{q} + R\dot{q} + \frac{1}{C}q = v_1 - v_2 \tag{4.51}$$

It should be noted that Kirchhoff's voltage law is the electrical domain counterpart of Newton's second law of motion for mechanical systems. Indeed, forcing and voltage are similar, and voltages across inductors, resistors, and capacitors are similar to inertia, damping, and spring forces, respectively—more on these similarities (analogies) in Chapter 10.

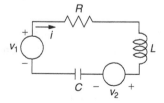

FIGURE 4.18

Single-Mesh Electrical Circuit to Illustrate Kirchhoff's Voltage Law.

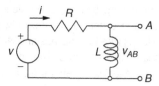

FIGURE 4.19

Electrical Circuit as a Single-DOF System.

4.2.2 Configuration, Degrees of Freedom

The configuration of an electrical system is determined by means of a *minimum number* of *independent physical parameters* (they can be currents, charges, voltages, voltage rates, etc.) that are sufficient to define the configuration of the electrical system as well as the values of all other variables of interest. These independent parameters are named *system coordinates*, and they actually are *degrees of freedom* (DOFs), similar to the DOFs introduced for mechanical systems in Chapter 3.

Figure 4.19 illustrates a circuit formed of a voltage source, a resistor, and an inductor, all assembled in one loop.

Assuming the values of the voltage v, resistance R, and inductance L are known, the current value needs to be determined by means of Kirchhoff's second law. The current is the only parameter necessary to fully define the configuration of the circuit at any moment in time; therefore, this system is a single-DOF one. Once the current was determined, the voltage v_{AB}, for instance, which falls across the inductor L, can be calculated. It can be considered as well that the independent parameter (DOF) of the circuit is the source voltage v, as knowledge of v suffices to enable calculation of any other parameter (variable) of the electrical system, but then the current i has to be known (specified).

Example 4.8

Determine the configuration (number of DOFs) of each of the three electrical circuits shown in Figure 4.20.

Solution

For the circuit of Figure 4.20(a), the branch currents i_1, i_2, and i_3 are DOF candidates. However, the three currents are connected at an adjacent node (either A or B) by means of Kirchhoff's first law:

$$i_1 = i_2 + i_3 \tag{4.52}$$

Equation (4.52) plays the role of a constraint on i_1, i_2, and i_3. It was shown in Chapter 3 that the number of DOFs of a mechanical system is the number of apparent DOF

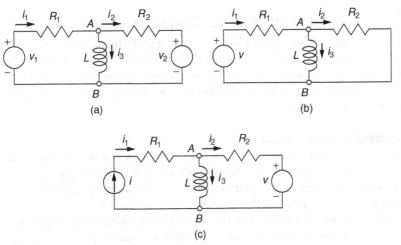

FIGURE 4.20

Two-Mesh Electrical Circuits with (a) Two Voltage Sources; (b) One Voltage Source; (c) One Voltage Source and One Current Source.

parameters (or candidate DOFs) minus the number of connection equations between these parameters. This rule is also valid for electrical systems, and as a consequence, the number of DOFs for the electrical system of Figure 4.20(a), the currents i_1 and i_2, for instance, can be the DOFs. For this system, the voltage sources v_1 and v_2 can also be the independent coordinates (DOFs) or any other voltage-current combination that is formed of two components (for instance, v_1 and i_2).

The same reasoning applies to the circuit of Figure 4.20(b), which needs two independent parameters to define its configuration and is therefore a two-DOF system. Although the voltage v is sufficient to generate all currents in the circuit, the system is still a two-DOF one, not a single-DOF system, as one might be tempted to conclude, because one single parameter is insufficient to fully define the configuration of the system.

Similarly, the electric circuit of Figure 4.20(c) needs two independent parameters to fully define its configuration at any time. They can be, for instance, i and i_1 or i and v. Two equations (formulated by means of Kirchhoff's laws) therefore are necessary to express the remaining variables in terms of the independent ones (the DOFs). ∎

A rule of thumb quickly provides the number of coordinates (DOFs) of an electrical network: The number of DOFs is equal to the number of meshes for which KVL equations are formulated. According to this rule, it can be seen that the electrical networks of Figure 4.20 are formed of two meshes; therefore, they should be two DOF systems, as demonstrated in the previous solved Example 4.8.

4.2.3 **Methods for Electrical Systems Modeling**

One of the most utilized procedures in modeling electrical networks is the *mesh analysis method*, and another method is the *node analysis*; both methods are briefly presented in this section. The *equivalent resistance method* is designed to model networks that are formed solely of resistors; this technique is presented in the companion website Chapter 4.

The Mesh Analysis Method

The *mesh analysis method* is normally used for electrical networks where the input is provided by voltage sources. The voltages supplied by the sources are known, as well as the parameters defining the electrical components making up the network. In mesh analysis, therefore, Kirchhoff's second law is applied to express voltage balances for each mesh, aided by Kirchhoff's first law to relate the currents converging at common nodes. In such a network, the unknowns are usually the currents or the charges. For circuits containing only resistors, the mathematical model consists of an algebraic equations system. When the circuits also contain capacitors or inductors, the mathematical model consists of differential (or differential-integral) equations.

Example 4.9

Determine the currents set up in the circuit of Figure 4.21 using the mesh analysis method. Known are $R_1 = 20\ \Omega$, $R_2 = 30\ \Omega$, $R_3 = 10\ \Omega$, $R_4 = 40\ \Omega$, $R_5 = 5\ \Omega$, and $v = 40$ V.

Solution

The circuit of Figure 4.21 indicates the currents and their arbitrary directions, as well as the arbitrary positive directions for each of the two meshes (the curved arrowed lines inside each mesh). The following equations are obtained through application of Kirchhoff's second law:

$$\begin{cases} R_1 i_1 + R_4 i_3 + R_5 i_1 = v \\ R_2 i_2 + R_3 i_2 - R_4 i_3 = 0 \end{cases}$$ (4.53)

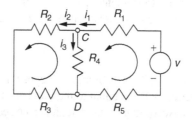

FIGURE 4.21

Electrical Network Comprising Resistors and a Voltage Source.

Kirchhoff's first law is applied at node C (or node D), which results in

$$i_1 = i_2 + i_3 \tag{4.54}$$

Equations (4.53) and (4.54) are rearranged in a system of three algebraic equations where the unknowns are the currents i_1, i_2, and i_3:

$$\begin{cases} (R_1 + R_5)i_1 + R_4 i_3 = v \\ (R_2 + R_3)i_2 - R_4 i_3 = 0 \\ i_1 - i_2 - i_3 = 0 \end{cases} \tag{4.55}$$

The equations system (4.55) can be written in vector-matrix form as

$$\begin{bmatrix} R_1 + R_5 & 0 & R_4 \\ 0 & R_2 + R_3 & -R_4 \\ 1 & -1 & -1 \end{bmatrix} \begin{Bmatrix} i_1 \\ i_2 \\ i_3 \end{Bmatrix} = \begin{Bmatrix} v \\ 0 \\ 0 \end{Bmatrix} \tag{4.56}$$

and the unknown vector containing the currents is determined by matrix algebra as

$$\begin{Bmatrix} i_1 \\ i_2 \\ i_3 \end{Bmatrix} = \begin{bmatrix} R_1 + R_5 & 0 & R_4 \\ 0 & R_2 + R_3 & -R_4 \\ 1 & -1 & -1 \end{bmatrix}^{-1} \begin{Bmatrix} v \\ 0 \\ 0 \end{Bmatrix} \tag{4.57}$$

The solution to Eqs. (4.57) is obtained using the symbolic calculation capabilities of MATLAB® by means of the following code:

```
>> syms r1 r2 r3 r4 r5 v
>> a = [r1+r5,0,r4;0,r2+r3,-r4;1,-1,-1];
>> f = [v;0;0];
>> i = inv(a)*f
```

After some algebraic conditioning, the returned currents are

$$\begin{cases} i_1 = \dfrac{(R_2 + R_3 + R_4)v}{R_1(R_2 + R_3 + R_4) + (R_2 + R_3)(R_4 + R_5) + R_4 R_5} \\[2ex] i_2 = \dfrac{R_4 v}{R_1(R_2 + R_3 + R_4) + (R_2 + R_3)(R_4 + R_5) + R_4 R_5} \\[2ex] i_3 = \dfrac{(R_2 + R_3)v}{R_1(R_2 + R_3 + R_4) + (R_2 + R_3)(R_4 + R_5) + R_4 R_5} \end{cases} \tag{4.58}$$

Numerically, the following values are obtained: $i_1 = 0.89$ A, $i_2 = i_3 = 0.44$ A.   ■

The Node Analysis Method

In the *node analysis method*, voltages are associated with each node of the network where current change occurs, then nodal equations are formulated by using

Kirchhoff's node law. With this method, we select one node to be the reference node, and all the voltages are expressed in terms of the reference node's voltage. Usually, it is computationally preferable to select as the reference node the one with the largest number of element branches connected to it. It is also customary to consider the voltage of the reference node to be zero.

Example 4.10

Calculate the currents produced by the current and voltage sources in the circuit sketched in Figure 4.22 employing the node analysis method. Known are $R_1 = 50\ \Omega$, $R_2 = 70\ \Omega$, $R_3 = 60\ \Omega$, $R_4 = 40\ \Omega$, $i = 0.1$ A, and $v = 80$ V.

Solution

For the circuit of Figure 4.22, if node B is considered to be the reference node with zero voltage, it follows that the entire line BD is grounded; therefore, the voltage of node D is also zero. Using Ohm's law, according to which a current is equal to the voltage difference across it divided by resistance, and Kirchhoff's node law, two node equations can be written when the voltages v_A and v_B are associated with nodes A and B, respectively. The equations are

$$\begin{cases} i = i_1 + i_2; \quad \text{or} \quad i = \dfrac{v_A - 0}{R_1} + \dfrac{v_A - v_C}{R_2} \\[2mm] i_2 = i_3 + i_4; \quad \text{or} \quad \dfrac{v_A - v_C}{R_2} = \dfrac{v_C - 0}{R_3} + \dfrac{v_C - v}{R_4} \end{cases} \tag{4.59}$$

The unknowns v_A and v_C are calculated symbolically from Eq. (4.59) using MATLAB® as:

$$\begin{cases} v_A = \dfrac{R_1(R_2R_3 + R_3R_4 + R_4R_2)}{(R_1 + R_2)(R_3 + R_4) + R_3R_4} i + \dfrac{R_1R_3}{(R_1 + R_2)(R_3 + R_4) + R_3R_4} v \\[3mm] v_C = \dfrac{R_1R_3R_4}{(R_1 + R_2)(R_3 + R_4) + R_3R_4} i + \dfrac{R_3(R_1 + R_2)}{(R_1 + R_2)(R_3 + R_4) + R_3R_4} v \end{cases} \tag{4.60}$$

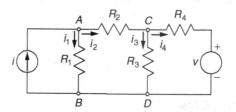

FIGURE 4.22

Electrical Network Comprising Resistors, a Voltage Source, and a Current Source.

The numerical values of the voltages of Eqs. (4.60) are $v_A = 19.93$ V and $v_C = 40.83$ V. The four currents are determined from Eqs. (4.59) as $i_1 = 0.40$ A, $i_2 = -0.30$ A, $i_3 = 0.68$ A, $i_4 = -0.98$ A. The minus signs of i_2 and i_4 indicate that these currents have directions opposite to the ones arbitrarily chosen in Figure 4.22. ■

4.2.4 Free Response

This section studies the natural response and the free damped response of electrical systems, both describing the behavior of electrical networks in the absence of voltage or current sources.

Natural Response

Electrical systems have a natural response when no voltage or current source is involved and no energy dissipation occurs (which translates in the absence of resistors). Electrical circuits that contain only capacitors and inductors are conservative systems; they display a natural response consisting of one or more natural frequencies, depending on the number of DOFs. The differential equation(s) defining the natural response of an electrical system can be derived using Kirchhoff's laws or applying the energy method, similarly to mechanical systems. MATLAB® can also be utilized to determine the eigen-frequencies and eigenvectors associated to the natural response. These methods are discussed next for single- and multiple-DOF electrical systems.

Single-DOF Conservative Electrical Systems

For single-DOF conservative electrical systems, the natural frequency is calculated by searching for sinusoidal (harmonical) solutions of the mathematical model differential equation, as introduced in Chapter 2.

Example 4.11
Consider the *LC* (resonant) circuit sketched in Figure 4.23(a), which is formed of an inductor and a capacitor. Derive its mathematical model using Kirchhoff's second law and also by the energy method. Determine the natural frequency of this electrical system for $L = 1$ H and $C = 4\ \mu F$.

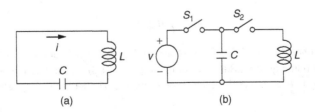

(a)   (b)

FIGURE 4.23

LC (Inductor-Capacitor) Circuit: (a) Actual Circuit; (b) Schematic for Charging the Circuit.

Solution

The single-DOF electrical system shown in Figure 4.23(a) is a conservative one, as no voltage or current sources input energy into the system and no resistors draw energy from the system. We can assume that the capacitor is charged separately from a voltage source (when switch S_1 is closed and switch S_2 is open) then disconnected from the source and connected to the inductor (by opening switch S_1 and closing S_2), as shown in Figure 4.23(b), so that the capacitor discharges on the inductor and a current i is produced through the circuit of Figure 4.23(a).

Application of Kirchhoff's second law to the electrical circuit of Figure 4.23(a) yields

$$L\frac{di(t)}{dt} + \frac{1}{C}\int i(t)dt = 0 \tag{4.61}$$

which can be written in terms of the charge q as

$$L\ddot{q} + \frac{1}{C}q = 0 \tag{4.62}$$

or

$$\ddot{q} + \frac{1}{LC}q = 0 \tag{4.63}$$

The electrical energy totaled by the capacitor and inductor in the circuit of Figure 4.23(a) is

$$E = \frac{1}{2}Li^2 + \frac{1}{2}\frac{q^2}{C} = \frac{1}{2}L\dot{q}^2 + \frac{1}{2}\frac{q^2}{C} \tag{4.64}$$

Because the total electrical energy is conserved, the time derivative of the energy is zero; therefore, the following equation results from Eq. (4.64):

$$\dot{q}\left(L\ddot{q} + \frac{1}{C}q\right) = 0 \tag{4.65}$$

The condition posed by Eq. (4.65) should be valid at all times, but the charge rate (the current) is not zero at all times; therefore, the only way that Eq. (4.65) is satisfied is when

$$L\ddot{q} + \frac{1}{C}q = 0 \tag{4.66}$$

which is identical to Eq. (4.62): They represent the mathematical model of the electrical system shown in Figure 4.23(a). By comparing Eq. (4.63) with the generic equation that modeled the free undamped response of a single-DOF mechanical system and was of the form

$$\ddot{x} + \omega_n^2 x = 0 \tag{4.67}$$

it follows that the natural frequency ω_n of the electrical system of Figure 4.23(a) is

$$\omega_n = \frac{1}{\sqrt{LC}} \tag{4.68}$$

and its numerical value is $\omega_n = 500\,\text{rad/s}$. The free response of the electrical system consists of a harmonic (sinusoidal or cosinusoidal) vibration at the natural frequency, quite similar to the case of a spring-mass mechanical system.   ∎

Multiple-DOF Conservative Electrical Systems

Kirchhoff's voltage law or the energy method (both in conjunction with Kirchhoff's node law) can be applied to derive the mathematical models of conservative electrical systems with configurations involving more than one DOF to determine their natural frequencies. The natural frequencies, together with their corresponding modes, can be evaluated analytically or using MATLAB® specialized commands, as discussed in Chapter 3.

Analytical Method The following example illustrates the analytical method of calculating the natural frequencies and the corresponding modes of a multiple-DOF electrical system.

Example 4.12

Derive the mathematical model of the two-mesh electrical system sketched in Figure 4.24 using Kirchhoff's laws and the energy method. Determine the natural frequencies and the modes (eigenvectors) of this system for $L_1 = 0.5\,\text{H}$, $L_2 = 0.3\,\text{H}$, and $C = 0.02\,\text{F}$.

Solution

Application of KVL to the two meshes of the electrical circuit of Figure 4.24 generates the equations

$$\begin{cases} L_1 \dfrac{di_1(t)}{dt} + \dfrac{1}{C}\int [i_1(t) - i_2(t)]dt = 0 \\[2mm] L_2 \dfrac{di_2(t)}{dt} - \dfrac{1}{C}\int [i_1(t) - i_2(t)]dt = 0 \end{cases} \tag{4.69}$$

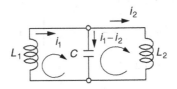

FIGURE 4.24

Two-Mesh Electrical Circuit with Energy Conservation.

which can be written in charge form as

$$\begin{cases} L_1 \ddot{q}_1 + \dfrac{1}{C}(q_1 - q_2) = 0 \\ L_2 \ddot{q}_2 + \dfrac{1}{C}(q_2 - q_1) = 0 \end{cases} \tag{4.70}$$

The energy collected by the three electrical components of the circuit is

$$E = \frac{1}{2}L_1 \dot{q}_1^2 + \frac{1}{2}\frac{(q_1 - q_2)^2}{C} + \frac{1}{2}L_2 \dot{q}_2^2 \tag{4.71}$$

The energy being constant, its time derivative is zero, which leads to

$$\dot{q}_1 \left[L_1 \ddot{q}_1 + \frac{1}{C}(q_1 - q_2) \right] + \dot{q}_2 \left[L_2 \ddot{q}_2 + \frac{1}{C}(q_2 - q_1) \right] = 0 \tag{4.72}$$

Equation (4.72) has to be valid at all times, but the charge rates cannot be zero at all times; therefore, compliance with the condition of Eq. (4.72) results in Eqs. (4.70), which have been obtained by means of Kirchhoff's voltage law: They represent the mathematical model of the conservative electrical system of Figure 4.24.

The free response of a conservative system requires solution of the harmonic type:

$$\begin{cases} q_1 = Q_1 \sin(\omega t) \\ q_2 = Q_2 \sin(\omega t) \end{cases} \tag{4.73}$$

Substitution of Eqs. (4.73) into Eqs. (4.70) yields the following algebraic equations system:

$$\begin{cases} \left(\dfrac{1}{C} - \omega^2 L_1 \right) Q_1 - \dfrac{1}{C} Q_2 = 0 \\ -\dfrac{1}{C} Q_1 + \left(\dfrac{1}{C} - \omega^2 L_2 \right) Q_2 = 0 \end{cases} \tag{4.74}$$

For the equations system (4.74) to have nontrivial solutions in the amplitudes Q_1 and Q_2, the determinant of the system needs to be zero:

$$\begin{vmatrix} \dfrac{1}{C} - \omega^2 L_1 & -\dfrac{1}{C} \\ -\dfrac{1}{C} & \dfrac{1}{C} - \omega^2 L_2 \end{vmatrix} = 0 \tag{4.75}$$

which produces the following algebraic equation (characteristic equation) in ω:

$$\omega^2 \left[\omega^2 L_1 L_2 - \frac{1}{C}(L_1 + L_2) \right] = 0 \tag{4.76}$$

whose solution consists of the natural frequencies

$$\begin{cases} \omega_{n1} = 0 \\ \omega_{n2} = \sqrt{\dfrac{L_1 + L_2}{C L_1 L_2}} \end{cases} \tag{4.77}$$

The numerical value of the nonzero natural frequency of Eqs. (4.77) is $\omega_{n,2} = 16.33$ rad/s.

Similar to mechanical systems, modes and eigenvectors can be expressed for electrical systems. The following ratio is obtained from Eqs. (4.74):

$$\frac{Q_1}{Q_2} = \frac{1}{1 - \omega^2 L_1 C} \tag{4.78}$$

which, for the nontrivial natural frequency of Eq. (4.77), becomes

$$\left(\frac{Q_1}{Q_2}\right)_{\omega = \sqrt{\frac{L_1 + L_2}{CL_1 L_2}}} = -\frac{L_2}{L_1} = -0.6 \tag{4.79}$$

Equation (4.79) indicates that the mode consists of two charge amplitudes that have opposite signs (which means the corresponding currents circulate in opposite directions at ω_{n2}). The magnitude of Q_2 is larger than that of Q_1. One eigenvector, corresponding to the nonzero natural frequency and the ratio of Eq. (4.79), is obtained by considering that $Q_2 = 1$, for instance, which leads to

$$\{Q\}_{\omega = \omega_{n2}} = \{Q\}_2 = \left\{ \begin{array}{c} -\dfrac{L_2}{L_1} \\ 1 \end{array} \right\} = \left\{ \begin{array}{c} -0.6 \\ 1 \end{array} \right\} \tag{4.80}$$

■

Using MATLAB® to Calculate Natural Frequencies, the Eigenvalue Problem Similar to mechanical systems, the electrical vibrations corresponding to the natural frequencies of a multimesh electrical system (which is the counterpart of a multiple-DOF mechanical system) can be expressed in vector-matrix form as an eigenvalue problem. The mathematical model of Example 4.12, Eqs. (4.70), can be written as

$$[L]\{\ddot{q}\} + [C]\{q\} = \{0\} \tag{4.81}$$

where

$$[L] = \begin{bmatrix} L_1 & 0 \\ 0 & L_2 \end{bmatrix}; [C] = \begin{bmatrix} \dfrac{1}{C} & -\dfrac{1}{C} \\ -\dfrac{1}{C} & \dfrac{1}{C} \end{bmatrix} \tag{4.82}$$

are the inductance and capacitance matrices and

$$\{q\} = \{q_1 \quad q_2\}^t \tag{4.83}$$

By following a development similar to the one applied to mechanical systems in Chapter 3 (more derivation details are in the companion website Chapter 4), the following equation is obtained when a sinusoidal solution is sought for $\{q\}$ in Eq. (4.81):

$$\det\left([L]^{-1}[C] - \lambda[I]\right) = 0 \tag{4.84}$$

where $\lambda = \omega^2$ are the eigenvalues, $[I]$ is the identity matrix and $[L]^{-1}[C] = [D]$ is the *dynamic matrix* of the conservative electrical system.

The MATLAB® command [V, D] = eig(D), which was introduced and utilized in Chapter 3, returns the modal matrix V, whose columns are the eigenvectors, and the diagonal matrix D, whose diagonal elements are the eigenvalues. Simply using the command eig(D) produces the eigenvalues solely.

Example 4.13

Derive the mathematical model for the electrical circuit of Figure 4.25, and then determine the natural frequencies (eigenvalues) and the eigenvectors by using the vector-matrix formulation method and MATLAB®. Consider $L_1 = L_2 = L = 50$ mH, $C_1 = C_2 = C = 800$ μF.

Solution

The electric energy corresponding to the elements of the circuit is

$$E = \frac{1}{2}L_1\dot{q}_1^2 + \frac{1}{2}L_2(\dot{q}_1 - \dot{q}_2)^2 + \frac{1}{2}\frac{q_1^2}{C_1} + \frac{1}{2}\frac{q_2^2}{C_2} \tag{4.85}$$

The system being conservative, the time derivative of the electric energy is zero; therefore, the following two differential equations are obtained by annulling the multipliers of the two charge first derivatives:

$$\begin{cases} (L_1 + L_2)\ddot{q}_1 - L_2\ddot{q}_2 + \dfrac{1}{C_1}q_1 = 0 \\[2mm] -L_2\ddot{q}_1 + L_2\ddot{q}_2 + \dfrac{1}{C_2}q_2 = 0 \end{cases} \tag{4.86}$$

The inductance and capacitances matrices, $[L]$ and $[C]$, result from comparing Eqs. (4.81) and (4.86):

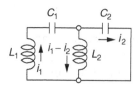

FIGURE 4.25

Conservative Two-Mesh Electrical Circuit.

$$[L] = \begin{bmatrix} L_1 + L_2 & -L_2 \\ -L_2 & L_2 \end{bmatrix}; [C] = \begin{bmatrix} \dfrac{1}{C_1} & 0 \\ 0 & \dfrac{1}{C_2} \end{bmatrix} \tag{4.87}$$

Using MATLAB® with its symbolic calculation capability, the following natural frequencies are obtained:

$$\omega_{n1} = \frac{0.618}{\sqrt{LC}}; \omega_{n2} = \frac{1.618}{\sqrt{LC}} \tag{4.88}$$

whose numerical values are ω_{n1} = 97.71 rad/s and ω_{n2} = 255.83 rad/s. The unit-norm eigenvectors corresponding to the two natural frequencies are also found by means of MATLAB® as:

$$\{Q\}_{\omega = \omega_{n1}} = \begin{Bmatrix} 0.53 \\ 0.85 \end{Bmatrix}; \{Q\}_{\omega = \omega_{n2}} = \begin{Bmatrix} -0.85 \\ 0.53 \end{Bmatrix} \tag{4.89}$$

Equations (4.89) indicate that, during the first modal motion, the charges q_1 and q_2 have identical flow directions. The amplitude of q_2 is larger than that of q_2. During the second modal motion, the charges vibrate about opposing directions and the magnitude of q_1 is larger than the magnitude of q_2. ▪

Free Damped Response

Let us consider an *RLC* series circuit without a source, as sketched in Figure 4.26.
The dynamic equation is obtained by means of Kirchhoff's voltage law as

$$L\frac{di(t)}{dt} + Ri(t) + \frac{1}{C}\int i(t)\,dt = 0 \tag{4.90}$$

which can be written in terms of charge as

$$\ddot{q} + \frac{R}{L}\dot{q} + \frac{1}{LC}q = 0 \tag{4.91}$$

It has been shown for a similar mechanical system in Chapter 2 that the ratio multiplying q in Eq. (4.91) represents the square of the system's natural frequency ω_n.

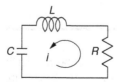

FIGURE 4.26

Single-Mesh Electrical Circuit with Resistor, Inductor, and Capacitor.

In addition to the natural frequency, let us introduce another parameter, ξ, the *electric damping ratio*:

$$\begin{cases} \omega_n^2 = \dfrac{1}{LC} \\ 2\xi\omega_n = \dfrac{R}{L} \end{cases} \tag{4.92}$$

such that Eq. (4.91) becomes

$$\ddot{q} + 2\xi\omega_n\dot{q} + \omega_n^2 q = 0 \tag{4.93}$$

Equations (4.93) and Eq. (2.71), which define the free damped vibrations of a single-DOF mechanical system, are similar. More about similarity (and analogy) between systems of different kinds is discussed in Chapter 10, but it should be mentioned that the electrical damping ratio was introduced in a way that is similar to the mechanical damping ratio. The dissipative nature of ξ is highlighted by the presence of R (which accounts for energy dissipation in an electrical system), and the value of ξ can be determined directly in terms of the actual system parameters by combining the two Eqs. (4.92) in

$$\xi = \frac{R}{2}\sqrt{\frac{C}{L}} \tag{4.94}$$

It has been shown in mechanical systems that for $0 < \xi < 1$ there is underdamping, for $\xi > 1$ there is overdamping, and for $\xi = 1$ the damping is critical. The same definitions apply for electrical damping.

For $0 < \xi < 1$, the solution to Eq. (4.93) is

$$q = Q_1 e^{\sigma t}\cos(\omega_d t) + Q_2 e^{\sigma t}\sin(\omega_d t) = Q e^{\sigma t}\sin(\omega_d t + \varphi) \tag{4.95}$$

with

$$\begin{cases} \sigma = -\xi\omega_n \\ \omega_d = \omega_n\sqrt{1 - \xi^2} \end{cases} \tag{4.96}$$

and Q_1, Q_2 (or Q and φ) are found from the initial conditions. The cases of critical damping ($\xi = 1$) and overdamping are presented in the companion website Chapter 4.

Example 4.14

In the electrical circuit of Figure 4.26, the resistor has a resistance $R = 250\ \Omega$. It is determined experimentally that the system's damping ratio is 0.45 and its natural frequency is 1200 Hz. What changes need to be applied to the system's capacitance C and inductance L such that the damping ratio is reduced by 25% and the natural frequency is increased by 30% when the same resistance is used? Plot the system's response in its latter configuration when the capacitor is charged by a voltage source of $v = 100$ V, which is subsequently removed.

Solution
Equations (4.92) enable expressing the inductance and capacitance as

$$L = \frac{R}{2\xi\omega_n}; \ C = \frac{2\xi}{R\omega_n} \tag{4.97}$$

With the original-design numerical values, these parameters are $L_1 = 0.0368$ H and $C_1 = 4.77 \times 10^{-7}$ F. The changed values of the damping ratio and natural frequency are

$$\xi_2 = 0.75\xi_1; \ \omega_{n2} = 1.3\omega_{n1} \tag{4.98}$$

which numerically are $\xi_2 = 0.3375$ and $\omega_{n2} = 1560$ Hz. With these values, Eqs. (4.92) are used again, generating the altered values of the inductance and capacitance: $L_2 = 0.0378$ H and $C_2 = 2.75 \times 10^{-7}$ F. Equations (4.95) and (4.96) give the charge equation as a function of time, and the constants Q_1 and Q_2 are determined by using the initial conditions:

$$q(0) = vC; \ \frac{dq(t)}{dt}\bigg|_{t=0} = 0 \tag{4.99}$$

stating that the initial charge is furnished by the voltage v applied to the capacitor and the initial current through the circuit is zero. The charge is plotted as a function of time in Figure 4.27; it can be seen that the charge settles to a steady-state value of zero in approximately 2 ms.

The MATLAB® dsolve command is employed here to solve a second-order differential equation. For the present example the command line that generates $q(t)$ is

```
>> dsolve('D2q+2*csi*omn*Dq+omn^2*q=0','q(0)=v*c',
   'Dq(0)=0')
```

The charge is returned symbolically, so the pertinent numerical values have to be subsequently substituted, and the result plotted as shown in Figure 4.27. ■

4.2.5 Operational Amplifier Circuits
As mentioned previously, circuits with operational amplifiers can realize various functions, such as inversion, amplification, addition, integration, differentiation, and filtering, to mention a few capabilities. We discuss next some of these functions.

Inverting Amplifier Circuits
In a circuit with two resistors, such as the one of Figure 4.28, the operational amplifier can change the sign of the input voltage at the output as shown in the following.

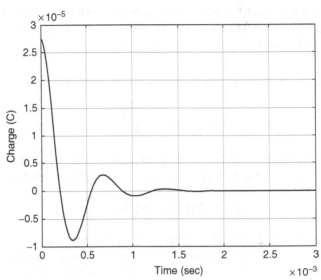

FIGURE 4.27

Charge as a Function of Time.

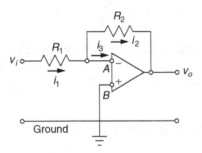

FIGURE 4.28

Basic Operational Amplifier in a Negative Feedback Circuit.

Based on Kirchhoff's node law, the currents identified in Figure 4.28 are connected as

$$i_1 = i_2 + i_3 \tag{4.100}$$

However, it has been shown that the input impedance of an ideal op amp is infinity; therefore, no current is entering the op amp, that is, $i_3 = 0$, which means that $i_1 = i_2$. The two currents can be expressed by means of Ohm's law as

$$i_1 = \frac{v_i - v_A}{R_1}; \; i_2 = \frac{v_A - v_o}{R_2} \tag{4.101}$$

where $v_i = v_1$ because $v_2 = 0$ in Eq. (4.46). The input-output voltage relationship introduced in Eq. (4.46) becomes

$$v_o = -Kv_A \tag{4.102}$$

Combining Eqs. (4.100), (4.101), and (4.102) yields

$$\frac{v_i}{R_1} + \frac{v_o}{KR_1} = -\frac{v_o}{R_2} - \frac{v_o}{KR_2} \tag{4.103}$$

Regularly, the gain K of an op amp is high (larger than 10^5); therefore, the two terms in Eq. (4.103) with K to the denominator are very small and can be neglected. As a consequence, Eq. (4.103) changes to

$$\frac{v_o}{v_i} = -\frac{R_2}{R_1} \tag{4.104}$$

Equation (4.104) indicates that the input and output voltages have different signs, which proves the *inverting effect* of this particular op amp circuit. In addition, if $R_2 > R_1$, there is an *amplification effect* as well, which justifies the name of *inverting amplifier*. The effect of multiplying and changing the sign of the input voltage at the output can be used as a logical operation in analog computing, as is shown in the companion website Chapter 4.

If we now analyze Eq. (4.101) in conjunction with Eq. (4.104) by taking into consideration that $i_1 = i_2$, we conclude that $v_A = 0$. At the same time, as Figure 4.28 shows, the positive input terminal is connected to the ground; therefore, its voltage is also zero. It follows that the voltages at the two input terminals, v_A and v_B, are zero. Actually, this property can be extended for the op amp circuits where the output is fed back to the negative input terminal, the case known as *negative feedback*, when the negative and positive input voltages are always equal.

Mathematical Operations with Operational Amplifier Circuits

We study here addition and integration; other operational amplifier examples are presented in the companion website Chapter 4.

Example 4.15

Determine the relationship between the output voltage v_o and the two input voltages v_1 and v_2 for the op amp circuit of Figure 4.29; establish the function produced by this circuit.

Solution

The op amp circuit has negative feedback, and because the positive input voltage is zero, it follows that the negative input terminal voltage is zero as well. No current is drawn into the op amp; therefore,

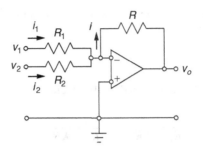

FIGURE 4.29

Operational Amplifier in a Negative Feedback Circuit with Two Input Resistors.

$$i_1 + i_2 = i \tag{4.105}$$

The currents of Eq. (4.105) are expressed by means of Ohm's law as

$$i_1 = \frac{v_1 - 0}{R_1}; \ i_2 = \frac{v_2 - 0}{R_2}; \ i = \frac{0 - v_o}{R} \tag{4.106}$$

By substituting Eqs. (4.105) into Eq. (4.106), the following relationship is obtained between the output and input voltages:

$$v_o = -\left(\frac{R}{R_1}v_1 + \frac{R}{R_2}v_2\right) \tag{4.107}$$

In the case where the three resistors are identical, that is, $R_1 = R_2 = R$, Eq. (4.107) reduces to

$$v_o = -(v_1 + v_2) \tag{4.108}$$

which indicates the function of the op amp circuit of Figure 4.29 is that of an *inverting adder*. The particular case with two input resistors can be generalized. When *n* resistors of corresponding resistances $R_1, R_2, ..., R_n$ are connected to the negative input terminal of Figure 4.29, having $v_1, v_2, ..., v_n$, the corresponding voltages with respect to the ground, the output voltage becomes

$$v_o = -\left(\frac{R}{R_1}v_1 + \frac{R}{R_2}v_2 + \cdots + \frac{R}{R_n}v_n\right) \tag{4.109}$$

For identical resistances, $R_1 = R_2 = \cdots = R_n = R$, Eq. (4.109) reduces to

$$v_o = -(v_1 + v_2 + \cdots + v_n) \tag{4.110}$$

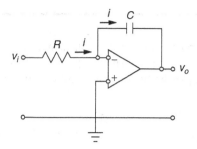

FIGURE 4.30

Operational Amplifier in a Negative Feedback Circuit with Input Resistor and Feedback Capacitor.

Example 4.16

Formulate the relationship between the output and input voltages for the op amp circuit of Figure 4.30 and determine the equation of the output voltage v_o as a function of R, C, and the input voltage v_i.

Solution

This circuit, too, has negative feedback and the positive input connected to the ground; as a consequence, the two input ports have zero voltages applied to them. The current i can be expressed in two ways, for the input branch and for the feedback branch:

$$i = \frac{v_i - 0}{R}; \; i = C\frac{d}{dt}(0 - v_o) \tag{4.111}$$

Equation (4.111) is rewritten as

$$\frac{dv_o}{dt} = -\frac{v_i}{RC} \tag{4.112}$$

Integration of Eq. (4.112) yields

$$v_o = -\frac{1}{RC}\int v_i dt \tag{4.113}$$

which indicates that the role of the op amp circuit of Figure 4.30 is that of an *integrator*. ▪

4.2.6 Forced Response with Simulink®

Simulink® can be utilized to graphically model and solve differential equations pertaining to electrical systems dynamic models, which can incorporate nonlinear effects, such as saturation. A couple of Simulink® examples are studied next.

Example 4.17

Identify an electrical circuit whose mathematical model is the differential equation $dx(t)/dt + 10x(t) = \cos(\omega t)$. Use Simulink® to solve this differential equation for $x(0) = 0$ and $\omega = 100$ rad/s, and plot the system response as a function of time.

Solution

Consider the electrical circuit of Figure 4.31.

The following equation is obtained by using Kirchhoff's second law:

$$Ri(t) + \frac{1}{C}\int i(t)\,dt = v(t) \tag{4.114}$$

which can be written in terms of charge as

$$\dot{q}(t) + \frac{1}{RC}q(t) = \frac{1}{R}v(t) \tag{4.115}$$

Equation (4.115) is, indeed, of the type indicated in this example with $R = 1\,\Omega$, $C = 0.1$ F, and $v = \cos(100t)$. The initial condition of the problem is $q(0) = 0$, which means there is no initial charge in the circuit.

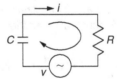

FIGURE 4.31

Electrical Circuit with Resistor, Capacitor, and Alternating Voltage Source.

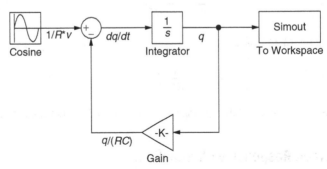

FIGURE 4.32

Simulink® Diagram for Calculating the Time-Response Corresponding to a First-Order Differential Equation Defining an Electrical System.

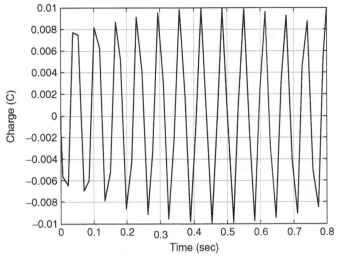

FIGURE 4.33

Charge as a Function of Time for the Electrical Circuit of Figure 4.32.

Solving differential equations using Simulink® was introduced in Chapter 2 in connection with single-DOF mechanical systems, and a similar procedure needs to be applied here. Equation (4.115) is reformulated as

$$\dot{q}(t) = -\frac{1}{RC}q(t) + \frac{1}{R}v(t) \tag{4.116}$$

The two signals in the right-hand side of Eq. (4.116) are the inputs to the summing point of Figure 4.32, and the result is the charge derivative in the left-hand side of Eq. (4.116). After integration, the charge $q(t)$ is obtained. It is known that $\cos(100t) = -\sin(100t - \pi/2)$; therefore, the `Sine Wave` source parameters are `Amplitude: -1`, `Frequency (rad/sec): 100`, `Phase (rad): pi/2`. Under `Simulation`, select `Configuration Parameters`; and under `Solver`, select a `Stop time` of 0.8 s. The charge plot as a function of time is shown in Figure 4.33. ◼

Example 4.18

The amplifier circuit of Figure 4.34, with $R_2 = 10R_1$ and $R_4 = 5R_3$, has a sinusoidal input voltage $v_i = 18 \sin(5t)$ V. Knowing that the output voltage saturates for $v_i < -10$ V and $v_i > 15$ V, use Simulink® to plot the output voltage, and compare it to the output voltage when the amplifier is ideal (without saturation).

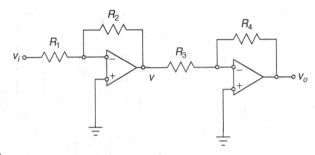

FIGURE 4.34

Two-Stage Operational Amplifier System.

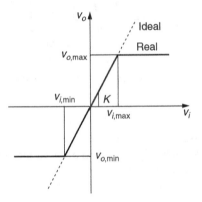

FIGURE 4.35

Saturation-Type Nonlinearity in an Amplifier.

Solution

The output voltage v_o, input voltage v_i, and interstage voltage v can be related according to

$$K = \frac{v_o}{v_i} = \frac{v}{v_i} \times \frac{v_o}{v} = \left(-\frac{R_2}{R_1}\right) \times \left(-\frac{R_4}{R_3}\right) = \frac{R_2 R_4}{R_1 R_3} \tag{4.117}$$

which results in a gain (amplification) of $K = 50$. For an *ideal amplifier*, the relationship between the output voltage and the input voltage is a linear one, as pictured in Figure 4.35. For input voltages exceeding some threshold values, say $v_{i,min}$ and $v_{i,max}$, the output voltage no longer increases, due to *saturation*—this situation is also illustrated in Figure 4.35 and was introduced in Chapter 1. Saturation itself does not occur in reality at such precise limits, and there is a transition between the unsaturated and saturated regions. However, for the ideal saturated amplifier, the output voltage is defined in terms of the input voltage as

$$
v_o = \begin{cases} v_{o,\min}, & v_i < v_{i,\min} \\ Kv_i, & v_{i,\min} \le v_i \le v_{i,\max} \\ v_{o,\max}, & v_i > v_{i,\max} \end{cases} \tag{4.118}
$$

The nonlinear output voltage profile can be modeled in Simulink®, as shown next. The block diagram needed to graphically solve this example is shown in Figure 4.36. The `Saturation` block is selected from the `Discontinuities` library. Under `Main`, an `Upper limit` of 750 V and a `Lower limit` of −500 V need to be specified. These limits are obtained by multiplying the limits of the input voltage by the gain of 50. The saturated and nonsaturated outputs v_o are mixed into a `Mux` block (the black rectangle is found in the `Commonly Used Blocks` library), which further transmits the two signals to the `Scope` to be plotted.

Figure 4.37 shows the plot of the simulation; the result of saturation is a truncation of the ideal sinusoidal output: The peak regions of the ideal sinusoid are chopped in the saturated response between −500 V and 750 V.

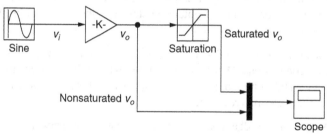

FIGURE 4.36

Simulink® Diagram for Plotting the Output Voltage from an Operational Amplifier with and without Saturation.

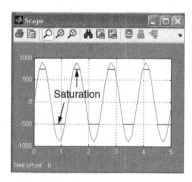

FIGURE 4.37

Simulink® Plot of Unsaturated and Saturated Output Voltages.

SUMMARY

This chapter discusses the electrical elements of voltage and current source, resistor, capacitor, inductor, and operational amplifier. It also introduces the methods of electrostatic actuation and sensing in microelectromechanical systems. Ohm's law, Kirchhoff's laws, the energy method, the mesh analysis method, and the node analysis method are applied to derive mathematical models for electrical systems. Examples are studied of how to apply MATLAB® for symbolic and numerical calculation of the natural, free, and forced responses of electrical systems. Simulink® is used to model and solve differential equations for electrical systems that might include nonlinearities, such as saturation. The next chapter studies, in a similar manner, the modeling of fluid and thermal systems.

PROBLEMS

4.1 Starting from the operating principle of a translation potentiometer (Figure 4.3), sketch a rotation (angular) potentiometer and find the relationship between voltage and angular displacement.

4.2 Five identical resistors ($R = 120\ \Omega$) need to be combined using both series and parallel connections. Determine the combinations that produce the maximum resistance and minimum resistance, respectively, between two end points; calculate the respective resistances.

4.3 A Wheatstone bridge consists of five resistors, as shown in Figure 4.38. Determine the equivalent resistance between nodes a and b, as well as the one between nodes c and d for $R_1 = R_3 = 210\ \Omega$, $R_2 = R_4 = 180\ \Omega$, $R_5 = 130\ \Omega$.

4.4 Determine the equivalent resistance between nodes a and b of the electrical system shown in Figure 4.39 for $R_1 = 50\ \Omega$, $R_2 = 80\ \Omega$, $R_3 = 40\ \Omega$, $R_4 = 60\ \Omega$, $R_5 = 30\ \Omega$, $R_6 = 10\ \Omega$.

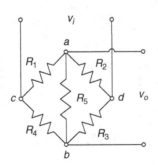

FIGURE 4.38

Wheatstone Bridge with Resistances.

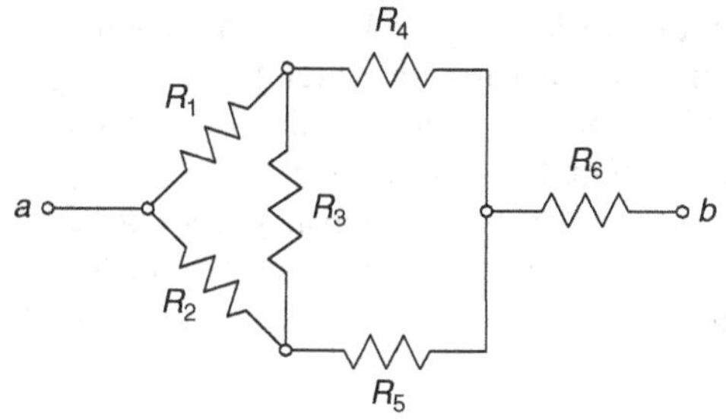

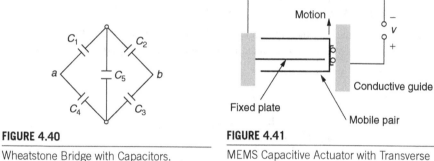

FIGURE 4.39

Electrical Resistances in a Mixed Connection.

FIGURE 4.40

Wheatstone Bridge with Capacitors.

FIGURE 4.41

MEMS Capacitive Actuator with Transverse Displacement.

4.5 Calculate the equivalent capacitance between nodes a and b that correspond to the capacitor Wheatstone bridge shown in Figure 4.40 for $C_1 = C_2 = 20$ µF, $C_3 = 15$ µF, $C_4 = 30$ µF, and $C_5 = 40$ µF. Hint: See companion website Chapter 4 for conversion between triangle and reversed-Y capacitance connections.

4.6 A MEMS capacitive actuator with one fixed plate and two mobile plates is sketched in Figure 4.41. Evaluate the resulting force acting on the mobile pair as a function of the changing gap when an external voltage $v = 20$ µV is applied between the fixed and the mobile plates. Initially, the fixed plate is placed symmetrically between the mobile plates. The distance between the two mobile plates is $d = 5$ µm, the plate overlapping area is 8000 µm², and $\varepsilon_0 = 8.8 \times 10^{-12}$ F/m. Also calculate the resulting force corresponding to a minimum gap $g_{min} = d/500$.

4.7 Compare the sensing performance of a transverse-motion capacitive unit with that of a longitudinal-motion one for MEMS displacement detection in terms of the initial gap g_0 and overlap x_0. Assume the two physical units are identical (same initial gap, plate areas, and dielectric properties).

4.8 The motion of the block of Figure 4.42 is sensed both longitudinally and transversely separately by two capacitive circuits. Knowing $g_0 = 10$ μm, $x_0 = 35$ μm, $v_b = 2$ μV, $\Delta v_t = 0.8$ μV (reading by the transverse unit), and $\Delta v_l = 0.2$ μV (reading by the longitudinal unit), calculate the corresponding block displacement. If the result indicates a discrepancy, consider the transverse unit is the precise one, indicate what design parameter of the longitudinal unit was erroneously evaluated, and determine its exact value. The motion starts as indicated in Figure 4.42.

4.9 A longitudinal MEMS capacitor is replaced with a transverse one to obtain a 10-fold increase in the actuation force. Consider that the two actuators' dimensions and operation conditions are identical, the initial gap of the transverse actuator is equal to the longitudinal actuator gap, and the maximum distance traveled is one third of that gap. For a capacitor plate that is square with a side length of 80 μm, calculate the initial gap of the transverse actuator.

4.10 Demonstrate that the equations allowing conversion of an inductor Δ connection into a reversed-Y connection (as sketched in Figure 4.43) are the ones

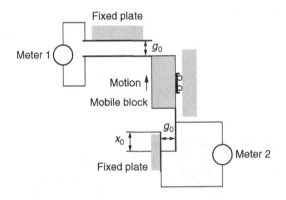

FIGURE 4.42

MEMS Capacitive Sensor with Transverse and Longitudinal Pickup.

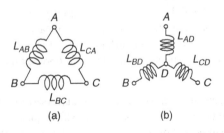

FIGURE 4.43

Three-Inductor Connections: (a) Δ (Delta) Connection; (b) Reversed-Y Connection.

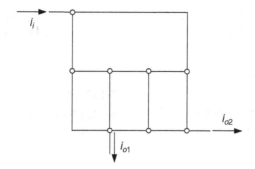

FIGURE 4.44

Electrical Network with Currents.

derived in Example 4.1 where the symbol R for resistance has to be substituted with the symbol L for inductance (see Figure 4.5). Calculate L_{AD}, L_{BD}, and L_{CD} for $L_{AB} = 80$ mH, $L_{BC} = 100$ mH, $L_{CA} = 70$ mH.

4.11 An electrical network is formed of five meshes enclosed in a rectangular area. Determine the number of corresponding loops and the configurations (degrees of freedom) for three geometric arrangements. Draw the electrical diagram of such a circuit that contains one voltage source and six resistors. Demonstrate using Kirchhoff's first law that five independent currents define the configuration of the electrical system.

4.12 Using Kirchhoff's current law,

(**a**) Demonstrate that the configuration of the electrical network schematized in Figure 4.44 is four.

(**b**) Demonstrate that $i_i = i_{o1} + i_{o2}$.

4.13 A single-mesh (single-DOF) circuit comprises series and parallel combinations of three identical inductors and two identical capacitors. Identify the two electrical systems for which the natural frequencies ratio is maximum and calculate that ratio.

4.14 What is the ratio of the natural frequencies for the electrical system of Figure 4.45?

4.15 Use the energy method to determine the mathematical model for the electrical circuit of Figure 4.46. Find the system's natural frequencies and the corresponding eigenvectors for $L_1 = 80$ mH, $L_2 = 120$ mH, $C_1 = 250$ μF, and $C_2 = 280$ μF.

4.16 Use the matrix eigenvalue method and MATLAB® to solve Problem 4.15.

4.17 For an electrical circuit containing a resistor, an inductor, and a capacitor connected in series it is known that the natural frequency is 3000 Hz and the

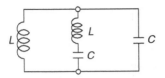

FIGURE 4.45

Two-Mesh Electrical Circuit with Coupling Inductor and Capacitor.

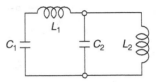

FIGURE 4.46

Two-Mesh Electrical Circuit with Coupling Capacitor.

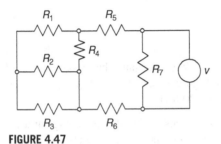

FIGURE 4.47

Electrical Circuit with Resistors and Voltage Source.

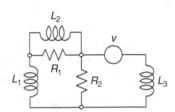

FIGURE 4.48

Three-Mesh Electrical Circuit.

system's period is 0.00035 s. It is also known that the resistance correspond-ing to critical damping is $R_{cr} = 200 \, \Omega$. Calculate the unknown parameters R, L, and C.

4.18 A solenoid is connected to a voltage source. Derive the mathematical mod-els of the electrical system when the solenoid is considered ideal (without electrical resistance) and when the solenoid resistance is taken into account. Calculate the two currents and plot them as functions of time. Known are the voltage $v = 80$ V, the wire diameter $d = 0.2$ mm, the coil median diameter $D = 50$ mm, the number of turns $N = 2000$, the electric resistivity $\rho = 17 \times 10^{-9} \, \Omega$m, and the magnetic permittivity $\mu = 0.01$ H/m.

4.19 Apply the mesh analysis method to derive the mathematical model for the elec-trical circuit of Figure 4.47. Find the currents in the circuit for $R_1 = 35 \, \Omega$, $R_2 = 50 \, \Omega$, $R_3 = 60 \, \Omega$, $R_4 = 20 \, \Omega$, $R_5 = 45 \, \Omega$, $R_6 = 10 \, \Omega$, and $R_7 = 25 \, \Omega$. The voltage is $v = 100$ V.

4.20 Use the mesh analysis method to determine the mathematical model for the electrical circuit of Figure 4.48. Solve for the currents and plot them using MATLAB® for $L_1 = 5$ H, $L_2 = 8$ H, $L_3 = 6$ H, $R_1 = 100 \, \Omega$, $R_2 = 150 \, \Omega$, $v = 20\sin(6t)$ V.

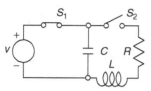

FIGURE 4.49

Two-Mesh Electrical Circuit with Two
Switches Swapping States.

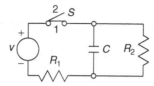

FIGURE 4.50

Two-Mesh Electrical Circuit with Switch.

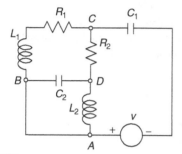

FIGURE 4.51

Electrical Network Comprising Resistors,
Capacitors, Inductors, and a Voltage
Source.

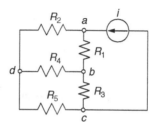

FIGURE 4.52

Electrical Circuit with Resistors and Current
Source.

4.21 Derive the mathematical model of the circuit shown in Figure 4.49. Consider
that, in the first phase, the switch S_1 is closed while switch S_2 is open; and in the
second phase, the two switches change their states. Plot the current through R for
the second phase. Known are $R = 120\ \Omega$, $C = 50\ \mu F$, $L = 0.6$ H, and $v = 80$ V.

4.22 Derive the mathematical model of the circuit shown in Figure 4.50 when the
switch is closed (position 1), then when the switch is opened. Plot the current
through R_2 when the switch in the open position. Consider $R_1 = R_2 = 200\ \Omega$,
$C = 20\ \mu F$, and $v = 100$ V.

4.23 Derive the mathematical model of the electrical circuit of Figure 4.51 using the
mesh analysis method.

4.24 Use Simulink® to model the circuit shown in Figure 4.51 of Problem 4.23. Plot
the branch currents in terms of time when the initial charge on capacitor C_1 is
$q_1(0) = 0.001$ C; known also are $R_1 = 350\ \Omega$, $R_2 = 220\ \Omega$, $L_1 = 1.2$ H, $L_2 = 2$ H,
$C_1 = 3.5$ mF, $C_2 = 2.8$ mF, $v = 110$ V.

4.25 Determine the DOFs of the electrical circuit sketched in Figure 4.52. Using
the node analysis method and MATLAB®, calculate the currents through the
branches for $R_1 = 50\ \Omega$, $R_2 = 20\ \Omega$, $R_3 = 40\ \Omega$, $R_4 = 40\ \Omega$, $R_5 = 60\ \Omega$, and
$i = 3$ mA.

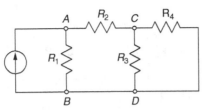

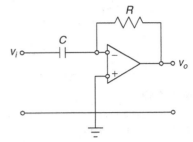

FIGURE 4.53

Electrical Network Comprising Resistors
and a Current Source.

FIGURE 4.54

Operational Amplifier in a Negative
Feedback Circuit with Input Capacitor and
Feedback Resistor.

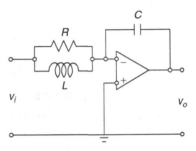

FIGURE 4.55

Electrical Circuit with Operational Amplifier.

4.26 Calculate the currents produced by the current source in the circuit sketched in Figure 4.53 by employing the nodal analysis method and MATLAB® for $i = 1$ mA, $R_1 = 30$ Ω, $R_2 = 50$ Ω, $R_3 = 10$ Ω, and $R_4 = 70$ Ω.

4.27 Determine the relationship between the output and input voltages for the op amp circuit of Figure 4.54 and establish the function produced by this circuit. Plot the output voltage for $v_i = 80t$ V, $R = 180$ Ω, and $C = 35$ mF.

4.28 Derive the mathematical model for the operational amplifier circuit shown in Figure 4.55. Use MATLAB® to find the output voltage for $R = 35$ Ω, $L = 0.5$ H, $C = 95$ μF, and an input voltage defined as $v = 40/(1 + 0.01t)$, where t is time.

4.29 In the Wheatstone bridge shown in Figure 4.38 of Problem 4.3, the input voltage is $v_i = 40 \sin(10t)$ V and the resistor R_1 saturates when the voltages $v_{min} = -20$ V and $v_{max} = 30$ V fall across it. Use Simulink® to plot the output voltage.

4.30 Solve Problem 4.20 based on Figure 4.48 utilizing Simulink® and the numerical values given in Problem 4.20.

Suggested Reading

R. C. Dorf and J. A. Svoboda, *Introduction to Electric Circuits*, 4th Ed. John Wiley & Sons, New York, 2006.

K. C. A. Smith and R. E. Alley, *Electrical Circuits—An Introduction*, Cambridge University Press, Cambridge, UK, 1992.

A. E. Fitzgerald, D. E. Higginbotham, and A. Grabel, *Basic Electrical Engineering*, 5th Ed. McGraw-Hill, 1981, New York.

W. H. Hyatt, Jr. and J. E. Kemmerly, *Engineering Circuit Analysis*, McGraw-Hill, New York, 1978.

S. B. Hammond and D. K. Gehmlich, *Electrical Engineering*, McGraw-Hill, New York, 1970.

Fluid and Thermal Systems

5

OBJECTIVES

The focus of this chapter is on modeling fluid and thermal systems, and the main addressed topics are

- Fluid (liquid and pneumatic) elements—inertance, capacitance, resistance, and fluid energy sources.
- Thermal capacitance and resistance elements.
- Mathematical modeling of fluid systems and formulation of the natural and forced responses.
- Mathematical modeling of thermal system forced response.
- Use of MATLAB® in symbolic and numerical calculations, and in evaluation of fluid systems eigenvalues and eigenvectors.
- Application of Simulink® to graphically model and plot the time response of forced fluid systems with linear and nonlinear properties.

INTRODUCTION

This chapter is dedicated to modeling the dynamics of fluid (liquid and pneumatic) systems as well as thermal systems. These systems are modeled using elements similar to those for mechanical and electrical systems: inertance, capacitance, and resistance. System models are derived when these components are coupled in various systems. In case only inertance and capacitance properties are present and no resistive losses occur, the natural response of fluid systems is studied. In many applications, the inertia properties of liquids and gases can be neglected, and the resulting mathematical models are based on only the capacitive and resistive properties used to formulate the forced response of fluid and thermal first-order systems. The use of MATLAB® and Simulink® in solving for the natural and forced responses of fluid and thermal systems is illustrated by several solved examples.

DOI: 10.1016/B978-0-240-81128-4.00005-9

5.1 LIQUID SYSTEMS MODELING

In liquid (or hydraulic) systems, the medium of energy transmission is a liquid. We review a few basic liquid laws first, such as Bernouli's law and the law of mass conservation. Next, we discuss the basic liquid elements of inertance, capacitance, and resistance, together with the sources generating liquid system motion. You will learn how to formulate the natural (free) response of inertance-capacitance liquid systems as well as the forced response of liquid systems containing capacitance and resistance components.

The notion of flow rate is utilized in liquid and pneumatic systems with different meanings. In *liquid systems*, the *volume flow rate* is employed, which is denoted here by q_v; whereas *pneumatic systems* use the *mass flow rate*, denoted by q_m. The two amounts are connected by means of the mass density:

$$q_v = \frac{\Delta V}{\Delta t} = \frac{\Delta x A}{\Delta t} = vA; \quad q_m = \frac{\Delta m}{\Delta t} = \rho \frac{\Delta V}{\Delta t} = \rho q_v \tag{5.1}$$

where V is volume, m is mass, v is the fluid velocity, x is distance travelled by fluid, and A is area perpendicular on flow direction.

5.1.1 Bernoulli's Law and the Law of Mass Conservation

An important instrument in modeling liquid dynamics is *Bernoulli's law*. For a conservative liquid (with no energy losses or gains) flowing in a pipe and when the liquid is considered incompressible (and therefore its mass density is constant), Bernoulli's law states that

$$p_2 + \frac{\rho v_2^2}{2} + \rho g h_2 = p_1 + \frac{\rho v_1^2}{2} + \rho g h_1 \tag{5.2}$$

which is based on Figure 5.1, and where p is the static pressure, h is the vertical distance (also named *head*) from a reference line, and v is the liquid velocity at the center point of a cross-section. When losses are accounted for (of a viscous nature) and energy is input into the system (such as by pumps or hydraulic actuators), Eq. (5.2) changes to

$$p_2 + \frac{\rho v_2^2}{2} + \rho g h_2 = p_1 + \frac{\rho v_1^2}{2} + \rho g h_1 + \rho w - \rho g h_f \tag{5.3}$$

where w is the specific work produced by a hydraulic source (it is energy per unit mass, being measured in N-m-kg^{-1} in SI) and h_f is the lost head (it is due mainly to viscous friction). Equations (5.2) and (5.3) are the pressure-form equations of Bernoulli's law.

The law of *volume/mass conservation* states that, if there is no accumulation or loss of liquid between points 1 and 2 of Figure 5.1, then the volume flow rate will not change:

$$q_{v1} = q_{v2} \tag{5.4}$$

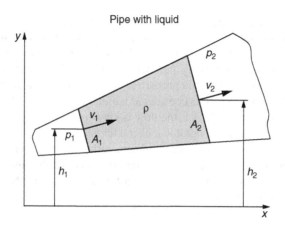

FIGURE 5.1

Liquid Column Traveling through Variable Cross-Section Pipe in a Vertical Plane.

or, based on Eq. (5.1)

$$v_1 A_1 = v_2 A_2 \tag{5.5}$$

For a pipe of length l, the lost head is calculated as

$$h_f = \frac{lf}{d_h} \times \frac{v^2}{2g} \tag{5.6}$$

where f is the *Moody friction factor* and d_h is the *hydraulic diameter*, which is calculated as

$$d_h = \frac{4A}{P_w} \tag{5.7}$$

with P_w being the *wetted perimeter* of the pipe internal cross-section. For laminar flow (which is defined shortly), the Moody friction factor is determined as

$$f = \frac{64}{Re} \tag{5.8}$$

The *Reynolds number*, Re, is the ratio of the inertia to viscous friction effects associated with the relative motion between a solid and a fluid. The mathematical expression of the Reynolds number is

$$Re = \frac{\rho v d_h}{\mu} = \frac{v d_h}{\nu} \tag{5.9}$$

where d_h is the hydraulic diameter (for a sphere moving in a fluid, d_h is the diameter; similarly, for fluid flowing in a pipe, d_h is the pipe inner diameter), μ is the *dynamic viscosity*, and ν is the *kinematic viscosity*. *Laminar flow* has a smooth, linear character and is defined by generally parallel streamlines. It is characterized by large viscosity effects and small inertia properties and occurs for low Reynolds numbers, for instance, the flow in a pipe is considered laminar when Re < 2000. For very small Reynolds number, such as Re < 1, the flow is known as *creep* or *Stokes flow*. On the contrary, *turbulent flow* has a nonlinear character and is defined by a chaotic motion containing vortices and eddies. In pipelines, turbulent flow occurs for Reynolds numbers larger than 4000 and where inertia effects are predominant. When 2000 < Re < 4000, the flow is considered to be of a *transition* nature, because it combines laminar and turbulent traits.

Example 5.1

A pump is used to send liquid vertically through a pipe of circular cross-section with inner diameter $d = 0.02$ m. Assume the pipe, whose height is $l = 6$ m, is open at its end opposite to the pump; also assume the flow is laminar and the incompressible liquid has a mass density $\rho = 1000$ kg/m³ and dynamic viscosity $\mu = 0.00001$ N-s/m². Known also is the input volume flow rate $q_{vi} = 0.0001$ m³/s. Calculate the specific work of the pump w that is necessary to send liquid to the top of the pipe by considering the friction losses. The pressure at the pump intake and at the pipe's free end is atmospheric.

Solution

Figure 5.2 shows schematically the pump and vertical pipe system, where point 0 is at the intake (input) to the pump, point 1 is at the outtake (output) of the pump, and point 2 is at the end of the vertical pipe segment.

Application of Bernoulli's law between points 0 and 1, which are assumed at the same height, leads to

$$p_1 + \frac{\rho v_1^2}{2} = p_0 + \frac{\rho v_0^2}{2} + \rho w \qquad (5.10)$$

The law of mass conservation gives

$$q_{vi} = v_0 A = v_1 A \qquad (5.11)$$

which assumed that the pump intake and outtake areas are identical; the result is

$$v_0 = v_1 = \frac{4 q_{vi}}{\pi d^2} \qquad (5.12)$$

Because $v_0 = v_1$, Eq. (5.10) simplifies to

$$p_1 = p_0 + \rho w \qquad (5.13)$$

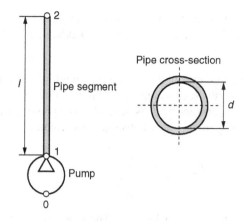

FIGURE 5.2

Pump with Vertical Pipe Segment.

Bernoulli's law is now applied between points 1 and 2:

$$P_a + \frac{\rho v_2^2}{2} + \rho g l = P_1 + \frac{\rho v_1^2}{2} - \rho g h_f \quad \text{or} \quad \rho g l = \rho w - \rho g h_f \qquad (5.14)$$

since $v_2 = v_1$ because of mass conservation and $p_0 = p_a$. The friction head is found, using Eqs. (5.6) through (5.9), as

$$h_f = \frac{32\mu l v_1}{\rho g d^2} \qquad (5.15)$$

which takes into consideration that the hydraulic diameter is equal to the actual inner pipe diameter d. Combining Eqs. (5.12), (5.13), (5.14), and (5.15), the specific work of the pump is

$$w = g l + \frac{128\mu l q_{vi}}{\pi \rho d^4} \qquad (5.16)$$

The numerical value of the pump specific work is $w = 58.8$ m²/s². ■

5.1.2 Liquid Elements

Similar to mechanical or electrical systems, which are formed of elements with iner-tia, storage capacity, losses, and energy input, liquid systems can be defined by such elements. Inertance elements portray liquid inertia effects, whereas capacitances and resistances characterize liquid storage and loss features, respectively. The pressure or the difference of level (head) among various components of a liquid system represent source elements that set the liquid into motion. As a consequence of the dual manner of generating liquid motion, the elements' definitions can be provided by either using the pressure or the head in addition to other amounts of interest. The subscript p (for pressure) or h (for head) accompanies the letter l (standing for liquid) in the following.

Inertance

The *inertance* quantifies the inertia effects in liquid systems and is particularly important in long conduits such as pipes. For laminar flow and in terms of pressure, the inertance (which is denoted by I) is defined as the pressure difference necessary to produce a unit change in the rate of change of the volume flow rate:

$$I_{l,p} = \frac{\Delta p}{\dot{q}_v} = \frac{\Delta p}{\dfrac{dq_v}{dt}} = \frac{\Delta p dt}{dq_v} \tag{5.17}$$

The SI unit of $I_{l,p}$ is N-s^2-m^{-5} (or kg-m^{-4}). The head-related inertance definition is similar to that of Eq. (5.17):

$$I_{l,h} = \frac{\Delta h}{\dot{q}_v} = \frac{\Delta h}{\dfrac{dq_v}{dt}} = \frac{\Delta h dt}{dq_v} \tag{5.18}$$

The SI unit of $I_{l,h}$ is kg-N^{-1}-m^{-1} (or s^2-m^{-2}). By taking into account that a static pressure difference is connected to the head difference as

$$\Delta p = \rho g \Delta h \tag{5.19}$$

it follows that the two inertance definitions of Eqs. (5.17) and (5.18) are related as

$$I_{l,p} = \rho g I_{l,h} \tag{5.20}$$

The kinetic energy that corresponds to inertance is very similar to the kinetic energy of mechanical and electrical systems, defined as

$$T_{l,p} = \frac{1}{2} I_{l,p} \left[\frac{dV(t)}{dt}\right]^2 = \frac{1}{2} I_{l,p} q_v^2 = \frac{1}{2} \frac{\Delta p}{\dot{q}_v} q_v^2 \tag{5.21}$$

The pressure-definition of Eq. (5.21) is kept here (whose SI unit is J or N-m), as the head-defined energy results in a quantity that is not actually an energy (the interested reader might want to check the units of $T_{l,h} = \frac{1}{2}(\Delta h/\dot{q}_v)q_v^2$).

Example 5.2

A microchannel used in a microfluidic application has the shape and dimensions indicated in Figure 5.3. Calculate the pressure-defined inertance of the liquid that flows through this channel segment, considering that known are the end widths w_1 and w_2, the thickness h, the length l, and the liquid mass density ρ. Obtain the inertance for the particular design with $w_1 = w_2$.

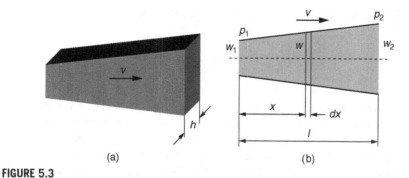

FIGURE 5.3

Tapered Rectangular Cross-Section Pipe with Laminar Flow: (a) Three-Dimensional View; (b) Side View.

Solution

The width w of Figure 5.3(b) can be expressed in terms of x as

$$w = w(x) = w_1 + \frac{w_2 - w_1}{l}x \tag{5.22}$$

The inertance of an elementary parallelepiped of width w, thickness h, and length dx is first determined. The total inertance is calculated afterward by summing all similar elementary prisms (which means integration over the tapered pipe length l). Newton's second law of motion is used for the prismatic element of length dx by expressing the external force as area times pressure difference:

$$dm\frac{dv}{dt} = A(x)[p(x) - p(x + dx)] = A(x)dp(x) \tag{5.23}$$

where it has been considered that $p(x) = p(x + dx) + dp(x)$. Knowing that $\dot{q}_v = A[dv(t)/dt]$, Eq. (5.23) enables expressing the inertance of the element as

$$dI_{l,p} = \frac{dp(x)}{\dot{q}_v} = \frac{dm\dfrac{dv(t)}{dt}}{A(x)^2\dfrac{dv(t)}{dt}} = \frac{\rho A(x)\,dx\dfrac{dv(t)}{dt}}{A(x)^2\dfrac{dv(t)}{dt}} = \frac{\rho dx}{A(x)} = \frac{\rho dx}{w(x)h} \tag{5.24}$$

MATLAB®'s Symbolic Math Toolbox™ is used to integrate the elementary inertance of Eq. (5.24) using the following code:

```
>> syms w1 w2 h l x rho
>> w = w1+(w2-w1)/l*x;
>> in =limit(int(1/w,x)...,x,l,'left')-limit(int(1/w,x)...,
   x,0,'right');
>> inertance = rho/h*in
```

The last MATLAB® command in the previous sequence returns:

$$I_{l,p} = \int dI_{l,p} = \frac{\rho l}{h(w_2 - w_1)} \times \ln\frac{w_2}{w_1} \tag{5.25}$$

It can be seen that the variable x is integrated between the limits of 0 and l in two steps: first the indefinite integral of $1/w(x)$ is calculated, then the definite integral is evaluated as the difference between the upper and lower limits of the indefinite integral; on attempting to directly calculate the definite integral by means of the command $\mathtt{int(1/w,x,0,1)}$, an error message is returned, as MATLAB® cannot determine whether $w(x)$ is between the limits 0 and l.

For $w_1 = w_2 = w$, the trapezoidal prism becomes a parallelepiped with an inertance of

$$I_{l,p}^* = \frac{\rho l}{hw} = \frac{\rho l}{A} \tag{5.26}$$

which is obtained using the MATLAB® command $\mathtt{limit(inertance,w1,w2,'left')}$—this command calculates the limit of $I_{l,p}$ when w_1 reaches w_2 from the left. The full MATLAB® code can be found on the companion website. Equation (5.26), where A is the inner cross-sectional area, is valid for any constant cross-section pipe. ▮

Capacitance

The *capacitance* in the liquid domain reflects the storage capacity by a tank-type device. The capacitance can be defined in terms of static pressure as the ratio between the volume flow rate and the rate of pressure variation:

$$C_{l,p} = \frac{q_v(t)}{\dfrac{dp(t)}{dt}} = \frac{q_v(t)\,dt}{dp(t)} = \frac{\dfrac{dV(t)}{dt}\,dt}{dp(t)} = \frac{dV}{dp} \tag{5.27}$$

In the International System (SI) of units system, the pressure-defined liquid capacitance is measured in m^5-N^{-1} (or m^4-s^2-kg^{-1}). In terms of head, the capacitance quantifies the volume flow rate necessary to change the head rate variation by one unit:

$$C_{l,h} = \frac{q_v(t)}{\dfrac{dh(t)}{dt}} = \frac{q_v(t)\,dt}{dh(t)} = \frac{\dfrac{dV(t)}{dt}\,dt}{dh(t)} = \frac{dV}{dh} \tag{5.28}$$

Due to the connection between head and static pressure, Eq. (5.19), the two capacitances of Eqs. (5.27) and (5.28) are related as

$$C_{l,h} = \rho g C_{l,p} \qquad (5.29)$$

The SI unit of $C_{l,h}$ is m².

Similar to springs in mechanical systems and capacitors in electrical systems, the liquid capacitance is incorporated into potential pressure-form energy as

$$U_{l,p} = \frac{1}{2} C_{l,p}(\Delta p)^2 = \frac{1}{2} \times \frac{(\Delta V)^2}{C_{l,p}} = \frac{1}{2}(\Delta V)(\Delta p) \qquad (5.30)$$

The hydraulic energy, according to Eq. (5.30), is measured in N-m.

Example 5.3
a. Determine the capacitance of the variable cross-section cylindrical vessel sketched in Figure 5.4(a).
b. Use the result to calculate the capacitance of the conical segment of Figure 5.4(b). Known are the end diameters d_1 and d_2, the height h, as well as the liquid mass density ρ and the gravitational acceleration g.

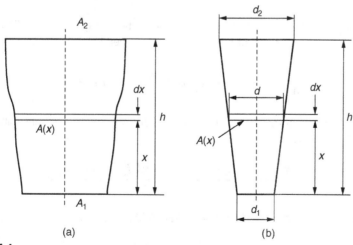

(a) (b)

FIGURE 5.4

(a) Variable Cross-Section Vertical Tank; (b) Conical-Segment Vertical Tank.

Solution

a. The pressure-defined capacitance can be expressed as

$$C_{l,p} = \frac{dV}{dp} = \frac{\dfrac{dV}{dh}}{\dfrac{dp}{dh}} \tag{5.31}$$

The pressure at the bottom of the tank is

$$p = \rho g h \tag{5.32}$$

The volume occupied by the liquid in a variable cross-section tank with a head (height) of h is calculated as

$$V = \int_0^h A(x)\,dx \tag{5.33}$$

where $A(x)$ is the tank variable cross-sectional area at a distance x measured from the bottom of the tank. By taking derivatives of p and V in Eqs. (5.32) and (5.33) with respect to h and substituting these derivatives into Eq. (5.31), the pressure-defined and head-defined capacitances become

$$C_{l,p} = \frac{1}{\rho g} \times \frac{dV}{dh} = \frac{\dfrac{d}{dh}\displaystyle\int_0^h A(x)\,dx}{\rho g}; \quad C_{l,h} = \frac{d}{dh}\int_0^h A(x)\,dx \tag{5.34}$$

b. The diameter defining the variable area $A(x)$ is the one given in Eq. (5.22) with d instead of w and h instead of l. The tank volume is therefore expressed as

$$V = \frac{\pi}{4}\int_0^h \left(d_1 + \frac{d_2 - d_1}{h}x\right)^2 dx = \frac{\pi h}{12}\left(d_1^2 + d_1 d_2 + d_2^2\right) \tag{5.35}$$

Calculating the derivative of V with respect to h in Eq. (5.35) and substituting it into Eqs. (5.34) yields

$$C_{l,p} = \frac{\pi\left(d_1^2 + d_1 d_2 + d_2^2\right)}{12\rho g}; \quad C_{l,h} = \frac{\pi\left(d_1^2 + d_1 d_2 + d_2^2\right)}{12} \tag{5.36}$$

For a cylindrical tank with $d_1 = d_2 = d$, Eqs. (5.36) change to

$$C_{l,p} = \frac{\pi d^2}{4\rho g}; \quad C_{l,h} = \frac{\pi d^2}{4} \tag{5.37}$$

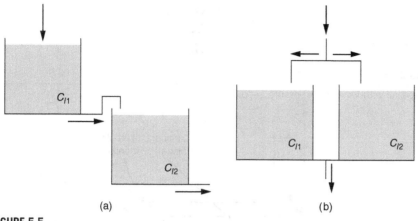

FIGURE 5.5

Liquid Storage Tanks Connected in (a) Series; (b) Parallel.

Liquid capacitances can be connected in series and in parallel in the same way as electrical capacitances. The companion website Chapter 5 gives the derivation of the equivalent series and parallel capacitances, based on Figure 5.5:

$$\begin{cases} \dfrac{1}{C_{ls}} = \dfrac{1}{C_{l1}} + \dfrac{1}{C_{l2}} \\ C_{lp} = C_{l1} + C_{l2} \end{cases} \tag{5.38}$$

Resistance

The liquid motion through pipes or conduits encounters resistance from changes in the conduit direction, the existence of valves, or other constrictions, which act as energy dissipaters. The dissipative action is quantified by means of *resistances*, very similar to electrical systems. Figure 5.6 shows the portion of a pipeline with a valve on it. The valve changes the area of flow and the net result of it is a pressure drop from p_1 to p_2.

Mathematically, the liquid resistance is defined as the ratio of the pressure drop to the volume flow rate:

$$R_{l,p} = \frac{\Delta p}{q_v} = \frac{p_1 - p_2}{q_v} \tag{5.39}$$

The SI unit for $R_{l,p}$ is N-s-m^{-5} (or kg-s^{-1}-m^{-4}). In terms of equivalent head, the liquid resistance is defined as

$$R_{l,h} = \frac{h}{q_v} \tag{5.40}$$

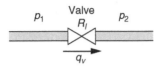

FIGURE 5.6

Pipeline with Valve and Pressure Drop.

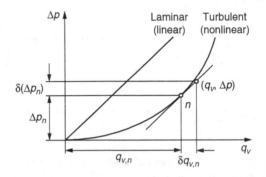

FIGURE 5.7

Laminar- and Turbulent-Regime Relationships between Pressure Drop and Volume Flow Rate.

$R_{l,h}$ is measured in s-m^{-2} in SI units. Again, when the static pressure-head relationship is considered, the two resistances of Eqs. (5.39) and (5.40) are connected as

$$R_{l,p} = \frac{\Delta p}{q_v} = \frac{\rho g h}{q_v} = \rho g R_{l,h} \tag{5.41}$$

Both definitions assume a linear relationship between pressure (or head) variation and volume flow rate. While this linearity covers the laminar flow domain, for turbulent flow, the relationship between these variables becomes nonlinear and it is accepted that pressure varies with the square of the volume flow rate:

$$\Delta p = k q_v^2 \tag{5.42}$$

where k is a constant determined experimentally. Figure 5.7 shows the variation of the pressure in terms of volume flow rate for the laminar and turbulent regimes.

Assuming the pressure varies nonlinearly with the volume flow rate, this relationship can be linearized using a Taylor series expansion about a given *nominal* (or *operational*) point, such as the one indicated with the letter n in Figure 5.7. The Taylor series expansion can keep the first two (linear) terms:

$$\Delta p = \Delta p_n + \frac{\partial(\Delta p)}{\partial q_v}\bigg|_{q_v=q_{v,n}} (q_v - q_{v,n}) \tag{5.43}$$

As seen in Figure 5.7,

$$\Delta p - \Delta p_n = \delta(\Delta p_n); \quad q_v - q_{v,n} = \delta q_{v,n} \tag{5.44}$$

represent small variations in pressure and volume flow rate. As a consequence, Eq. (5.43) can be written as

$$\delta(\Delta p_n) = \left.\frac{\partial(\Delta p)}{\partial q_v}\right|_{q_v=q_{v,n}} \delta q_{v,n} \tag{5.45}$$

Equation (5.45) suggests the linearized resistance about the nominal point is

$$R_{l,p} = \left.\frac{\partial(\Delta p)}{\partial q_v}\right|_{q_v=q_{v,n}} = \frac{\delta(\Delta p_n)}{\delta q_{v,n}} \tag{5.46}$$

Equation (5.46) indicates that the linearized liquid resistance is the ratio of a small variation of the pressure variation to the small variation of the volume flow rate and can be calculated as the partial derivative of the pressure variation in terms of the volume flow rate at the nominal point. By taking into account Eq. (5.42), the linearized resistance defined in Eq. (5.46) becomes

$$R_{l,p} = \left.\frac{\partial(\Delta p)}{\partial q_v}\right|_{q_v=q_{v,n}} = 2kq_{v,n} = 2\left(\frac{\Delta p}{q_v}\right)_{q_v=q_{v,n}} \tag{5.47}$$

In other words, the linearized liquid resistance for turbulent flow is twice the liquid resistance of laminar (linear) flow.

Example 5.4

The pressure variation in a pipe with turbulent liquid flow is expressed in terms of volume flow rate as $\Delta p = q_v + 3q_v^2$. Compare the linearized hydraulic resistance corresponding to this relationship to the linear resistance, which can be defined as $R_{l,p}^* = \Delta p/q_v$ by plotting the two-resistance ratio in terms of q_v. Calculate this ratio for $q_v = 0.01$ m³/s and $q_v = 2$ m³/s.

Solution

According to Eq. (5.47), the linearized hydraulic resistance is obtained as

$$R_{l,p} = \frac{\partial(\Delta p)}{\partial q_v} = 1 + 6q_v \tag{5.48}$$

At the same time, the resistance $R_{l,p}^*$ is calculated as

$$R_{l,p}^* = \frac{\Delta p}{q_v} = 1 + 3q_v \tag{5.49}$$

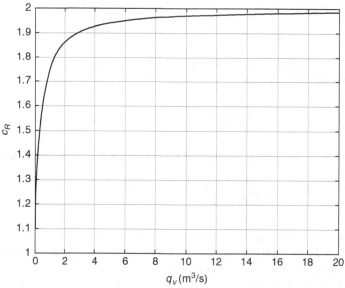

FIGURE 5.8

Linear-to-Nonlinear Hydraulic Resistance Ratio as a Function of Volume Flow Rate.

The following ratio can be used to compare the resistances of Eqs. (5.48) and (5.49):

$$c_R = \frac{R_{l,p}}{R_{l,p}^*} = \frac{1 + 6q_v}{1 + 3q_v} = 2 - \frac{1}{1 + 3q_v} \tag{5.50}$$

which indicates the linearized resistance is almost twice the resistance defined in Eq. (5.49). Equation (5.50) also shows that the ratio increases with the increasing volume flow rate, and the limit is

$$c_{R,\max} = \lim_{q_v \to \infty} c_R = 2 \tag{5.51}$$

The numerical values of the resistance ratio of Eq. (5.50) are $c_R = 1.029$ for $q_v = 0.01$ m³/s and $c_R = 1.857$ for $q_v = 2$ m³/s. Figure 5.8 is the plot of c_R as a function of q_v. ▮▮

In the companion website Chapter 5, the *Hagen-Poiseuille equation* is demonstrated, which gives the resistance of a cylindrical pipe of length l and internal diameter d for laminar flow as

$$R_{l,p} = \frac{\Delta p}{q_v} = \frac{128 \mu l}{\pi d^4} \tag{5.52}$$

Example 5.5

Determine the hydraulic resistance of a tapered pipe having a length of l and end diameters of d_1 and d_2, as indicated in Figure 5.9. Consider laminar flow and calculate the resistance numerical value for $d_1 = 0.5$ m, $d_2 = 0.3$ m, $l = 10$ m, and $\mu = 0.001$ N-s/m^2.

Solution

If an elementary portion of length dx of the tapered pipe is studied, the respective portion is approximately a cylinder of diameter d, as shown in Figure 5.9; therefore, its pressure difference is given by the Hagen-Poiseuille equation as

$$d(\Delta p) = \frac{128\mu dx}{\pi d^4} q_v \tag{5.53}$$

The diameter d is expressed geometrically in terms of x as

$$d = d_2 + (d_1 - d_2)\frac{l - x}{l} \tag{5.54}$$

Substitution of Eq. (5.54) into Eq. (5.53) and integration between the limits of 0 and l with respect to x yields the pressure difference between the input and the output of the pipe:

$$p_1 - p_2 = \Delta p = \int d(\Delta p) = \frac{128\mu l\left(d_1^2 + d_1 d_2 + d_2^2\right)}{3\pi d_1^3 d_2^3} q_v \tag{5.55}$$

which indicates that the fluid resistance of the tapered pipe of Figure 5.8 is

$$R_{l,p} = \frac{128\mu l\left(d_1^2 + d_1 d_2 + d_2^2\right)}{3\pi d_1^3 d_2^3} \tag{5.56}$$

Obviously, when $d_1 = d_2$, and therefore the taper is zero (the tapered pipe becomes a cylindrical one), Eq. (5.56) reduces to the classical Hagen-Poiseuille equation resistance, Eq. (5.52), which corresponds to a constant cross-section cylindrical pipe. The solution to this problem has been obtained using the MATLAB® Symbolic Math Toolbox™; the code is included in the companion website. For the numerical parameters of this example, the hydraulic resistance is $R_{l,p} = 19.72$ N-s/m^5. ∎

Similar to dampers in mechanical systems and resistors in electrical systems, energy is lost through liquid resistances, and this loss can be expressed in pressure form as

$$U_{dl} = \frac{1}{2}R_{l,p}q_v^2 = \frac{1}{2}\frac{(\Delta p)^2}{R_{l,p}} = \frac{1}{2}\Delta p q_v \tag{5.57}$$

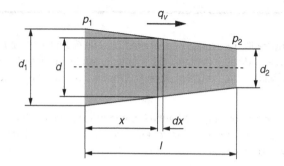

FIGURE 5.9

Tapered Pipe with Laminar Flow.

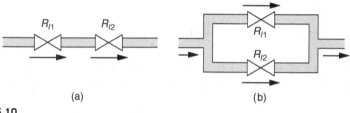

(a)   (b)

FIGURE 5.10

Liquid Resistances Connected in (a) Series; (b) Parallel.

It should be mentioned that Eq. (5.57) actually expresses power: The pressure is measured in N/m², hydraulic resistance is measured in N-s/m⁵, and therefore the unit of U_{dl} is N-m/s, which is the unit for power.

Liquid resistances can be connected in series (as shown in Figure 5.10(a)) or in parallel (as illustrated in Figure 5.10(b)), and the equivalent resistances R_{ls} and R_{lp}, which are derived in the companion website Chapter 5, are

$$\begin{cases} R_{ls} = R_{l1} + R_{l2} \\ \dfrac{1}{R_{lp}} = \dfrac{1}{R_{l1}} + \dfrac{1}{R_{l2}} \end{cases} \tag{5.58}$$

Example 5.6

A microfluidic channel system is shown in Figure 5.11. Calculate the equivalent hydraulic resistance between points 1 and 2 by considering that the liquid losses are produced according to the Hagen-Poiseuille equation. Also calculate the power lost in the microsystem. Known are $l = 100\ \mu m$, $\mu = 0.0005$ N-s/m², $d = 20\ \mu m$ (pipe diameter), and $\Delta p = p_1 - p_2 = 10^3$ N/m².

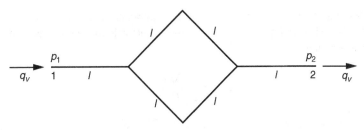

FIGURE 5.11

Six-Component Pipeline System with Flowing Liquid.

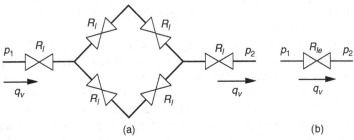

FIGURE 5.12

(a) Actual Microchannel System with Liquid Resistances; (b) Equivalent, One-Resistance Liquid System.

Solution

Figure 5.12(a) shows the microchannel system of Figure 5.11 with the corresponding hydraulic resistances, which are identical. The aim is to obtain the equivalent, one-resistance system of Figure 5.12(b). To achieve that, the resistance, which is equivalent to the four actual resistances in the middle of the system of Figure 5.12(a), can be calculated by combining in parallel two groups of series connected resistances, which yields

$$R_{l1} = \frac{(2R_l)(2R_l)}{(2R_l) + (2R_l)} = R_l \tag{5.59}$$

The equivalent resistance is formed by connecting in series the end resistances of the original system of Figure 5.12(a) to the middle resistance of Eq. (5.59):

$$R_{le} = R_l + R_{l1} + R_l \tag{5.60}$$

Substituting Eq. (5.59) into Eq. (5.60) yields

$$R_{le} = 3R_l = 3 \times 128 \frac{\mu l}{\pi d^4} = 384 \frac{\mu l}{\pi d^4} \tag{5.61}$$

With the numerical values of this example, the equivalent resistance is found to be $R_{le} = 3.8197 \times 10^{13}$ N-s/m^5. Equation (5.57) is used to calculate the power dissipated through the pipeline system:

$$U_{dl} = \frac{1}{2}\frac{(\Delta p)^2}{R_{le}} \tag{5.62}$$

Numerically, the dissipated power is $U_{dl} = 1.31 \times 10^{-8}$ W.   ■

Sources of Hydraulic Energy

For *liquid-level systems*, as is discussed shortly in this chapter, the liquid motion is generated through the head difference among various components, such as tanks and piping. *Hydraulic actuators* are components that convert high input fluid pressure into kinetic energy at the output.

In many liquid applications, the energy necessary to generate flow in a liquid network is provided by *pumps*, which transform the input electric energy into output liquid work, generally manifested as flow rate or equivalent head. Pumps can be of several configurations, such as centrifugal, axial, rotary, or reciprocating in regular-scale applications, as well as diaphragm (or membrane) in micro- and nano-applications. The characteristic centrifugal pump, a widely used configuration, shows the head variation as a function of the flow rate at the output is generally nonlinear, as sketched in Figure 5.13. Its equation is

$$h = h_g - K_h q_v^2 \tag{5.63}$$

where h_g, the *geometric head*, is the maximum head-type energy a pump produces when no energy is lost through friction and K_h is a coefficient related to the energetic losses. A point along the characteristic curve indicates that a specified head h corresponds to a given value of the volume flow rate and that, as the flow rate increases (together with the corresponding losses), the head of the pump

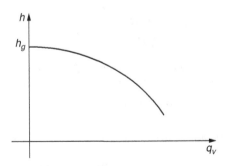

FIGURE 5.13

Head-Flow Rate Characteristic of a Generic Centrifugal Pump.

decreases. An equation similar to Eq. (5.63) can be written when using pressure instead of head:

$$p = p_0 - K_p q_v^2 \qquad (5.64)$$

where p_0 is the maximum attainable pressure at zero flow rate and K_p is a constant depending on the pump construction.

 ### 5.1.3 Liquid Systems

Assembling several of the liquid elements presented thus far in this section leads to the formation of liquid systems. When only inertance and capacitances are present in a liquid system and no forcing source is considered, the natural response can be formulated similarly to mechanical and electrical systems. When external action or energy is applied in any form in a liquid system, the response is forced. Both liquid system responses are studied next.

Natural Response

We study the natural response of free lossless liquid systems that are described by one single variable (single-DOF systems) as well as for systems whose response needs to be formulated in terms of more than one liquid-system variable (multiple-DOF systems). The natural frequencies and corresponding modes (eigenvectors) of multiple-DOF systems can be calculated analytically or by MATLAB®, as shown in previous chapters.

Single-DOF Conservative Liquid Systems

Consider a lumped-parameter liquid system that is defined by inertance I_l and capacitance C_l. This system possesses a natural frequency, which can easily be found using the energy method, similar to the modality used to determine the natural frequencies of single-DOF mechanical or electrical systems. The demonstration of the following natural frequency

$$\omega_n = \frac{1}{\sqrt{I_l C_l}} \qquad (5.65)$$

is given in the companion website Chapter 5. This natural frequency is very similar to the natural frequency of an electrical system formed of an impedance L and a capacitance C, as discussed in Chapter 4. It can be checked that ω_n of Eq. (5.65) is measured in s^{-1}, which is identical to rad-s^{-1}, the unit of natural frequency.

Example 5.7

A lossless pipe segment of given length l and inner diameter d through which a liquid of known properties flows needs its hydraulic natural frequency reduced by 20%. What design changes can be made to achieve that goal?

Solution

Based on Eqs. (5.26) and (5.37), the pressure-defined liquid inertance and capacitance (where the subscript p has been dropped) are

$$\begin{cases} I_l = \dfrac{\rho L}{A} = \dfrac{4\rho l}{\pi d^2} \\ C_l = \dfrac{\pi d^2}{4\rho g} \end{cases} \tag{5.66}$$

By substituting Eqs. (5.66) into Eq. (5.65), the hydraulic natural frequency of the pipe segment becomes

$$\omega_n = \sqrt{\dfrac{g}{l}} \tag{5.67}$$

As a side note, the particular natural frequency of Eq. (5.67) is identical to the natural frequency of a simple pendulum of length l under the action of gravity.

Equation (5.67) indicates that changes in the natural frequency can be operated by using length alterations. The requirement is that the new natural frequency be

$$\omega_n^* = \omega_n - 0.2\omega_n = 0.8\omega_n \tag{5.68}$$

which, based on Eq. (5.67), is

$$\omega_n^* = \sqrt{\dfrac{g}{l^*}} \tag{5.69}$$

A combination of Eqs. (5.67), (5.68), and (5.69) yields the new length:

$$l^* = \dfrac{l}{0.64} = 1.56l \tag{5.70}$$

which shows that an increase of 56% in the pipe length is necessary to produce a reduction of 20% in the natural frequency. ■

Multiple-DOF Conservative Liquid Systems

The natural response of multiple-DOF conservative liquid systems is studied employing the analytical approach and MATLAB®. Schematic liquid circuits can be drawn and analyzed similarly to electrical systems, as shown in the following example. It should be mentioned that, while relative agreement exists in terms of the symbols used for hydraulic elements such as pumps, actuators, or resistances, there is no consensus on the graphical representation of liquid inertances and capacitances. Due to the similitude between electrical and hydraulic systems, the symbol used for electrical inductances is used for hydraulic inertances, whereas hydraulic capacitances are symbolized similar to electrical capacitances.

Analytical Approach The analytical approach to finding the natural frequency of multiple-DOF liquid systems is similar to the approach used for mechanical and electrical systems in Chapters 3 and 4 and is illustrated by an example.

Example 5.8

Utilize the energy method to derive the mathematical model for the liquid system whose circuit is sketched in Figure 5.14. Calculate its natural frequencies and determine the corresponding modes (eigenvectors) by the analytical approach. Consider $I_{l1} = I_{l2} = I_l = 2 \times 10^6 \, \text{kg/m}^4$ and $C_{l1} = C_{l2} = C_l = 3 \times 10^{-8} \, \text{m}^4\text{-s}^2\text{-kg}^{-1}$.

Solution

Using the volume flows indicated in Figure 5.14 and based on the energy Eq. (5.21) and Eq. (5.30), the total energy corresponding to the four liquid components is

$$E_l = \frac{1}{2} I_{l1} \dot{v}_1^2 + \frac{(v_1 - v_2)^2}{2C_{l1}} + \frac{1}{2} I_{l2} \dot{v}_2^2 + \frac{1}{2C_{l2}} v_2^2 \tag{5.71}$$

with v indicating volume (not to be confounded with translatory velocity, which uses the same symbol). This energy is constant, therefore its time derivative is zero, which leads to

$$\dot{v}_1 \left[I_{l1} \ddot{v}_1 + \frac{1}{C_{l1}} (v_1 - v_2) \right] + \dot{v}_2 \left[I_{l2} \ddot{v}_2 + \frac{1}{C_{l1}} (v_2 - v_1) + \frac{1}{C_{l2}} v_2 \right] = 0 \tag{5.72}$$

Equation (5.72) needs to be satisfied at all times, and since the volume flow rates cannot be zero at all times, the only way of validating Eq. (5.72) is when

$$\begin{cases} I_{l1} \ddot{v}_1 + \frac{1}{C_{l1}} v_1 - \frac{1}{C_{l1}} v_2 = 0 \\ I_{l2} \ddot{v}_2 - \frac{1}{C_{l1}} v_1 + \left(\frac{1}{C_{l1}} + \frac{1}{C_{l2}} \right) v_2 = 0 \end{cases} \tag{5.73}$$

The solution to Eq. (5.73) is harmonic:

$$\begin{cases} v_1 = V_1 \sin(\omega t) \\ v_2 = V_2 \sin(\omega t) \end{cases} \tag{5.74}$$

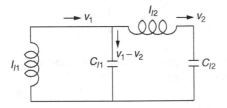

FIGURE 5.14
Two-Mesh Conservative Liquid System.

where V_1 and V_2 are volume amplitudes. Equations (5.74) are substituted into Eqs. (5.73) and the result is the following algebraic equations system, as $\sin(\omega t)$ cannot be zero at all times:

$$
\begin{cases}
\left(\dfrac{1}{C_{l1}} - \omega^2 I_{l1}\right)V_1 - \dfrac{1}{C_{l1}}V_2 = 0 \\[2mm]
-\dfrac{1}{C_{l1}}V_1 + \left(\dfrac{1}{C_{l1}} + \dfrac{1}{C_{l2}} - \omega^2 I_{l2}\right)V_2 = 0
\end{cases}
\tag{5.75}
$$

The homogeneous equations system (5.75) has nontrivial solutions in V_1 and V_2 when the determinant of the system is zero:

$$
\begin{vmatrix}
\dfrac{1}{C_{l1}} - \omega^2 I_{l1} & -\dfrac{1}{C_{l1}} \\[3mm]
-\dfrac{1}{C_{l1}} & \dfrac{1}{C_{l1}} + \dfrac{1}{C_{l2}} - \omega^2 I_{l2}
\end{vmatrix} = 0
\tag{5.76}
$$

which, by also using the specific values of the inductances and capacitances of the problem, results in the characteristic equation

$$
I_l^2 \omega^4 - 3\frac{I_l}{C_l}\omega^2 + \frac{1}{C_l^2} = 0
\tag{5.77}
$$

The solution of this equation in ω provides the two natural frequencies of the hydraulic system:

$$
\omega_{n1} = \sqrt{\frac{3 - \sqrt{5}}{2I_l C_l}}; \quad \omega_{n2} = \sqrt{\frac{3 + \sqrt{5}}{2I_l C_l}}
\tag{5.78}
$$

With the numerical data of the problem, the natural frequencies are $\omega_{n1} = 2.5$ rad/s and $\omega_{n2} = 6.6$ rad/s.

The following amplitude ratio is obtained from the first Eq. (5.75):

$$
\frac{V_1}{V_2} = \frac{1}{1 - \omega^2 I_l C_l}
\tag{5.79}
$$

Using ω_{n1} in Eq. (5.79) and requiring that the corresponding eigenvector is unit-norm, the following equations are obtained:

$$
\left(\frac{V_1}{V_2}\right)_{\omega=\omega_{n1}} = \frac{V_{11}}{V_{12}} = 1.6; \quad \sqrt{V_{11}^2 + V_{12}^2} = 1
\tag{5.80}
$$

The eigenvector corresponding to Eqs. (5.80) is

$$\{V\}_{\omega=\omega_{n1}} = \begin{Bmatrix} V_{11} \\ V_{12} \end{Bmatrix} = \begin{Bmatrix} 0.85 \\ 0.53 \end{Bmatrix} \tag{5.81}$$

As seen in Eq. (5.81), the two flows v_1 and v_2 have identical directions and the amplitude V_1 is larger than the amplitude V_2 during the modal motion at the resonant frequency ω_{n1}. The second natural frequency of Eq. (5.78) is now substituted into the amplitude ratio of Eq. (5.79) and a unit-norm eigenvector is again sought, which results in the following equations:

$$\left(\frac{V_1}{V_2}\right)_{\omega=\omega_{n2}} = \frac{V_{21}}{V_{22}} = -0.62; \quad \sqrt{V_{21}^2 + V_{22}^2} = 1 \tag{5.82}$$

whose corresponding eigenvector is

$$\{V\}_{\omega=\omega_{n2}} = \begin{Bmatrix} V_{21} \\ V_{22} \end{Bmatrix} = \begin{Bmatrix} -0.53 \\ 0.85 \end{Bmatrix} \tag{5.83}$$

Inspection of Eq. (5.83) shows that, during the second modal motion at ω_{n2}, the two flows have opposite directions (this is indicated by the minus sign) and the amplitude V_1 is smaller than the amplitude V_2. ■

Using MATLAB® to Calculate Natural Frequencies, the Eigenvalue Problem Similar to mechanical and electrical systems, the natural response of a multiple-DOF hydraulic system can be formulated in vector-matrix form as an eigenvalue problem, and MATLAB® can be employed to solve for eigenvalues and eigenvectors. The equation describing the free vibrations of a multiple-DOF hydraulic system can be written as

$$[I_l]\{\ddot{v}\} + [C_l]\{v\} = \{0\} \tag{5.84}$$

where $[I_l]$ is the *inertance matrix*, $[C_l]$ is the *capacitance matrix*, and $\{v\}$ is the *volume vector*. It can be shown by following a development similar to the one applied to mechanical systems in Chapter 3, that when sinusoidal solution of the type $\{v\} = \{V\} \sin(\omega t)$ is sought for Eq. (5.84), the following equation is obtained:

$$\det([I_l]^{-1}[C_l] - \lambda[I]) = 0 \tag{5.85}$$

where $\lambda = \omega^2$ are the eigenvalues, $[I]$ is the identity matrix and $[I_l]^{-1}[C_l] = [D_l]$ is the *liquid dynamic matrix*.

Once $[D_l]$ has been determined, the MATLAB® command $[V, D_l] = eig(D_l)$, which was utilized in Chapters 3 and 4, returns the modal matrix V, whose columns are the eigenvectors, and the diagonal matrix D_l, whose diagonal elements are the eigenvalues.

Example 5.9

Calculate the eigenvalues and the eigenvectors corresponding to the hydraulic system of Example 5.8 using the eigenvalue method and MATLAB®.

Solution

Equations (5.75) can be arranged in the matrix-vector form of Eq. (5.84) with

$$[I_l] = \begin{bmatrix} I_{l1} & 0 \\ 0 & I_{l2} \end{bmatrix}; \quad [C_l] = \begin{bmatrix} \dfrac{1}{C_{l1}} & -\dfrac{1}{C_{l1}} \\ -\dfrac{1}{C_{l1}} & \dfrac{1}{C_{l1}} + \dfrac{1}{C_{l2}} \end{bmatrix} \tag{5.86}$$

By taking into consideration that the inertances are identical and the conductances are also identical, the dynamic matrix is calculated using MATLAB® symbolic calculation:

$$[D_l] = [I_l]^{-1}[C_l] = \frac{1}{I_l C_l} \begin{bmatrix} 1 & -1 \\ -1 & 2 \end{bmatrix} \tag{5.87}$$

As expected, the eigenvalues returned by MATLAB® are the squares of the natural frequencies of Eqs. (5.78) and the eigenvectors are

$$\{V\}_{\omega=\omega_{n1}} = \begin{Bmatrix} -0.85 \\ -0.53 \end{Bmatrix}; \quad \{V\}_{\omega=\omega_{n2}} = \begin{Bmatrix} -0.53 \\ 0.85 \end{Bmatrix} \tag{5.88}$$

which are essentially the eigenvectors obtained in Example 5.8 (the minus signs in the first eigenvector of Eq. (5.88) shows that the two components move in the same direction).

Forced Response of Liquid-Level Systems

As mentioned previously in this chapter, one modality of generating the forced response of liquid systems is by means of tanks, where flow connects in and out with pipe lines, for instance. Often, in such liquid systems, known as *liquid-level systems*, hydraulic calculations can be performed without considering the inertia effects; as a consequence, such systems consist of only capacitances and resistances.

Example 5.10

A liquid-level system is formed of a tank of capacitance C_l communicating with a pipe segment equipped with a valve of resistance R_l, as sketched in Figure 5.15.

a. Derive a mathematical model for this system by connecting the output (either the flow rate q_o or the head h) to the input flow rate q_i. Assume that the pressure at the input and output ports is zero.

b. For a unit ramp input, find the solution $q_o(t)$ and plot it against time considering that $R_{l,p} = 5 \times 10^4$ N-s/m⁵ and $C_{l,p} = 2 \times 10^{-6}$ m⁵/N. The slope of the ramp input is $Q = 10^{-4}$ m³/s.

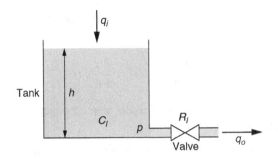

FIGURE 5.15

Liquid-Level System with Tank and Valve.

Solution

a. If we use the pressure definitions of resistances and capacitances, the following equations can be written:

$$\begin{cases} p(t) = R_{l,p}q_o(t) \\ q_i(t) - q_o(t) = C_{l,p}\dfrac{dp(t)}{dt} \end{cases} \tag{5.89}$$

where the pressure is at the bottom of the tank just before the valve. Taking the time derivative of the first Eq. (5.89) and considering the resistance is constant, the following equation results:

$$\frac{dp(t)}{dt} = R_{l,p}\frac{dq_o(t)}{dt} \tag{5.90}$$

which, substituted into the second Eq. (5.89), produces

$$R_{l,p}C_{l,p}\frac{dq_o(t)}{dt} + q_o(t) = q_i(t) \tag{5.91}$$

Equation (5.91) represents the mathematical model of the liquid-level system of Figure 5.15, where q_i is the input and q_o is the output. Since the order of the differential equation is one, the physical system is a first-order one.

An alternate model can be obtained by considering that the pressure before the resistance is actually

$$p(t) = \rho g h(t) \tag{5.92}$$

It is simple to see that

$$\frac{dp(t)}{dt} = \rho g \frac{dh(t)}{dt} \tag{5.93}$$

Equations (5.92) and (5.93) are used in conjunction with Eq. (5.89) to obtain the first-order differential equation:

$$R_{l,p}C_{l,p}\frac{dh(t)}{dt} + h(t) = \frac{R_{l,p}}{\rho g}q_i(t) \tag{5.94}$$

which represents another mathematical model, where the output is the head h and the input is the flow rate q_i.

b. When the input flow rate is of a ramp form, namely, $q_i = Qt$ (where Q is a constant), the solution to the differential Eq. (5.91) is

$$q_o(t) = \left[t + \left(e^{-\frac{t}{\tau}} - 1\right)\tau\right]Q \tag{5.95}$$

which can be found using the MATLAB® command dsolve, as introduced in Chapter 4. The time constant, $\tau = R_{l,p}C_{l,p}$ has a value of 0.1 s for the numerical values of this example. Due to its ramp nature, the input flow rate grows to infinity when time goes to infinity, and this is shown in the plot of Figure 5.16. The following code generates Eq. (5.95):

```
>> dsolve('R*C*Dqo + qo = q*t','qo(0) = 0')
```

after defining the symbolic variables.

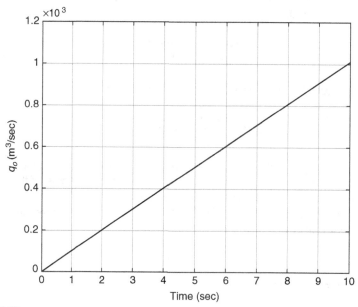

FIGURE 5.16

Output Flow Rate as a Function of Time.

Example 5.11

Derive the mathematical model of the liquid system sketched in Figure 5.17 by considering the inputs to the system are the pressure p_1 and the flow rate q_2, whereas the output is the flow rate q_o.

Solution

Based on the hydraulic capacitance and resistance definitions, as well as on Figure 5.17, the following relationships can be formulated:

$$R_{l1} = \frac{p_1(t) - p_2(t)}{q_i(t)}; R_{l2} = \frac{p_2(t) - p_3(t)}{q_1(t)}; R_{l3} = \frac{p_3(t) - p_a}{q_o(t)};$$

$$C_{l1} = \frac{q_i(t) - q_1(t)}{\dfrac{dp_2(t)}{dt}}; C_{l2} = \frac{q_1(t) + q_2(t) - q_o(t)}{\dfrac{dp_3(t)}{dt}} \tag{5.96}$$

The third Eq. (5.96) allows expressing p_3 as

$$p_3(t) = R_{l3}q_o(t) + p_a \tag{5.97}$$

which can be used to obtain q_1 from the fifth Eq. (5.96):

$$q_1(t) = R_{l3}C_{l2}\frac{dq_o(t)}{dt} + q_o(t) - q_2(t) \tag{5.98}$$

Combining now the second Eq. (5.96) with Eqs. (5.97) and (5.98) enables expressing the pressure p_2:

$$p_2(t) = R_{l2}R_{l3}C_{l2}\frac{dq_o(t)}{dt} + (R_{l2} + R_{l3})q_o(t) - R_{l2}q_2(t) + p_a \tag{5.99}$$

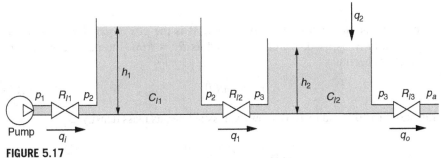

FIGURE 5.17

Liquid-Level System with Pump, Two Tanks, and Three Valves.

The first Eq. (5.96) and Eq. (5.99) yield

$$q_i(t) = \frac{1}{R_{l1}}\left[p_1(t) - R_{l2}R_{l3}C_{l2}\frac{dq_o(t)}{dt} - (R_{l2} + R_{l3})q_o(t) + R_{l2}q_2(t) - p_a\right] \quad (5.100)$$

Eventually, the fourth Eq. (5.96) is used in conjunction with Eqs. (5.98) and (5.100) to produce the following second-order differential equation:

$$
\begin{aligned}
& R_{l2}R_{l3}C_{l1}C_{l2}\frac{d^2q_o(t)}{dt^2} + \left[(R_{l2} + R_{l3})C_{l1} + \frac{R_{l2}R_{l3}C_{l2}}{R_{l1}} + R_{l3}C_{l2}\right]\frac{dq_o(t)}{dt} \\
& + \frac{R_{l1} + R_{l2} + R_{l3}}{R_{l1}}q_o(t) = R_{l2}C_{l1}\frac{dq_2(t)}{dt} + \frac{R_{l1} + R_{l2}}{R_{l1}}q_2(t) + \frac{p_1(t)}{R_{l1}} - \frac{p_a}{R_{l1}}
\end{aligned}
\quad (5.101)
$$

Equation (5.101), which can be solved independently for q_o, is the mathematical model of the liquid system of Figure 5.17.

5.2 PNEUMATIC SYSTEMS MODELING

In pneumatic systems, the motion agent is a gas, most often air. Gases are compressible, particularly at large velocities; therefore, pneumatic systems produce responses that are slower than liquid systems, but for velocities that are smaller than the sound velocity, they are nearly incompressible. We discuss some basic gas laws then introduce the pneumatic elements and modeling of pneumatic systems.

5.2.1 Gas Laws

Gas laws describe either the *state* or the *transformation* (*process*) between different states of a gaseous substance. The *perfect* (or *ideal*) *gas law* postulates that, for a given gas state that is defined by pressure p, volume V, and absolute (Kelvin-scale) temperature θ, the following relationship applies for a gas mass of m:

$$pV = \frac{m}{M}R\theta \quad (5.102)$$

where R is the *universal gas constant* and M is the *gas molecular mass*. If the gas constant R_g is used, which is defined as $R_g = R/M$, the perfect gas law of Eq. (5.102) becomes

$$pV = mR_g\theta \quad (5.103)$$

Equations (5.102) and (5.103) allow expressing the gas mass density as

$$\rho = \frac{m}{V} = \frac{pM}{R\theta} = \frac{p}{R_g\theta} \quad (5.104)$$

Transformation or process gas laws connect two states using specific conditions. The *polytropic transformation* is defined by the equation

$$p = a\rho^n = a\left(\frac{m}{V}\right)^n \tag{5.105}$$

where a is a constant and n is the *polytropic exponent*. Several particular transformations, relating to actual physical conditions, can be derived from the general polytropic transformation, each defined by a specific exponent n. All assume the gas mass m is constant. The *adiabatic transformation*, which considers no heat exchange between the gas and its surroundings, has an exponent defined as

$$n = \frac{c_p}{c_v} \tag{5.106}$$

where c_p is the *constant-pressure specific heat* and c_v is the *constant-volume specific heat*. The specific heat is defined as the heat (energy) Q necessary to raise the temperature of the mass unit by one degree:

$$c = \frac{Q}{m\Delta\theta} \tag{5.107}$$

Transformations that keep the temperature constant are called *isothermal*, and for such processes, the exponent is $n = 1$. That can easily be checked as follows: Eq. (5.102) shows that $pV = $ constant, a condition that also results from Eq. (5.105) when $n = 1$. *Constant-pressure transformations* (also called *isobaric*) satisfy the condition that the ratio V/θ is a constant, as it results from Eq. (5.102). This means the polytropic exponent needs to be $n = 0$, as can be checked in Eq. (5.105). Eventually, *constant-volume transformations* result in equations of the type $p/\theta = $ constant, as seen in Eq. (5.102). It can also be checked out that such processes imply $n \to \infty$ because $V = $ constant when $n \to \infty$, Eq. (5.105).

5.2.2 Pneumatic Elements

The pneumatic elements are defined similarly to liquid systems, particularly the inertance and the resistance, but as mentioned in the introduction to this chapter, the mass flow rate is used instead of the volume flow rate that operates for liquids. The pneumatic capacitance is discussed in terms of the specific gas transformation.

Inertance

A column of gas moving in a duct possesses kinetic energy; therefore, its *inertance* is defined as

$$I_g = \frac{\Delta p}{\dot{q}_m} = \frac{\Delta p}{\rho \dot{q}_v} = \frac{1}{\rho}I_l \tag{5.108}$$

where I_l is the pressure-defined inertance of the liquid studied at the beginning of this chapter. The SI unit of I_g is m^{-1}. The kinetic energy of a gas is

$$T_g = \frac{1}{2}I_g q_m^2 = \frac{1}{2} \times \frac{I_l}{\rho}\rho^2 q_v^2 = \rho T_l \tag{5.109}$$

where T_l is the liquid kinetic energy; the gas energy's SI unit is kg^2-m^{-1}-s^{-2}.

Capacitance

For constant volume, the pneumatic capacitance of a container is defined as

$$C_g = \frac{q_m(t)}{\dfrac{dp(t)}{dt}} = \frac{dm(t)}{dp(t)} = \rho\frac{q_v(t)}{\dfrac{dp(t)}{dt}} = \rho C_l \tag{5.110}$$

where C_l is the capacitance of a liquid. The SI unit for gas capacitance is m-s^2. For a polytropic process, where the pressure is defined as in Eq. (5.105),

$$dp = n\frac{p}{\rho}d\rho \tag{5.111}$$

which indicates the capacitance of Eq. (5.110) can be written for constant volume as

$$C_g = \frac{dm}{dp} = \frac{Vd\rho}{dp} = \frac{V}{nR_g\theta} \tag{5.112}$$

where Eq. (5.104) has been used. It can be seen that, for a constant-pressure transformation where the polytropic coefficient is $n = 0$, the pneumatic capacitance is infinity; whereas for a constant-volume process with $n \to \infty$, the pneumatic capacitance is equal to zero. In general, the pneumatic capacitance is variable, as it depends on temperature; for an isothermal transformation only (where the temperature is constant), the pneumatic capacitance is constant and equal to

$$C_g = \frac{V}{R_g\theta} \tag{5.113}$$

The energy stored by a pneumatic capacitive element is

$$U_g = \frac{1}{2}C_g p^2 = \frac{1}{2}\rho C_l p^2 = \rho U_l \tag{5.114}$$

and its SI unit is N-kg-m^{-2} (or kg^2-m^{-1}-s^{-2}).

Resistance

The pneumatic resistance is defined in terms of pressure variation and mass flow rate as

$$R_g = \frac{\Delta p}{q_m} = \frac{\Delta p}{\rho q_v} = \frac{1}{\rho}R_l \tag{5.115}$$

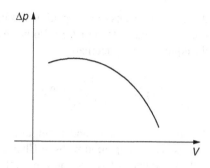

FIGURE 5.18

Characteristic Curve of a Fan.

where R_l is the resistance of a liquid. Gas resistance is measured in m^{-1}-s^{-1} in the International System of units. The energy dissipated through pneumatic resistance is

$$U_{dg} = \frac{1}{2}R_g q_m^2 = \frac{1}{2} \times \frac{R_l}{\rho}\rho^2 q_v^2 = \rho U_{dl} \qquad (5.116)$$

and the SI energy unit is N^2-s-m^{-3} (or kg^2-m^{-1}-s^{-3}).

Sources of Pneumatic Energy

Similar to liquid systems, energy needs to be supplied to a pneumatic system to allow operation. The main pneumatic energy sources are the *fans* or *blowers*, for which the delivered pressure is a parabolic function of the air volume; this is expressed by means of an equation that also includes the energy losses, such as those due to impeller friction or shock:

$$\Delta p = c_1\left(\frac{c_2}{V} - 1\right)^2 \qquad (5.117)$$

whose characteristic curve is shown in Figure 5.18; c_1 and c_2 are constants, depending on the type of pneumatic source and performance.

 ### 5.2.3 **Pneumatic Systems**

The natural response of conservative pneumatic systems is briefly discussed next, followed by an example illustrating the forced response of a pneumatic system with losses.

Natural Response

Pneumatic systems with no losses, therefore defined by only inertance and capacitance properties, can be analyzed in terms of their natural response, similarly to liquid, mechanical, or electrical systems. For a single-DOF pneumatic system, the natural frequency is calculated by means of an equation similar to Eq. (5.65), where the subscript g (gas) should be used instead of l, for liquid. Also using Eqs. (5.108) and (5.110), the natural frequency of a pneumatic system is

$$\omega_{n,g} = \frac{1}{\sqrt{I_g C_g}} = \frac{1}{\sqrt{I_l C_l}} = \omega_{n,l} \tag{5.118}$$

therefore, a single-DOF pneumatic system and a single-DOF liquid system have identical natural frequencies. The companion website Chapter 5 includes examples of calculating the natural frequencies of single- and multiple-DOF pneumatic systems using the energy method and MATLAB®, but the procedure is identical to that used for the natural response of liquid systems and is not pursued here. However, a few problems dedicated to this topic are proposed at end of this chapter.

Forced Response

When pneumatic sources are included in a pneumatic system, the dynamic response is forced. We study the forced response of a two-DOF pneumatic system with resistance losses and negligible inertia.

Example 5.12

A fan is used to pressurize the container of capacitance C_{g2} as in Figure 5.19, where another vessel of capacitance C_{g1} is connected to the target vessel and the fan. Derive the mathematical model of this pneumatic system that connects the output pressure p_o to the input pressure created by the fan, p_i, by also considering the duct losses R_{g1} and R_{g2}.

Solution

The following equations are written for the four pneumatic components:

$$R_{g1} = \frac{p_i(t) - p(t)}{q_{mi}(t)}; \ R_{g2} = \frac{p(t) - p_o(t)}{q_{mo}(t)}; \ C_{g1} = \frac{q_{mi}(t) - q_{mo}(t)}{\dfrac{dp(t)}{dt}}; \ C_{g2} = \frac{q_{mo}(t)}{\dfrac{dp_o(t)}{dt}} \tag{5.119}$$

The mass flow rate of the second Eq. (5.119) is substituted into the fourth Eq. (5.119), which enables expressing the pressure p as

$$p(t) = p_o(t) + R_{g2} C_{g2} \frac{dp_o(t)}{dt} \tag{5.120}$$

and this results in

$$\frac{dp(t)}{dt} = \frac{dp_o(t)}{dt} + R_{g2} C_{g2} \frac{d^2 p_o(t)}{dt^2} \tag{5.121}$$

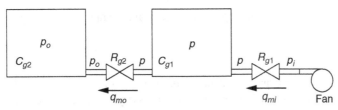

FIGURE 5.19

Pneumatic System with Fan, Two Containers, and Two Valves.

The first and the third Eqs. (5.119) are used in conjunction with Eqs. (5.120) and (5.121), which yields the following differential equation:

$$R_{g1}R_{g2}C_{g1}C_{g2}\frac{d^2 p_o(t)}{dt^2} + (R_{g1}C_{g1} + (R_{g1} + R_{g2})C_{g2})\frac{dp_o(t)}{dt} + p_o(t) = p_i(t) \quad (5.122)$$

Equation (5.122), which can independently be solved for p_o, and Eq. (5.120), which can subsequently be solved for p, form the mathematical model of the pneumatic system of Figure 5.19. ◼

5.3 THERMAL SYSTEMS MODELING

In thermal systems, the focus is on heat and mass exchange among various states of a medium or different media. The analysis in this section is restricted to heat exchange, but more advanced notions can be learned from texts specializing in heat and mass transfer. We introduce the thermal elements of capacitance and resistance, followed by the mathematical modeling of thermal systems. Since thermal inertia can safely be neglected, thermal systems behave as first-order systems. To keep notation unitary with that used for electrical, fluid, and pneumatic systems, the symbol q_{th} or simply q is used here to indicate the *heat flow rate*, which is defined as the time derivative of the *heat flow*, or *thermal energy Q*:

$$q_{th}(t) = q(t) = \frac{dQ(t)}{dt} \quad (5.123)$$

Another notation used here is θ for *temperature*; this symbol is preferred to the symbol t, which has been reserved to denote time.

5.3.1 Thermal Elements

As inertia effects are negligible in thermal systems, the elements of interest are the *thermal capacitance* (involved with thermal energy storing) and *thermal resistance* (responsible for energy losses). These amounts are assumed to be of a lumped-parameter nature. The role of the electrical charge rate (current) in electrical systems or flow rate in fluid systems is played by the heat flow rate q in defining these thermal quantities.

Capacitance

Thermal capacitance is connected to the energy storage capacity and assumes no energy losses. It is defined as the heat flow necessary to change the temperature rate of a medium by one unit in one second:

$$C_{th} = \frac{q(t)}{\dfrac{d\theta(t)}{dt}} = \frac{\dfrac{dQ(t)}{dt}}{\dfrac{d\theta(t)}{dt}} = \frac{dQ}{d\theta} \tag{5.124}$$

The SI unit for thermal capacitance is N-m-K^{-1} (or J-K^{-1}). As known from physics, the heat quantity Q is related to a change of temperature θ in a medium (solid or fluid) of mass m as

$$Q = mc\theta \tag{5.125}$$

where c is the *specific heat*. The specific heat can be defined under process conditions of constant pressure, when denoted by c_p, or constant volume, when the symbol c_v is used. Applying the temperature derivative to the variables of Eq. (5.125) results in

$$\frac{dQ}{d\theta} = mc \tag{5.126}$$

Comparison of Eqs. (5.124) and (5.126) shows that

$$C_{th} = mc \tag{5.127}$$

Resistance

Thermal resistance is formulated with regard to energy losses and under the assumption of no storage capacity. Similar to electrical or fluid systems, thermal resistance is defined as the temperature variation produced by a unit heat flow rate:

$$R_{th} = \frac{\Delta\theta}{q} \tag{5.128}$$

To find a specific expression for R_{th} in terms of physical parameters, the types of heat transfer processes (*conduction*, *convection*, and *radiation*) need to be considered separately, as each process is governed by a specific law that connects the heat flow rate to temperature variation. Radiation, which involves emission of heat by electromagnetic waves, does not require the presence of a medium to transmit the energy, unlike conduction and convection, and is not studied here.

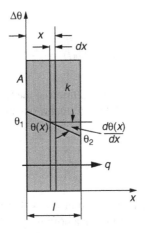

FIGURE 5.20

Conduction Heat Transfer through a Planar Wall.

Conduction

In *conduction*, the heat is transmitted through one single medium (solid or fluid) as energy released by particles that possess more energy to adjacent particles having less energy. *Fourier's law of heat conduction* governs the conduction process, which, based on the wall schematic of Figure 5.20, is expressed as

$$q = -kA\frac{d\theta(x)}{dx} \tag{5.129}$$

where k is the *thermal conductivity* of the wall material, A is the area of the surface normal to the heat flow direction, dx is the thickness of an elementary layer (of total thickness l), and $\theta(x)$ is the temperature at the surface determined by the abscissa x. The temperature to the left of the wall, θ_1, is assumed higher than the temperature to the right of the wall, θ_2; therefore, the heat flow direction is from left to right. The minus sign in Eq. (5.129) indicates that the temperature decreases in the wall from left to right; therefore, the slope of the temperature as a function of distance is negative. As a consequence, a positive heat flow rate going from the higher temperature to the lower one is possible only with a minus sign as in Eq. (5.129).

For constant heat flow rate, integration of Eq. (5.129) results in

$$qx = -kA\theta(x) + C \tag{5.130}$$

The integration constant is found using $\theta = \theta_1$ for $x = 0$, which results in $C = kA\theta_1$, whose value is substituted back in Eq. (5.130) to yield

$$\theta_1 - \theta(x) = \Delta\theta(x) = \frac{q}{kA}x \tag{5.131}$$

For $x = l$, Eq. (5.131) becomes

$$\theta_1 - \theta_2 = \Delta\theta(l) = \frac{q}{kA}l \qquad (5.132)$$

Comparison of Eq. (5.132) with the definition of the thermal resistance, Eq. (5.128), indicates that the conductance-related thermal resistance of a planar wall is (ignoring the minus sign)

$$R_{th} = \frac{l}{kA} \qquad (5.133)$$

The companion website Chapter 5 demonstrates that the thermal resistance corresponding to the radial heat flow through a hollow cylinder of internal radius r_i, external radius r_o, length l, and thermal conductivity k is

$$R_{th} = \frac{\ln\dfrac{r_o}{r_i}}{2\pi l k} \qquad (5.134)$$

Thermal resistances can be connected in series, in parallel, or in series/parallel combinations. A series combination is sketched in Figure 5.21(a), and a parallel one in Figure 5.21(b). The equivalent series ($R_{th,s}$) and parallel ($R_{th,p}$) resistances (which are derived in the companion website Chapter 5) are

$$\begin{cases} R_{th,s} = R_{th,1} + R_{th,2} \\ \dfrac{1}{R_{th,p}} = \dfrac{1}{R_{th,1}} + \dfrac{1}{R_{th,2}} \end{cases} \qquad (5.135)$$

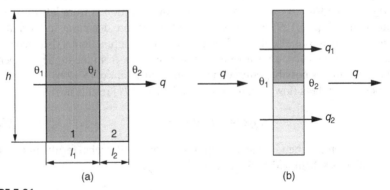

(a)   (b)

FIGURE 5.21

Connection of Planar Walls in (a) Series; (b) Parallel.

Example 5.13

Consider the heat conduction through the three-component wall sketched in Figure 5.22. The external layers are identical, and the middle layer has its thermal conductivity $k_2 = 4k_1$, where k_1 is the thermal conductivity of the external layers. Known are θ_i, θ_o $(\theta_i > \theta_o)$, these are the indoor and outdoor temperatures; l and A, wall component thicknesses and common surface area; and k_1, thermal conductivity of external layers.

a. Calculate the heat flow rate through the composite wall.
b. Calculate the temperature θ_i' at the interface between the inner wall layer and the middle wall layer.
   Numerical application: $l = 0.01$ m, $A = 20$ m², $\theta_i = 23°C$, $\theta_o = 5°C$, $k_1 = 0.15$ W/m-C.

Solution

a. If considering only heat flow through conduction from the inside toward the outside, as shown in Figure 5.22, then the three panels behave as three thermal resistors connected in series. The equivalent thermal resistance is therefore

$$R_{th,e} = 2R_{th,1} + R_{th,2} = \frac{2l}{k_1 A} + \frac{4l}{k_2 A} = \frac{3l}{k_1 A} \tag{5.136}$$

According to Eq. (5.128), the heat flow rate is

$$q = \frac{\theta_i - \theta_o}{R_{th,e}} = \frac{k_1 A}{3l}(\theta_i - \theta_o) \tag{5.137}$$

The heat flow rate's numerical value is $q = 1800$ W.

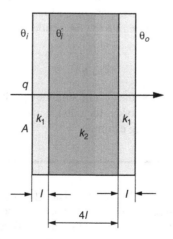

FIGURE 5.22

Three-Panel Composite Wall.

b. The same flow rate passes through the inner layer; therefore,

$$q = \frac{\theta_i - \theta_i'}{R_{th,1}} = \frac{k_1 A}{l}(\theta_i - \theta_i')$$
(5.138)

Equations (5.137) and (5.138) yield the unknown temperature

$$\theta_i' = \frac{2}{3}\theta_i + \frac{1}{3}\theta_o$$
(5.139)

Numerically, this temperature is equal to 17°C.   ■

Convection

In *convection*, the heat is transmitted between two media, one of them a fluid in motion and the second, in most of the cases, (the surface of) a solid. In general, the solid has a higher temperature; therefore, heat transfers from the solid to the moving fluid. The governing law defining the heat flow, known as *Newton's law of cooling*, is expressed as

$$q = hA\Delta\theta$$
(5.140)

where h is the *convection heat transfer coefficient*, A is the area of the surface of the solid in contact with the moving fluid, and

$$\Delta\theta = \theta_s - \theta_\infty$$
(5.141)

The temperature at the heat exchange surface is θ_s, and θ_∞ is the fluid temperature at a distance far enough from the surface. Values of h are generally obtained experimentally, as h depends on many factors, such as surface geometry, fluid and solid properties, and fluid velocity. Comparison of Eqs. (5.128) and (5.140) shows that the thermal resistance corresponding to convection is

$$R_{th} = \frac{1}{hA}$$
(5.142)

Conduction and convection may occur simultaneously, such as the case is with a wall exposed to wind on the outside, the outside and inside temperatures being different, as shown in the following example.

───

Example 5.14

Calculate the heat flow rate for the two-panel wall of Figure 5.21(a) when convection is considered both on the outside surface (film coefficient is h_o) and on the inside one (film coefficient is h_i). Also calculate the two surface temperatures. The two panels have dimensions of h (height), w (width), and thicknesses l_1 and l_2; and the thermal conductivities are k_1 and k_2, respectively. Consider it wintertime; therefore, the outdoor temperature θ_2 is lower than the indoor one θ_1. Numerical application: $h = 4$ m, $w = 6$ m, $l_1 = 0.1$ m, $l_2 = 0.004$ m, $k_1 = 0.1$ W/m-C, $k_2 = 0.0002$ W/m-C, $h_o = 25$ W/m²-C, $h_i = 0.1$ W/m²-C, $\theta_1 = 25°C$, $\theta_2 = -10°C$.

Solution

The equivalent thermal resistance is a series combination of the four resistances of the actual system:

$$R_{th,s} = R_{th,o}^{conv} + R_{th,2}^{cond} + R_{th,1}^{cond} + R_{th,i}^{conv} \tag{5.143}$$

By using Eqs. (5.133) and (5.142), the thermal resistance of Eq. (5.143) becomes

$$R_{th,s} = \frac{1}{A}\left(\frac{1}{h_o} + \frac{l_2}{k_2} + \frac{l_1}{k_1} + \frac{1}{h_i}\right) \tag{5.144}$$

Knowing the inside and outside temperatures allows finding the heat flow rate as

$$q = \frac{\theta_1 - \theta_2}{R_{th,s}} = \frac{(\theta_1 - \theta_2)A}{\dfrac{1}{h_o} + \dfrac{l_2}{k_2} + \dfrac{l_1}{k_1} + \dfrac{1}{h_i}} \tag{5.145}$$

The temperature on the outside wall surface is determined by considering only the outside convection:

$$\theta_{so} = \theta_2 + \frac{q}{h_o A} = \theta_2 + \frac{(\theta_1 - \theta_2)}{1 + h_0\left(\dfrac{l_2}{k_2} + \dfrac{l_1}{k_1} + \dfrac{1}{h_i}\right)} \tag{5.146}$$

Similarly, the temperature on the inside wall surface is calculated from convection:

$$\theta_{si} = \theta_1 - \frac{q}{h_i A} = \theta_1 - \frac{(\theta_1 - \theta_2)}{1 + h_i\left(\dfrac{1}{h_o} + \dfrac{l_2}{k_2} + \dfrac{l_1}{k_1}\right)} \tag{5.147}$$

Numerically, it is obtained that $A = hw = 24$ m², $q = 27.06$ W, $\theta_{so} = -9.95°C$, $\theta_{si} = 13.72°C$. ■

5.3.2 Thermal Systems

Thermal systems, as we saw in the few examples discussed thus far in this section, are used for heating or cooling (as parts of the more complex indoor air conditioning equipment or HVAC, heating, ventilating, and air conditioning), in sensing devices (such as thermometers), or in thermal homogenizing equipment. Systems formed of thermal capacitors and resistors are described by differential equations that are the mathematical models of the analyzed thermal systems. Let us analyze the following example.

Example 5.15

Two identical rooms are designed, as sketched in Figure 5.23, with the six identical external walls. A cooling system operates in only one of the rooms. Consider it is summertime and heat flow infiltrates from the outside. Heat also flows through the wall separating the two rooms. Knowing the following amounts, heat flow absorbed by the cooler q_i, the outside temperature θ_o as well as the walls thermal resistances and the two rooms thermal capacitances, find the mathematical model of this thermal system that expresses the two indoor temperatures θ_1 and θ_2.

Solution

The following equations can be written for the left room in terms of its thermal capacitance:

$$C_{th}\frac{d\theta_1(t)}{dt} = q_{o1}(t) - q(t) \tag{5.148}$$

Similarly, the capacitance-related equation for the right room is

$$C_{th}\frac{d\theta_2(t)}{dt} = q_{o2}(t) + q(t) - q_i(t) \tag{5.149}$$

The room walls behave as thermal resistances; therefore, the following relationships can be formulated:

$$R_{th,o} = \frac{\theta_o(t) - \theta_1(t)}{q_{o1}(t)} \tag{5.150}$$

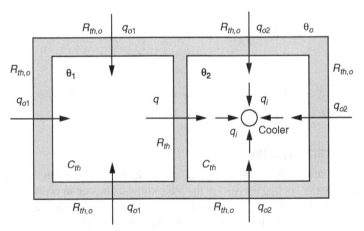

FIGURE 5.23

Two-Room Thermal System with Internal Cooling.

which corresponds to the three external walls on the left room, and

$$R_{th,o} = \frac{\theta_o(t) - \theta_2(t)}{q_{o2}(t)} \qquad (5.151)$$

which corresponds to the three external walls of the right room. For the intermediate wall, the thermal resistance is

$$R_{th} = \frac{\theta_1(t) - \theta_2(t)}{q(t)} \qquad (5.152)$$

The three heat flow rates are taken from Eqs. (5.150), (5.151), and (5.152) and substituted into Eqs. (5.148) and (5.149) adequately, such that the latter ones become

$$\begin{cases} C_{th}\dfrac{d\theta_1(t)}{dt} + \left(\dfrac{1}{R_{th,o}} + \dfrac{1}{R_{th}}\right)\theta_1(t) - \dfrac{1}{R_{th}}\theta_2(t) = \dfrac{1}{R_{th,o}}\theta_o(t) \\[3mm] C_{th}\dfrac{d\theta_2(t)}{dt} + \left(\dfrac{1}{R_{th,o}} + \dfrac{1}{R_{th}}\right)\theta_2(t) - \dfrac{1}{R_{th}}\theta_1(t) = \dfrac{1}{R_{th,o}}\theta_o(t) - q_i(t) \end{cases} \qquad (5.153)$$

The two differential Eqs. (5.153) form the mathematical model of the thermal system of Figure 5.23.

5.4 FORCED RESPONSE WITH SIMULINK®

Similar to mechanical and electrical systems, Simulink® can be used to model the forced response of fluid and thermal systems, as shown in the following examples.

Example 5.16
Plot the pressure p, which results from applying an input $q_i = 0.1 + 0.01\sin(10t)$ m³/s to the liquid system shown in Figure 5.15 of *Example 5.10*. The capacitance is $C_{l,p} = 2 \times 10^{-6}$ m⁵/N and the liquid resistance is nonlinear, defined by the pressure-output flow rate as $p = R_{l,p}q_o^2$ N/m². Compare this pressure with the pressure obtained when the resistance is linear with $R_{l,p} = 1000$ N-s/m⁵.

Solution
Equations (5.94) in conjunction with Eqs. (5.92) and (5.93) allow formulating the following differential equation corresponding to the linear liquid resistance

$$\frac{dp(t)}{dt} = -\frac{1}{R_{l,p}C_{l,p}}p(t) + \frac{1}{C_{l,p}}q_i(t) \qquad (5.154)$$

The two terms in the right-hand side of Eq. (5.154) are the inputs to a two-input summing point, whose output is the pressure derivative in the left-hand side of Eq. (5.154).

Figure 5.24 is the block diagram that integrates Eq. (5.154). The top part of Figure 5.24 is the block diagram that integrates Eq. (5.154), and the result is plotted by Scope 1, as shown in Figure 5.25(a). The second Eq. (5.89) can be written as

$$q_o(t) = q_i(t) - C_{l,p}\frac{dp(t)}{dt} \tag{5.155}$$

Using the nonlinear relationship between output flow rate q_o and the pressure p in conjunction with Eq. (5.155), yields the following differential equation corresponding to the nonlinear liquid resistance

$$\frac{dp(t)}{dt} = -\frac{1}{\sqrt{R_{l,p}}\,C_{l,p}}\sqrt{p(t)} + \frac{1}{C_{l,p}}q_i(t) \tag{5.156}$$

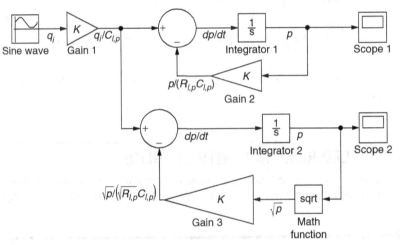

FIGURE 5.24

Simulink® Diagram for Integrating the Differential Eqs. (5.154) and (5.156).

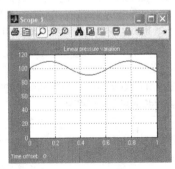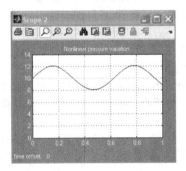

FIGURE 5.25

Time Variation of Tank Pressure (in N/m²) for Linear and Nonlinear Resistances.

The bottom part of the Simulink® diagram of Figure 5.24 realizes integration of the differential Eq. (5.156) and the resulting plot is illustrated in Figure 5.25(b) as displayed by Scope 2.

The input flow rate is modeled through a Sine Wave block from the Source Block library. Its Amplitude is 0.01, the Bias (or offset) is 0.1, and the Frequency is 10 rad/s. The value of the gains are Gain 1 = 5 x 10^5, Gain 2 = 500 and Gain 3 = 15811, as a result of the numerical values of parameters. The Math Function block generating the square root of the pressure p is found in the Math Operations library. As Figure 5.25 indicates, the linear-resistance pressure variation is approximately 10 times larger than the pressure variation corresponding to the nonlinear resistance. ◼

Example 5.17

Consider the following numerical values for the two-room thermal system of Example 5.15: R_{th} = 0.8889 K-s/J, $R_{th,o}$ = 1.2 K-s/J, C_{th} = 250,000 J/K, and q_i = 450 W. The outdoor temperature increases linearly from a value of 1,001 K (28°C) to a value of 1,013 K (40°C) in 1,000 s and remains constant afterwards. Use Simulink® to plot the two room temperatures over a time period of 4,000 s considering the initial temperatures in both rooms are 300 K (27°C).

Solution

The input (outdoors) temperature can be modeled in Simulink® by the Signal Builder block from the Sources library, which is shown in Figure 5.26. The time range is changed from the default value to 4,000 s by clicking Axes and then Change Time Range. The specific input profile is generated by using the boxes underneath the plot region to define the start and end values of time and temperature for two time intervals: one from 0 to 1,000 s and the next one from 1,000 to 4,000 s, as shown in Figure 5.26.

The mathematical model Eqs. (5.153) can be written as

$$\begin{cases} \dot{\theta}_1 = a_{11}\theta_1 + a_{12}\theta_2 + a_{13}\theta_o \\ \dot{\theta}_2 = a_{12}\theta_1 + a_{11}\theta_2 + a_{13}\theta_o + a_{21} \end{cases} \qquad (5.157)$$

where

$$a_{11} = -\frac{1}{C_{th}}\left(\frac{1}{R_{th,o}} + \frac{1}{R_{th}}\right); a_{12} = \frac{1}{R_{th}C_{th}}; a_{13} = \frac{1}{R_{th,o}C_{th}}; a_{21} = -\frac{q_i}{C_{th}} \qquad (5.158)$$

The parameters above have the following numerical values: a_{11} = -7.8×10^{-6}, a_{12} = 4.5×10^{-6}, a_{13} = 3.33×10^{-6}, and a_{21} = -0.0018. The Simulink® diagram that realizes integration of Eqs. (5.157) is shown in Figure 5.27, whereas the time-domain room temperatures are plotted in Figure 5.28. As Figure 5.28 indicates, the temperature in the room without a cooling unit slightly increases, whereas the temperature in the cooled room slightly decreases. ◼

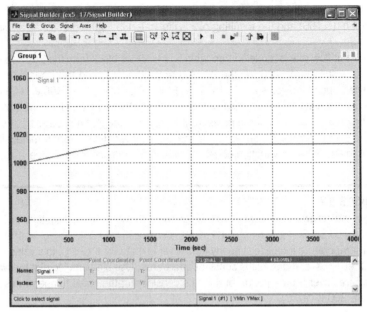

FIGURE 5.26

Input Temperature Profile.

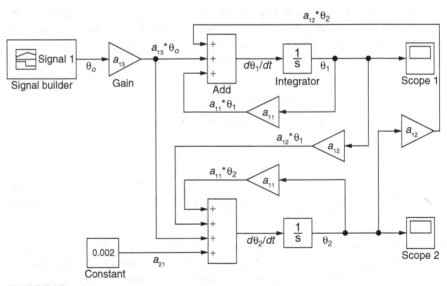

FIGURE 5.27

Simulink® Diagram for Integrating the Differential Eqs. (5.157).

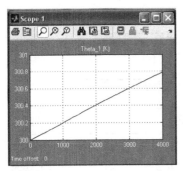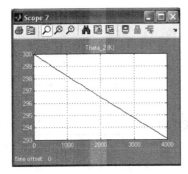

FIGURE 5.28

Room Temperatures versus Time.

SUMMARY

Using an approach similar to the one utilized in Chapters 2, 3, and 4 that focuses on modeling mechanical and electrical systems, this chapter introduces the fluid elements of inertance, capacitance, resistance, as well as the pumps and fans. The thermal elements of capacitance and resistance are also studied. Based on these elements, mathematical models are formulated for the natural and forced responses of fluid systems as well as to derive mathematical models for the forced response of thermal systems. MATLAB® is utilized to run symbolic and numerical calculations and evaluate the eigenvalues and eigenvectors of fluid systems; an example on how to apply Simulink® to model and solve for the time response of a fluid system example is also included. Subsequent chapters employ fluid and thermal system models, along with mechanical and electrical system models, to identify various responses in the time or frequency domains.

PROBLEMS

5.1 Find the pressure at point 4 of the vertical-plane pipeline drawn in Figure 5.29 under laminar flow conditions. The pipeline cross-section is circular with an inner diameter $d_i = 30$ mm. Known are the pump specific work $w = 2$ m²/s², $l = 10$ m, $h = 2$ m, $v_0 = 1$ m/s, $\rho = 1000$ kg/m³, $\mu = 0.001$ N-s/m², and the atmospheric pressure $p_a = 10^5$ N/m².

5.2 Determine the liquid inertance of the parabolic-shaped conduit segment defined by the end radii r_1 and r_2 and height h, as sketched in Figure 5.30. Find the inertance numerical value for $\rho = 800$ kg/m³, $h = 1.2$ m, $r_1 = 0.05$ m, and $r_2 = 0.09$ m.

5.3 Calculate algebraically the liquid inertance of the pipeline segment shown in Figure 5.31 and its numerical value for $l = 3$ m, inner diameters $d_1 = 20$ mm, $d_2 = 40$ mm, and fluid density $\rho = 1,800$ kg/m³.

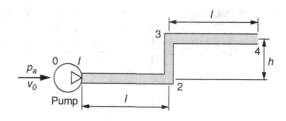

FIGURE 5.29

Vertical Liquid System with Pump and Pipeline.

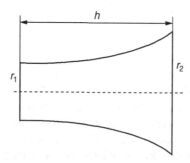

FIGURE 5.30

Parabolic-Profile Pipe Segment.

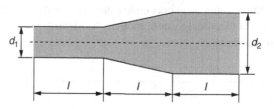

FIGURE 5.31

Variable Cross-Section Pipeline Segment.

5.4   Determine the hydraulic capacitance of a semi-spherical tank of inner radius R.

5.5   Calculate the capacitance of a cylindrical vertical tank whose cross-section varies according to a decreasing exponential: $r(x) = c_1 e^{-c_2 x}$ (x being the distance measured from the upper end of the tank). The tank has its maximum radius r_1 at the top and minimum radius r_2 at the bottom; its total height is h. Numerical application: $r_1 = 1.2$ m, $r_2 = 0.8$ m, $h = 5$ m.

5.6 Calculate the capacitance of the tank shaped as in Figure 5.32.

5.7 Using the Hagen-Poiseuille approach, determine the hydraulic resistance of the parabolic-shaped pipe segment shown in Figure 5.30 of Problem 5.2.

5.8 Answer the same question as in Problem 5.7 for a pipe segment having the exponential decreasing profile defined in Problem 5.5 and placed horizontally.

5.9 In turbulent flow, the volume flow rate is expressed in terms of head as $q = c\sqrt{h}$, where c is a constant. Determine the pressure-defined hydraulic resistance.

5.10 The kinetic energy loss is measured for a valve corresponding to three values of the flow rate, and the following values are obtained: $q_{v1} = 0.1$ m³/s, $U_{d1} = 2100$ W; $q_{v2} = 0.3$ m³/s, $U_{d2} = 2800$ W; $q_{v3} = 0.5$ m³/s, $U_{d3} = 3500$ W. Determine a relationship between the liquid resistance of the valve $R_{l,p}$ and the volume flow rate q.

5.11 Liquid with a volume flow rate $q_v = 1$ m³/s flows through the pipe system sketched in Figure 5.33. Knowing that $l = 2$ m, $d_i = 25$ mm (the inner diameter of the circular piping), and $\mu = 0.00005$ N-s/m², determine the loss of pressure between points 1 and 2. Hint: The transformation between triangle and reversed-Y liquid resistance connections is identical to the one of electrical resistances of Chapter 4.

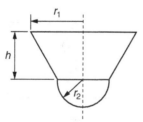

FIGURE 5.32

Vertical Tank with Tapered and Semi-Spherical Segments.

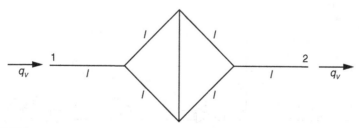

FIGURE 5.33

Pipeline System with Seven Lines Containing Flowing Liquid.

5.12 Calculate the hydraulic resistance of the pipeline shown in Figure 5.31 of Problem 5.3 and the energy that is lost. Known also are $q_v = 0.02$ m³/s and $\mu = 0.00001$ N-s/m².

5.13 Calculate the natural frequencies of the hydraulic system sketched in Figure 5.34. Is it possible that the amplitudes of the flow rates in the two meshes are equal during either of the two modal motions and the directions of these flows are identical (i.e., either clockwise or counterclockwise)?

Known are: $C_{l1} = 1.8 \times 10^{-6}$ m⁵/N, $C_{l2} = 2.2 \times 10^{-6}$ m⁵/N, $I_l = 3 \times 10^4$ N-s²/m⁵.

5.14 Water flows in a tank of diameter $d = 1$ m with an input volume flow rate $q_i = 0.1$ m³/s. At the base of the tank is a discharge pipe of length $l = 3$ m and inner diameter $d_i = 40$ mm. Find the output flow rate exiting the pipe after $t = 20$ s and the difference between the input and output flow rates as time goes to infinity (the steady-state error). Known are: $\mu = 0.001$ N-s/m² and $\rho = 1000$ kg/m³.

5.15 Derive a mathematical model for the liquid-level system shown in Figure 5.35 when the input to the system is the volume flow rate q_i and the output is the volume flow rate q_o.

5.16 Find the capacitance of a conical vessel of height h and base circle radius r with a polytropic gas (of coefficient n and gas constant R_g) in it for a specified temperature θ.

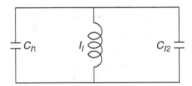

FIGURE 5.34

Two-Mesh Conservative Hydraulic Circuit.

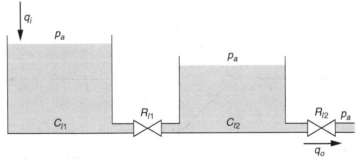

FIGURE 5.35

Liquid-Level System Consisting of Two Tanks and Two Valves.

5.17 Calculate the gas capacitance related to an incompressible gas that flows through the conduit segment of Figure 5.36.

5.18 Using the Hagen-Poiseuille approach, determine the resistance of an incompressible gas that flows through the conduit segment shown in Figure 5.36 of Problem 5.17.

5.19 Disregarding the pneumatic resistances, determine the natural frequency of the system sketched in Figure 5.37. The conduits are tubular of inner diameter $d_i = 20$ mm and $l = 1$ m. The vertical-plane tapered tank is defined by $d_1 = 0.2$ m and, $d_2 = 0.4$ m.

5.20 Neglect pneumatic resistances and determine the natural frequencies of the pneumatic system sketched in Figure 5.38. Use lumped-parameter inertances and capacitances. The middle conduit has an inner diameter d_2 that is twice the inner diameter of all other conduits, d_1. Known are $g = 9.8$ m/s^2 and $l = 3$ m.

5.21 Derive a mathematical model for the pneumatic system of Figure 5.39, which is formed of two tanks and a conduit. The input is the pressure p_i and the output is the pressure in the conical tank, p_o. Known are $d = 0.5$ m and $l = 3$ m; the

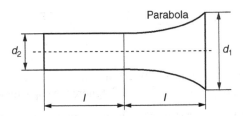

FIGURE 5.36

Parabolic-Profile Pipe Segment.

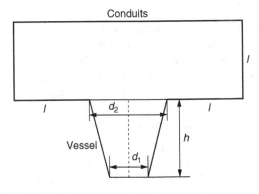

FIGURE 5.37

Pneumatic System with Tapered Vessel and Conduits.

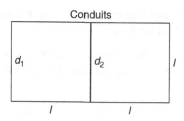

FIGURE 5.38

Conduit Pneumatic System.

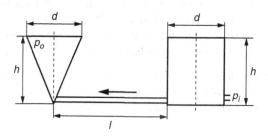

FIGURE 5.39

Pneumatic System with Two Containers and Conduit.

conduit inner diameter is $d_i = 0.005$ m. The pneumatic agent is a gas with $\rho = 1.2$ kg/m³ and $\mu = 1.3 \times 10^{-5}$ N-s/m². Ignore the capacitance of the conduit.

5.22 A structural layer is sandwiched between two identical insulation layers to reduce the thermal loss by 50%. Considering that the thermal conductivity of the insulation layers is 20 times smaller than the thermal conductivity of the structural layer, find the thickness of the insulation layer as a fraction of the structural layer thickness.

5.23 A wall is formed of two identical panels that enclose air space. The thickness of the panels is $l_p = 0.02$ m and that of the air space is $l_a = 0.04$ m. The thermal conductivity of the panel is $k_p = 2$ W/m-C and that of the air space is $k_a = 0.02$ W/m-C. It is calculated that during wintertime and for an outside temperature of $-10°C$, the inside temperature should be 23°C. However, the measured inside temperature is 25°C. Evaluate the outside convection coefficient considering combined heat conduction and convection.

5.24 The composite wall sketched in Figure 5.40 is formed of five panels. Sketch the thermal resistance connection for the outside-to-inside heat flow. Knowing that $l_1 = l_2 = l_3 = 0.02$ m, $A_1 = A_2 = A_3 = 25$ m², $A_4 = A_5 = 0.1$ m², $k_1 = k_5 = 0.001$ W/m-C, $k_2 = k_4 = 0.9$ W/m-C, $k_3 = 0.002$ W/-C, $\theta_o = 35°C$, and $\theta_i = 20°C$, find the heat flow rate penetrating from outside.

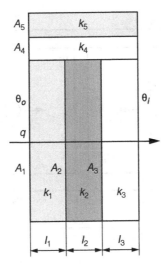

FIGURE 5.40

Five-Panel Composite Wall.

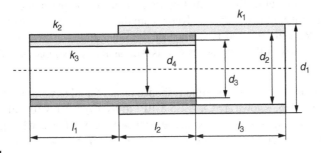

FIGURE 5.41

Longitudinal Section through Composite Cylinder Tubing.

5.25 A stainless steel pipe (k = 15 W/m-C) of length l = 10 m, inner diameter d_i = 0.5 m, and outer diameter d_o = 0.6 m is coated with an insulation material (k_i = 0.0001 W/m-C) to allow a total heat loss of q = 10 W when the outdoor temperature is −5°C and the temperature of the fluid inside the pipe is 30°C. What is the thickness of the insulating layer?

5.26 Several cylinders are combined as shown in the longitudinal section of Figure 5.41. Find the radial heat flow corresponding to a temperature difference of 40°C. Known are k_1 = 0.1 W/m-C, k_2 = 0.02 W/m-C, k_3 = 25 W/m-C, d_1 = 0.050 m, d_2 = 0.042 m, d_3 = 0.036 m, d_4 = 0.030 m, l_1 = 1 m, l_2 = 0.8 m, l_3 = 1.1 m.

5.27 Two identical rooms (as sketched in Figure 5.42) are heated differently during wintertime by means of two heat sources of known heat flow rates q_1 and q_2. The

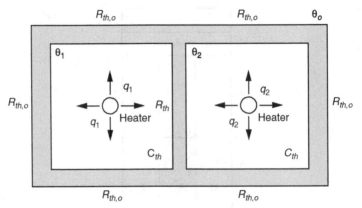

FIGURE 5.42

Two-Room Thermal System with Internal Heating.

rooms have identical walls separating the conditioned space from the outside. Heat also flows through the wall separating the two rooms, which is different from the outside walls. Knowing the outside temperature is θ_o, derive the mathematical model of this thermal system where the output amounts are the two room temperatures θ_1 and θ_2. Known are the thermal resistances of the outside walls $R_{th,o}$, the thermal resistance of the separating wall R_{th}, and the thermal capacitances of the two rooms C_{th}, as well as the two source heat flow rates q_1 and q_2.

5.28 Use Simulink® to plot the time response of the liquid system of Figure 5.35 in Problem 5.15. Known are $C_{l1} = 1.5 \times 10^{-6}$ m⁵/N, $C_{l2} = 2.1 \times 10^{-6}$ m⁵/N, $R_{l1} = 3 \times 10^4$ N-s/m⁵/, $q_i = 0.001$ m³/s; the resistance R_{l2} is nonlinear with the following relationship between pressure loss and flow rate: $p = 1100q^3$ N/m².

5.29 Plot the output pressure p_o in the conical vessel of Figure 5.39 in Problem 5.21 using Simulink®. Consider the input pressure is a ramp function defined as $p_i = 2000t$ N/m² for 10 minutes and then remains constant for 10 minutes after that.

5.30 Known are the following amounts for the thermal system of Figure 5.42 in Problem 5.27: $\theta_o = -10°C$, $R_{th} = 0.3$ K-s/J, $R_{th,o} = 1$ K-s/J, $C_{th} = 200,000$ J/K, $q_1 = 1900$ W, $q_2 = 1600$ W. Plot the two rooms' temperatures as a function of time using Simulink®.

Suggested Reading

I. Cochin, *Analysis and Design of Dynamic Systems*, Harper & Row, New York, 1980.

J. W. Brewer, *Control Systems: Analysis, Design and Simulation*, Prentice-Hall, Englewood Cliffs, NJ, 1974.

K. Ogata, *System Dynamics*, 4th Ed. Pearson Prentice Hall, Upper Saddle River, NJ, 2004.

W. J. Palm III, *System Dynamics*, 2nd Ed. McGraw-Hill, New York, 2009.

F. C. McQuiston, J. D. Parker, and J. D. Spitler, *Heating, Ventillating and Air Conditioning—Analysis and Design*, 6th Ed. John Wiley & Sons, New York, 2005.

N. Cheremisinoff, *Fluid Flow Pocket Handbook*, Gulf Publishing Company, Houston, 1984.

D. McCloy and H. R. Martin, *Control of Fluid Power: Analysis and Design*, 2nd Ed. Ellis Horwood Limited, Chichester, UK, 1980.

C. A. Belsterling, *Fluidic System Design*, Wiley Interscience, New York, 1971.

B. Eck, *Fans*, Pergamon Press, Oxford, UK, 1973.

Y. A. Cengel, *Heat Transfer: A Practical Approach*, 2nd Ed. McGraw-Hill, Boston, 2002.

H. Klee, *Simulation of Dynamic Systems with MATLAB® and Simulink®*, CRC Press, Boca Raton, FL, 2007.

The Laplace Transform

OBJECTIVES

The chapter introduces the Laplace transform, which can be used to solve the differential equations corresponding to mechanical, electrical, and fluid or thermal systems' mathematical models as derived in previous chapters. The Laplace transform is also instrumental in formulating other system dynamics models, such as the transfer function, the state space model, or frequency-domain analysis, as shown in Chapters 7, 8, and 9. The following topics are studied:

- Direct and inverse Laplace transforms.
- Laplace pairs and main Laplace-transform properties.
- Techniques of partial fraction expansion.
- Application of direct and inverse Laplace transforms to solving linear differential equations with constant and time-varying coefficients, integral equations, and integral-differential equations.
- Use of MATLAB® specialized commands for partial-fraction expansion and direct or inverse Laplace transform calculations.

INTRODUCTION

This chapter presents the main properties and applications of the Laplace transform. One important applicative strength of the Laplace method is the conversion of differential equations into algebraic equations by means of the direct Laplace transform, which enables calculation of the transformed solution with relative ease. Subsequent use of the inverse Laplace transform allows retrieving the original, time-dependent solution. Integral and integral-differential equations can also be solved by the same two-phase Laplace approach. The Laplace transform is also fundamental in the transfer function, state space, and frequency-domain approaches. While the Laplace transform is utilized mainly for constant-coefficient, linear differential equations, some equations with time-varying coefficients can also be solved through this methodology, a topic also discussed here. The chapter includes the application of

DOI: 10.1016/B978-0-240-81128-4.00006-4

the Laplace transform to multiple degree-of-freedom dynamic models, which are expressed in vector-matrix form. Time-domain system identification from Laplace-domain information also is discussed.

6.1 DIRECT LAPLACE AND INVERSE LAPLACE TRANSFORMATIONS

Transformations, also named *operators*, convert functions into other functions; this is schematically illustrated in Figure 6.1. Transformations change the original function, but they can either keep the independent variable or modify it. An operator that takes a function and multiplies it by a number *n*, for instance, preserves the independent variable of the original function. Such an operator can formally be expressed as

$$O[f(t)] = nf(t) \tag{6.1}$$

If, for example, $f(t)$ is $\cos(\omega t)$ in Eq. (6.1), the result, $n\cos(\omega t)$, depends on time t as well as on the input (original) function.

Another set of transformations (or simply, *transforms*) modify both the original function and its independent variable through their application. Such transforms are the *integral* ones, which change a function and its variable using integration, and the *Laplace transform* is one of them. Other integral transforms are the Fourier transform, the Melin transform, the Hankel transform, the Kantorovich transform, and the Mehler-Fock transform, all of which operate over an infinite domain. Still other transforms apply over a finite domain and are known as *finite transforms*, including specific techniques such as those of Sturm-Liouville, Legendre, and Tchebycheff.

The *direct Laplace transform* takes a function $f(t)$ that depends on time t and transforms into another function $F(s)$ that depends on the complex variable s, as sketched in Figure 6.2. The *inverse Laplace transform*, also indicated in Figure 6.2, operates on $F(s)$ to obtain $f(t)$. Mathematically, the Laplace transform is defined as

$$F(s) = \mathcal{L}[f(t)] = \int_0^\infty f(t)e^{-st}\,dt \tag{6.2}$$

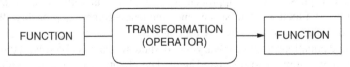

FIGURE 6.1

Relationships between Functions and Transformations (Operators).

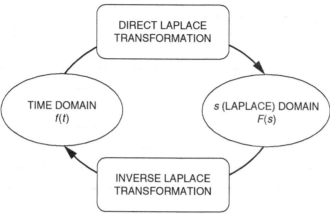

FIGURE 6.2

Direct and Inverse Laplace Transformations.

The integral defining the Laplace transform needs to be convergent for the transformation to exist, which requires the original function $f(t)$ to satisfy the following conditions:

1. $f(t)$ has to be *piecewise continuous* over the interval $0 < t < \infty$, which means it has to be continuous on any partition of that interval (provided the interval can be partitioned into a finite number of nonintersecting intervals), and it needs to have finite limits at the ends of each subinterval.
2. $f(t)$ needs to be of *exponential order*, which means is has to satisfy the following equality:

$$\lim_{t \to \infty} |f(t)| e^{-\sigma t} = 0 \qquad (6.3)$$

for values of the real constant σ that are larger than a threshold value σ_c, which is known as the *abscissa of convergence*.

6.1.1 Direct Laplace Transform and Laplace Transform Pairs

Formally, the inverse Laplace transform, which is often needed to return the original (generally unknown) time-domain function $f(t)$, is computed as

$$f(t) = \mathcal{L}^{-1}[F(s)] \qquad (6.4)$$

but the integration technique allowing direct calculation of $f(t)$ from $F(s)$ is a bit more involved for this introductory text. Suffice to say that application of the direct Laplace transform, Eq. (6.2), enables formulating $f(t) - F(s)$ pairs for elementary functions that satisfy the requirements for obtaining the Laplace transform. Table 6.1 contains several basic Laplace-transform pairs for quick translation between the two domains.

Table 6.1 Laplace Transform Pairs

$f(t)$	$F(s)$
$\delta(t)$	1
$1(t)$	$1/s$
t	$1/s^2$
t^n	$n!/s^{n+1}$
e^{-at}	$1/(s+a)$
$\sin(\omega t)$	$\omega/(s^2 + \omega^2)$
$\cos(\omega t)$	$s/(s^2 + \omega^2)$
$1/\sqrt{t}$	$\sqrt{\pi}/\sqrt{s}$
$t^n e^{-at}$	$n!/(s+a)^{n+1}$
$e^{at} - e^{bt}$	$(a-b)/[(s-a)(s-b)]$
$ae^{at} - be^{bt}$	$(a-b)s/[(s-a)(s-b)]$
$(e^{at} - e^{bt})/t$	$-\ln[(s-a)/(s-b)]$
$t - (1 - e^{-at})/a$	$a/[s^2(s+a)]$
$\sinh(\omega t)$	$\omega/(s^2 - \omega^2)$
$\cosh(\omega t)$	$s/(s^2 - \omega^2)$
$\sin(\omega t + \varphi)$	$[\omega \cos\varphi + (\sin\varphi)s]/(s^2 + \omega^2)$
$t \sin(\omega t)$	$2\omega s/(s^2 + \omega^2)^2$
$\sin(\omega t)/t$	$\tan^{-1}(\omega/s)$
$t \cos(\omega t)$	$(s^2 - \omega^2)/(s^2 + \omega^2)^2$
$t \sinh(\omega t)$	$2\omega s/(s^2 - \omega^2)^2$
$t \cosh(\omega t)$	$(s^2 + \omega^2)/(s^2 - \omega^2)^2$
$\sin(\omega_1 t)/\omega_1 - \sin(\omega_2 t)/\omega_2$	$\left(\omega_2^2 - \omega_1^2\right)/\left[\left(s^2 + \omega_1^2\right)\left(s^2 + \omega_2^2\right)\right]$
$1 - \cos(\omega t)$	$\omega^2/[s(s^2 + \omega^2)]$
$\omega t - \sin(\omega t)$	$\omega^3/[s^2(s^2 + \omega^2)]$
$\sin(\omega t) - \omega t \cos(\omega t)$	$2\omega^3/(s^2 + \omega^2)^2$
$\sin(\omega t) + \omega t \cos(\omega t)$	$2\omega s^2/(s^2 + \omega^2)^2$
$\sin(\omega t) \sinh(\omega t)$	$2\omega^2 s/[s^4 + (2\omega^2)^2]$
$\cos(\omega t) \cosh(\omega t)$	$s^3/[s^4 + (2\omega^2)^2]$
$\sin(\omega t) \cosh(\omega t)$	$\omega(s^2 + 2\omega^2)/[s^4 + (2\omega^2)^2]$
$\sinh(\omega t) - \sin(\omega t)$	$2\omega^3/(s^4 - \omega^4)$
$\cosh(\omega t) - \cos(\omega t)$	$2\omega^2 s/(s^4 - \omega^4)$
$\sinh(\omega t) + \sin(\omega t)$	$2\omega s^2/(s^4 - \omega^4)$
$\cosh(\omega t) + \cos(\omega t)$	$2s^3/(s^4 - \omega^4)$

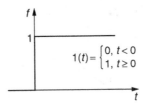

FIGURE 6.3

Unit-Step Function.

One of the simplest transforms is the one applied to the unit step function sketched in Figure 6.3. Instead of simply using a value of 1 (or the more generic symbol $u(t)$, as in other texts) to define it, the symbol $1(t)$ can be utilized with the function definition indicated in Figure 6.3, which highlights that the unit-step function is zero for $t < 0$. This special symbol allows truncating various other functions by adequate combination with $1(t)$, as will be shown shortly. The Laplace transform of $1(t)$ is

$$\mathcal{L}[1(t)] = \int_0^\infty e^{-st}\,dt = -\frac{e^{-st}}{s}\Big|_0^\infty = \frac{1}{s} \tag{6.5}$$

and the pair is included in Table 6.1.

Example 6.1

Calculate the Laplace transform of the unit-ramp function, which is sketched in Figure 6.4. Using the principle of mathematical induction, also derive the Laplace transform of the function $u(t) = t^n$, where n is an integer and $n > 1$.

Solution

The Laplace transform of the unit ramp function is calculated through integration by parts as

$$\mathcal{L}[t] = \int_0^\infty t e^{-st}\,dt = -t\frac{e^{-st}}{s}\Big|_0^\infty + \frac{1}{s}\int_0^\infty e^{-st}\,dt = -\frac{e^{-st}}{s^2}\Big|_0^\infty = \frac{1}{s^2} \tag{6.6}$$

which is also shown in Table 6.1. It can be demonstrated using L'Hospital rule in Eq. (6.6) that

$$\lim_{t\to\infty} -t\frac{e^{-st}}{s} = \lim_{t\to\infty}\frac{-1}{s^2 e^{st}} = 0 \tag{6.7}$$

To calculate the Laplace transform of t^2, a procedure similar to the one applied previously is used, which yields

$$\mathcal{L}[t^2] = \int_0^\infty t^2 e^{-st}\,dt = -t^2\frac{e^{-st}}{s}\Big|_0^\infty + \frac{2}{s}\int_0^\infty t e^{-st}\,dt = \frac{2}{s^3} \tag{6.8}$$

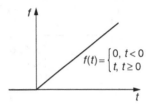

FIGURE 6.4

Unit-Ramp Function.

Equation (6.8) took into account both the result of Eq. (6.6) and the fact that

$$\lim_{t \to \infty} -t^2 \frac{e^{-st}}{s} = \lim_{t \to \infty} \frac{-2t}{s^2 e^{st}} = \lim_{t \to \infty} \frac{-2}{s^3 e^{st}} = 0 \tag{6.9}$$

Equations (6.6) and (6.8) suggest that

$$\mathcal{L}[t^n] = \frac{n!}{s^{n+1}} \tag{6.10}$$

but this has to be demonstrated. If Eq. (6.10) is valid, the following relationship needs to be valid as well:

$$\mathcal{L}[t^{n+1}] = \frac{(n+1)!}{s^{n+2}} \tag{6.11}$$

and, according to the principle of mathematical induction, Eq. (6.10) would be proven. This can be shown by using integration by parts again to calculate $\mathcal{L}[t^{n+1}]$:

$$\mathcal{L}[t^{n+1}] = \int_0^\infty t^{n+1} e^{-st} dt = -t^{n+1} \frac{e^{-st}}{s} \Big|_0^\infty + \frac{n+1}{s} \int_0^\infty t^n e^{-st} dt$$

$$= \frac{(n+1)n!}{ss^{n+1}} = \frac{(n+1)!}{s^{n+2}} \tag{6.12}$$

This proves that Eq. (6.11) is valid; therefore, Eq. (6.10) is valid as well, as also indicated in Table 6.1. The following limit is zero (as can be demonstrated through mathematical induction again):

$$\lim_{t \to \infty} -t^{n+1} \frac{e^{-st}}{s} = \lim_{t \to \infty} \frac{-(n+1)t^n}{s^2 e^{st}} = \lim_{t \to \infty} \frac{-(n+1)nt^{n-1}}{s^3 e^{st}}$$

$$= \dots \lim_{t \to \infty} \frac{-(n+1)n \times \dots \times 2 \times 1}{s^{n+1} e^{st}} = 0 \tag{6.13}$$

Example 6.2

Calculate the Laplace transform of the unit-amplitude sinusoidal function of Figure 6.5.

Solution

According to the definition, the Laplace transform of the unit-amplitude sinusoidal function is

$$F(s) = \mathcal{L}[\sin(\omega t)] = \int_0^\infty \sin(\omega t)e^{-st}dt \tag{6.14}$$

Integration by parts is applied to this equation, which yields

$$F(s) = -\frac{\sin(\omega t)}{s}e^{-st}\bigg|_0^\infty + \frac{\omega}{s}\int_0^\infty \cos(\omega t)e^{-st}dt = \frac{\omega}{s}\int_0^\infty \cos(\omega t)e^{-st}dt \tag{6.15}$$

because

$$-\frac{\sin(\omega t)}{s}e^{-st}\bigg|_0^\infty = \lim_{t\to\infty}\left(-\frac{\sin(\omega t)}{s}e^{-st}\right) - \lim_{t\to 0}\left(-\frac{\sin(\omega t)}{s}e^{-st}\right)$$

$$= 0 - 0 = 0 \tag{6.16}$$

Integration by parts is applied to the integral on the right-hand side of Eq. (6.15), which results in

$$F(s) = \frac{\omega}{s}\left[-\frac{\cos(\omega t)}{s}e^{-st}\bigg|_0^\infty - \frac{\omega}{s}F(s)\right] \tag{6.17}$$

By taking into account that

$$-\frac{\cos(\omega t)}{s}e^{-st}\bigg|_0^\infty = \lim_{t\to\infty}\left(-\frac{\cos(\omega t)}{s}e^{-st}\right) - \lim_{t\to 0}\left(-\frac{\cos(\omega t)}{s}e^{-st}\right)$$

$$= 0 + \frac{1}{s} = \frac{1}{s} \tag{6.18}$$

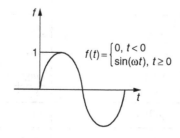

$$f(t) = \begin{cases} 0, & t < 0 \\ \sin(\omega t), & t \geq 0 \end{cases}$$

FIGURE 6.5

Unit-Amplitude Sinusoidal Function.

the Laplace transform of $\sin(\omega t)$ is found algebraically from Eq. (6.17) as

$$F(s) = \mathcal{L}[\sin(\omega t)] = \frac{\omega}{s^2 + \omega^2} \qquad (6.19)$$

6.1.2 Properties of the Laplace Transform

Table 6.2 synthesizes the main Laplace transform properties, which are demonstrated in the following. Several other properties and examples are included in the companion website Chapter 6.

Linearity

An important property with applications in solving differential equations and determining transfer functions is the linearity character of both the direct and inverse Laplace operators. Consider a linear combination of n time-dependent functions, $f_1(t), f_2(t), \dots, f_n(t)$ by means of n real constants, $a_1, a_2, \dots, a_n$. Application of the Laplace transform to this linear combination results in:

$$\mathcal{L}[a_1 f_1(t) + a_2 f_2(t) + \cdots + a_n f_n(t)] = a_1 F_1(s) + a_2 F_2(s) + \cdots + a_n F_n(s) \qquad (6.20)$$

This propriety can easily be verified by using the definition of the Laplace transform. Application of the inverse Laplace transform to Eq. (6.20) and consideration of

$$\mathcal{L}^{-1}[\mathcal{L}[f(t)]] = f(t) \qquad (6.21)$$

Table 6.2 Main Laplace Transforms Properties (Theorems)

Theorem	$f(t)$	$F(s)$
Linearity	$a_1 f_1(t) + a_2 f_2(t)$	$a_1 F_1(s) + a_2 F_2(s)$
Frequency shift	$e^{-at} f(t)$	$F(s + a)$
Time shift	$f(t - a) 1(t - a)$	$e^{-as} F(s)$
Time derivatives	$\dfrac{d}{dt} f(t)$	$s F(s) - f(0)$
	$\dfrac{d^2}{dt^2} f(t)$	$s^2 F(s) - s f(0) - \left. \dfrac{df(t)}{dt} \right|_{t=0}$
Time integral	$\displaystyle\int_0^t f(t)\,dt$	$\dfrac{F(s)}{s} + \dfrac{\left.\int f(t)\,dt\right|_{t=0}}{s}$
Final value	$\displaystyle\lim_{t \to \infty} f(t)$	$\displaystyle\lim_{s \to 0} s F(s)$
Initial value	$\displaystyle\lim_{\substack{t \to 0 \\ t > 0}} f(t)$	$\displaystyle\lim_{s \to \infty} s F(s)$

demonstrates the linear character of the inverse Laplace transform and results in

$$\mathcal{L}^{-1}[a_1 F_1(s) + a_2 F_2(s) + \cdots + a_n F_n(s)] = a_1 f_1(t) + a_2 f_2(t) + \cdots + a_n f_n(t) \quad (6.22)$$

Frequency-Shift Theorem

The *frequency shift theorem* (as shown in Table 6.2) yields the Laplace transform of the product between a function $f(t)$, which is Laplace transformable, and e^{-at}, where a is a real constant. It can be shown that

$$\mathcal{L}[e^{-at}f(t)] = \int_0^\infty e^{-at} f(t)e^{-st}dt = \int_0^\infty f(t)e^{-(s+a)t} dt = F(s + a) \quad (6.23)$$

The effect of multiplying the original function by an exponential function on applying the Laplace transform to this product is a translation (shift) of the variable s into the Laplace domain. Because the Laplace domain is closely connected to the frequency response, as we see in subsequent chapters, the theorem is known as the *frequency shift theorem*.

Example 6.3

The out-of-plane bending vibrations of a MEMS cantilever (see Fig. 3.1) are identified by the Laplace transform of the free end displacement as $Z(s) = 3/(s^2 + 0.2s + 30)$. Determine the time-domain coordinate $z(t)$.

Solution

In checking the functions in the right column of Table 6.1, the closest format to the one of the problem appears to be the pair corresponding to sin(ωt), but some modifications are needed before finding $z(t)$. By completing the square in the denominator, the function $Z(s)$ can be written as

$$Z(s) = \frac{3}{(s + 0.1)^2 + (\sqrt{29.99})^2} = \frac{3}{\sqrt{29.99}} \times \frac{\sqrt{29.99}}{(s + 0.1)^2 + (\sqrt{29.99})^2} \quad (6.24)$$

By considering now the linearity propriety, the frequency shift theorem, together with the line in Table 6.1 giving the Laplace transform of sin(ωt), it follows that

$$z(t) = \mathcal{L}^{-1}[Z(s)] = \mathcal{L}^{-1}\left[\frac{3}{\sqrt{29.99}} \times \frac{\sqrt{29.99}}{(s + 0.1)^2 + (\sqrt{29.99})^2}\right]$$

$$= \frac{3}{\sqrt{29.99}}e^{-0.1t}\sin(\sqrt{29.99}\,t) \quad (6.25)$$

Example 6.4

Use the frequency-shift theorem and the Laplace transforms of $\sin(\omega t)$ and $\cos(\omega t)$ to determine the Laplace transforms of the following functions: $f_1(t) = te^{-at}\sin(\omega t)$; $f_2(t) = te^{-at}\cos(\omega t)$.

Solution

The following results from the frequency-shift theorem:

$$\mathcal{L}\left[te^{-(a+j\omega)t}\right] = \frac{1}{(s+a+j\omega)^2} \tag{6.26}$$

The complex number of Eq. (6.26) is expressed in standard form as

$$
\begin{aligned}
\frac{1}{(s+a+j\omega)^2} &= \frac{1}{(s+a)^2 - \omega^2 + 2\omega(s+a)j} \\
&= \frac{(s+a)^2 - \omega^2 - 2\omega(s+a)j}{\left[(s+a)^2 - \omega^2\right]^2 + 4\omega^2(s+a)^2} \\
&= \frac{(s+a)^2 - \omega^2}{\left[(s+a)^2 + \omega^2\right]^2} - j\frac{2\omega(s+a)}{\left[(s+a)^2 + \omega^2\right]^2}
\end{aligned} \tag{6.27}
$$

At the same time, the original function of Eq. (6.26) can be formulated based on Euler's identity as

$$
\begin{aligned}
te^{-(a+j\omega)t} &= te^{-at}e^{-j\omega t} = te^{-at}\left[\cos(\omega t) - j\sin(\omega t)\right] \\
&= te^{-at}\cos(\omega t) - jte^{-at}\sin(\omega t)
\end{aligned} \tag{6.28}
$$

Applying the Laplace operator to Eq. (6.28) results in

$$\mathcal{L}\left[te^{-(a+j\omega)t}\right] = \mathcal{L}\left[te^{-at}\cos(\omega t)\right] - j\mathcal{L}\left[te^{-at}\sin(\omega t)\right] \tag{6.29}$$

Comparison of the real and imaginary parts of Eqs. (6.27) and (6.29) shows that

$$
\begin{cases}
\mathcal{L}[f_1(t)] = \mathcal{L}\left[te^{-at}\sin(\omega t)\right] = \dfrac{2\omega(s+a)}{\left[(s+a)^2 + \omega^2\right]^2} \\[4mm]
\mathcal{L}[f_2(t)] = \mathcal{L}\left[te^{-at}\cos(\omega t)\right] = \dfrac{(s+a)^2 - \omega^2}{\left[(s+a)^2 + \omega^2\right]^2}
\end{cases} \tag{6.30}
$$

Time-Shift Theorem

The previous theorem has a counterpart, which, as shown in Table 6.2, indicates that multiplication of the function $F(s)$ by an exponential e^{-as}, followed by application of the inverse Laplace transform, results in a time domain shift:

$$\mathcal{L}[f(t-a)1(t-a)] = e^{-as}F(s) \tag{6.31}$$

which is known as the *time-shift theorem*.

Before demonstrating this property, a brief preamble is needed to better understand the functions $f(t - a)$ and $1(t - a)$ on the left-hand side of Eq. (6.31). An original arbitrary function $f(t)$ is plotted in Figure 6.6(a), together with the function $f(t - a)$. As it can easily be verified, the function $f(t - a)$ is shifted (translated) horizontally to the right by a quantity a because $f(t - a) = f(t_1)$. A similar situation applies to the functions $1(t)$ and $1(t - a)$, as $1(t - a)$ translates to the right by the quantity a, as illustrated in Figure 6.6(b).

Let us analyze what happens with the product between an arbitrary function $f(t)$ and $1(t)$. This product is expressed as

$$f(t)1(t) = \begin{cases} f(t) \times 0 = 0, t < 0 \\ f(t) \times 1 = f(t), t \geq 0 \end{cases} \qquad (6.32)$$

In other words, the original function exists only for positive values of t and is zero (truncated) to the left of the origin. Multiplication of a function by $1(t)$ is omitted altogether in many applications, as this particular operation does not change the original function over the zero-to-infinity interval, where Laplace integration is being applied. It can simply be shown that multiplication of $f(t)$ by $1(t - a)$ results in a function that is zero to the left of a and is $f(t)$ to the right of that point. Let us analyze what is the result of multiplying $f(t - a)$ by $1(t - a)$. Without doing any math, the functions $f(t)$ and $1(t)$ are first translated to the right by a quantity a to yield the functions $f(t - a)$ and $1(t - a)$, respectively. The product of the two functions is zero, again, to the left of a and $f(t - a)$ to the right of a. The combined result is translating the product function $f(t)1(t)$ to the right by the quantity a:

$$f(t - a)1(t - a) = \begin{cases} f(t - a) \times 0 = 0, t < a \\ f(t - a) \times 1 = f(t - a), t > a \end{cases} \qquad (6.33)$$

Demonstrating the time-shift theorem, which is expressed in Eq. (6.31), can be done by starting from the Laplace transform definition:

$$\mathcal{L}[f(t - a)1(t - a)] = \int_0^\infty f(t - a)1(t - a)e^{-st}dt \qquad (6.34)$$

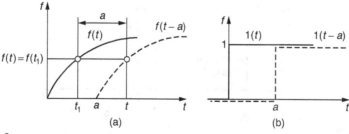

FIGURE 6.6

Horizontally Translated (a) Arbitrary Function; (b) Unit-Step Function.

By using the change of variables $t - a = t_1$, Eq. (6.34) can be written as

$$\mathcal{L}[f(t - a)1(t - a)] = \int_{-a}^{0} f(t_1)1(t_1)e^{-s(t_1+a)}dt_1 + \int_{0}^{\infty} f(t_1)1(t_1)e^{-s(t_1+a)}dt_1$$

$$= e^{-sa}\int_{0}^{\infty} f(t_1)1(t_1)e^{-st_1}dt_1 = e^{-sa}F(s) \qquad (6.35)$$

Equation (6.35) takes into account that

$$\int_{-a}^{0} f(t_1)1(t_1)e^{-s(t_1+a)}dt_1 = 0 \qquad (6.36)$$

because $f(t_1)1(t_1)$ is zero for t less than zero, as indicated in Eq. (6.32).

Example 6.5
Determine the Laplace transforms of the pulse and impulse functions plotted in Figure 6.7.

Solution
The *pulse function*, as can be seen in Figure 6.7(a), is nonzero only between 0 and τ, and it can be written as

$$p(t) = \begin{cases} \dfrac{A}{\tau}, & 0 \le t \le \tau \\ 0, & t > \tau \end{cases} \qquad (6.37)$$

The branch-form of this function is not amenable to using the definition of the Laplace transform over the zero-to-infinity time interval; therefore, reformulation in an adequate manner is needed. We saw that one convenient way of defining a function starting from a point on the time axis that is different from zero is by using the function $1(t - a)$ as a multiplier of the original function. As such, the pulse function can be considered as being made up of two step functions: one equal to A/τ, starting from zero and going to infinity;

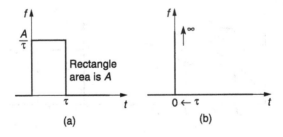

FIGURE 6.7

(a) Pulse Function; (b) Impulse Function.

and another one, equal to $-A/\tau$, applied only from $t = \tau$ to infinity. As a consequence, the pulse function can be rewritten as

$$p(t) = \frac{A}{\tau} - \frac{A}{\tau} 1(t - \tau) \tag{6.38}$$

where multiplication of the first term on the right-hand side with $1(t)$ has been omitted according to a previous discussion. By applying the linearity principle together with the time-shift theorem, the following is the Laplace transform of the original pulse function:

$$P(s) = \mathcal{L}\left[\frac{A}{\tau} - \frac{A}{\tau} 1(t - \tau)\right] = \frac{A}{\tau s}(1 - e^{-\tau s}) \tag{6.39}$$

The *impulse function* is a particular case of the pulse function occurring when the time interval over which the pulse is nonzero reduces to zero; in that case, the height of the needlelike rectangle must go to infinity to combine with a width that tends to zero, in order to keep a constant area A, as shown in Figure 6.7. Mathematically, the impulse is defined as

$$i(t) = \begin{cases} \lim\limits_{\tau \to 0} \dfrac{A}{\tau}, t = \tau \\ 0, t < \tau, t > \tau \end{cases} \tag{6.40}$$

Equations (6.38) and (6.40) indicate that

$$i(t) = \lim\limits_{\tau \to 0} p(t) \tag{6.41}$$

therefore,

$$I(s) = \mathcal{L}[i(t)] = \mathcal{L}\left[\lim\limits_{\tau \to 0} p(t)\right] = \lim\limits_{\tau \to 0} \mathcal{L}[p(t)] = \lim\limits_{\tau \to 0} \frac{A(1 - e^{-\tau s})}{\tau s}$$

$$= \lim\limits_{\tau \to 0} \frac{s A e^{-\tau s}}{s} = A \tag{6.42}$$

Interchangeability between the limit and the Laplace operators, as well as l'Hospital's rule, are applied in Eq. (6.42), because the original limit was of the form zero over zero. When $A = 1$, the impulse function becomes the unit-impulse function, also known as *Dirac delta function*, denoted by $\delta(t)$. ▇

Example 6.6
Determine the Laplace transform of the function sketched with thick solid line in Figure 6.8.

Solution
Figure 6.8 indicates that the function $f(t)$ is

$$f(t) = \begin{cases} f_1(t), 0 \le t \le a \\ f_1(t - a), a \le t \le 2a \\ 0, t \ge 2a \end{cases} = \begin{cases} t, 0 \le t \le a \\ t - a, a \le t \le 2a \\ 0, t \ge 2a \end{cases} \tag{6.43}$$

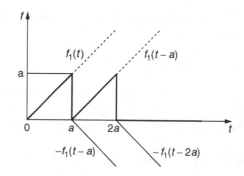

FIGURE 6.8

Truncated Train of Triangular Pulses.

The branch form of Eq. (6.43) can be rendered into the function expression

$$f(t) = t1(t) - a1(t - a) - a1(t - 2a) - (t - 2a)1(t - 2a) \qquad (6.44)$$

Let us analyze the specific form given in Eq. (6.44) and check whether this formula is adequate to the plot of Figure 6.8. The first term on the right-hand side of Eq. (6.44) is a ramp function that is defined from zero to infinity. The next term (the negative step) brings the function from the value of a at $t = a$ to a value of zero. However, the original function $f_1(t)$ continues to exist, which is necessary for the interval a to $2a$, where it actually is identical to the necessary function $t - a$. At $t = 2a$, another negative step function is applied, once more bringing the function to a value of zero from the value of a—this is achieved by the third (minus) term on the right-hand side of Eq. (6.44). The last negative term is used to cancel the effect of the remaining function $t - a$ when the time increases from a value of $2a$ to infinity. The Laplace transform applied to $f(t)$ of Eq. (6.44) results in

$$\mathcal{L}[f(t)] = \frac{1}{s}\left[\frac{1}{s} - ae^{-as}\left(e^{-as} + 1\right) - \frac{1}{s}e^{-2as}\right] \qquad (6.45)$$

Laplace Transform of Derivatives

The Laplace transform of the first derivative of a function $f(t)$ can be approached using the definition

$$\mathcal{L}\left[\frac{df(t)}{dt}\right] = \int_0^\infty \frac{df(t)}{dt}e^{-st}dt \qquad (6.46)$$

through integration by parts where $u = e^{-st}$ and the rest of the function in Eq. (6.46) being dv. In doing that, one obtains

$$\mathcal{L}\left[\frac{df(t)}{dt}\right] = \lim_{t\to\infty}\left[f(t)e^{-st}\right] - \lim_{t\to 0}\left[f(t)e^{-st}\right] + s\int_0^\infty f(t)e^{-st}dt = sF(s) - f(0) \qquad (6.47)$$

Equation (6.47) took into account the fact that $\lim_{t \to \infty}[f(t)e^{-st}] = 0$, because the function $f(t)$ is Laplace transformable and therefore is of exponential order.

Example 6.7

Find the Laplace transform of the function $f(t) = \sin(\omega t)$ by using the relationship

$$\mathcal{L}\left[\frac{d^2 f(t)}{dt^2}\right] = s^2 \mathcal{L}[f(t)] - sf(0) - \left.\frac{df(t)}{dt}\right|_{t=0}$$

Solution

The given relationship can be reformulated as

$$\mathcal{L}[f(t)] = \frac{1}{s^2}\left\{\mathcal{L}\left[\frac{d^2 f(t)}{dt^2}\right] + sf(0) + \left.\frac{df(t)}{dt}\right|_{t=0}\right\} \tag{6.48}$$

It is known that

$$\frac{d^2[\sin(\omega t)]}{dt^2} = -\omega^2 \sin(\omega t) \tag{6.49}$$

therefore,

$$\mathcal{L}\left[\frac{d^2 f(t)}{dt^2}\right] = -\omega^2 \mathcal{L}[f(t)] \tag{6.50}$$

for $f(t) = \sin(\omega t)$. Combining Eqs. (6.48) and (6.50) results in

$$\mathcal{L}[f(t)]\left(1 + \frac{\omega^2}{s^2}\right) = \frac{1}{s^2}\left\{sf(0) + \left.\frac{df(t)}{dt}\right|_{t=0}\right\} \tag{6.51}$$

Because

$$\left.\frac{df(t)}{dt}\right|_{t=0} = \omega \cos(0) = \omega \tag{6.52}$$

and $f(0) = 0$, Eq. (6.51) transforms to

$$\mathcal{L}[f(t)] = \frac{\omega}{s^2 + \omega^2} \tag{6.53}$$

which is the expected result.

Laplace Transform of Indefinite Integrals

There are two modalities of formulating indefinite integrals: one that specifies the limits of integration as zero and t (a generic time station), and another form that does not mention the limits of integration at all. As a consequence, there are two theorems related to the Laplace transformation of integrals. It will be shown next that

$$\mathcal{L}\left[\int_0^t f(t)\,dt\right] = \frac{F(s)}{s} \tag{6.54}$$

By applying the definition of the Laplace transform, it follows that

$$\mathcal{L}\left[\int_0^t f(t)\,dt\right] = \int_0^\infty \left(\int_0^t f(t)\,dt\right) e^{-st}\,dt \tag{6.55}$$

Integration by parts is used in Eq. (6.55) by taking $u = \int_0^t f(t)\,dt$ and $dv = e^{-st}dt$, which yields

$$\mathcal{L}\left[\int_0^t f(t)\,dt\right] = \lim_{t \to \infty}\left(-\frac{1}{s}e^{-st}\int_0^t f(t)\,dt\right) - \lim_{t \to 0}\left(-\frac{1}{s}e^{-st}\int_0^t f(t)\,dt\right)$$

$$+\frac{1}{s}\int_0^\infty f(t)e^{-st}\,dt = \frac{F(s)}{s} \tag{6.56}$$

because, as you can verify, both limits in Eq. (6.56) are zero; so Eq. (6.54) has been demonstrated.

We also demonstrate that

$$\mathcal{L}\left[\int f(t)\,dt\right] = \frac{F(s)}{s} + \frac{\int f(t)\,dt\Big|_{t=0}}{s} \tag{6.57}$$

By again applying integration by parts to the Laplace definition equation and selecting the variables u and v as we did previously, it can be shown that

$$\mathcal{L}\left[\int f(t)\,dt\right] = \lim_{t \to \infty}\left[-\frac{1}{s}e^{-st}\int f(t)\,dt\right] - \lim_{t \to 0}\left[-\frac{1}{s}e^{-st}\int f(t)\,dt\right] + \frac{1}{s}\int_0^\infty f(t)e^{-st}\,dt$$

$$= \frac{F(s)}{s} + \lim_{t \to 0}\left[\frac{1}{s}e^{-st}\int f(t)\,dt\right] = \frac{F(s)}{s} + \frac{\int f(t)\,dt\Big|_{t=0}}{s} \tag{6.58}$$

which took into account that $\lim_{t \to \infty}\left(-\frac{1}{s}e^{-st}\int f(t)\,dt\right) = 0$.

Initial-Value and Final-Value Theorems, use of MATLAB® to Calculate Limits

Both the initial- and final-value theorems can be demonstrated based on the theorem providing the Laplace transform of the first derivative of a function $f(t)$. Let us first calculate the limit:

$$\lim_{s \to \infty} \mathcal{L}\left[\frac{df(t)}{dt}\right] = \lim_{s \to \infty} \int_0^\infty \frac{df(t)}{dt} e^{-st} dt = \int_0^\infty \frac{df(t)}{dt} \left(\lim_{s \to \infty} e^{-st}\right) dt = 0 \qquad (6.59)$$

In this equation, advantage has been taken again of the interchangeability of the limit and integration operators with respect to the s variable. However, according to Eq. (6.47), it also follows that

$$\lim_{s \to \infty} \mathcal{L}\left[\frac{df(t)}{dt}\right] = \lim_{s \to \infty}\left[sF(s)\right] - f(0) = \lim_{s \to \infty}\left[sF(s)\right] - \lim_{t \to 0}\left[f(t)\right] \qquad (6.60)$$

Comparison of Eqs. (6.59) and (6.60) yields

$$\lim_{t \to 0}\left[f(t)\right] = \lim_{s \to \infty}\left[sF(s)\right] \qquad (6.61)$$

which is known as the *initial-value theorem*.

The final-value theorem can be established similarly by considering the following limit first:

$$\lim_{s \to 0} \mathcal{L}\left[\frac{df(t)}{dt}\right] = \lim_{s \to 0} \int_0^\infty \frac{df(t)}{dt} e^{-st} dt = \int_0^\infty \frac{df(t)}{dt} \left(\lim_{s \to 0} e^{-st}\right) dt = \int_0^\infty \frac{df(t)}{dt} dt$$
$$= \lim_{t \to \infty} f(t) - \lim_{t \to 0} f(t) \qquad (6.62)$$

It is also true that

$$\lim_{s \to 0} \mathcal{L}\left[\frac{df(t)}{dt}\right] = \lim_{s \to 0}\left[sF(s)\right] - f(0) = \lim_{s \to 0}\left[sF(s)\right] - \lim_{t \to 0}\left[f(t)\right] \qquad (6.63)$$

Because the algebraic sums in the right-hand sides of Eqs. (6.62) and (6.63) need to be equal, it is necessary that:

$$\lim_{t \to \infty}\left[f(t)\right] = \lim_{s \to 0}\left[sF(s)\right] \qquad (6.64)$$

which is the final-value theorem.

As mentioned in previous chapters, MATLAB® and its symbolic calculation capability can be used to evaluate such limits as the ones necessary in the initial- and final-value theorems. To calculate the limit $\lim_{t \to a, t < a}\left[f(t)\right]$, the following MATLAB® command is needed: `limit(f,t,a,'left')`, and to compute the limit $\lim_{t \to a, t > a}\left[f(t)\right]$, the MATLAB® command is `limit(f,t,a,'right')`.

The initial- and final-value theorems can be used to interrogate the behavior of a dynamic system only at the initial moment and the final one, without necessarily knowing the system's response over the whole time range, as shown in the following example.

Example 6.8

A single-loop electrical circuit is formed of a constant-voltage source of voltage v, an inductor of inductance L, and a resistor of resistance R, all connected, all in series. Calculate the initial value and the final value of the current in the mesh for $v = 120$ V and $R = 240$ Ω. Use MATLAB® to check the results.

Solution

According to Kirchhoff's second law, as shown in Chapter 4, the differential equation governing the dynamic behavior of this electrical circuit is

$$L\frac{di(t)}{dt} + Ri(t) = v \tag{6.65}$$

The Laplace transform is applied to Eq. (6.65), which results in

$$L[sI(s) - i(0)] + RI(s) = \frac{v}{s} \tag{6.66}$$

Solving for $I(s)$, the Laplace transform of the current $i(t)$, in Eq. (6.66) yields

$$I(s) = \frac{v}{s(Ls + R)} + \frac{Li(0)}{Ls + R} \tag{6.67}$$

The initial value of the current is determined by means of the initial-value theorem as

$$\lim_{t \to 0} i(t) = \lim_{s \to \infty} sI(s) = \lim_{s \to \infty}\left[\frac{v}{Ls + R} + \frac{sLi(0)}{Ls + R}\right] = i(0) \tag{6.68}$$

whereas the final (steady-state) value of the current is found using the final-value theorem as

$$\lim_{t \to \infty} i(t) = \lim_{s \to 0} sI(s) = \lim_{s \to 0}\left[\frac{v}{Ls + R} + \frac{sLi(0)}{Ls + R}\right] = \frac{v}{R} = 0.5A \tag{6.69}$$

The following MATLAB® code returns the results of Eqs. (6.68) and (6.69):

```
>> syms s i0 l r c v
>> I = v/(s*(l*s+r))+l*i0/(l*s+r);
>> limit(s*I,s,0,'right')
ans =
v/r
```

```
» limit(s*I,s,inf,'left')
  ans =
  10
```

Periodic Functions

Periodic functions have the property

$$f(t) = f(t + T) = f(t + 2T) = \cdots = f(t + nT) \tag{6.70}$$

for every value of the time variable t, where T is the period and n is any positive integer. Let us calculate the Laplace transform of such a periodic function. The zero-to-infinity time interval can be split in an infinite number of subintervals, namely: 0 to T, T to $2T$, ..., nT to $(n + 1)T$, ..., which means the Laplace transform of $f(t)$ can be written as

$$\int_0^\infty e^{-st} f(t)\,dt = \int_0^T e^{-st} f(t)\,dt + \int_T^{2T} e^{-st} f(t)\,dt + \cdots \int_{nT}^{(n+1)T} e^{-st} f(t)\,dt$$

$$+ \cdots = \sum_{n=0}^{\infty} \int_{nT}^{(n+1)T} e^{-st} f(t)\,dt \tag{6.71}$$

The following change of variable

$$t = t_1 + nT \tag{6.72}$$

transforms the limits of integration of the integral in Eq. (6.71) into 0 (the lower limit) and T (the upper limit), so that Eq. (6.71) becomes

$$\int_0^\infty e^{-st} f(t)\,dt = \sum_{n=0}^{\infty} \int_0^T e^{-s(t_1+nT)} f(t_1)\,dt_1 = \left(\sum_{n=0}^{\infty} e^{-nTs} \right) \int_0^T e^{-st_1} f(t_1)\,dt_1 \tag{6.73}$$

The infinite sum in Eq. (6.73) represents an infinite geometric series sum whose ratio is $r = e^{-Ts}$ and whose first term is $a = 1$. It is known that the sum of terms in a geometric series is calculated as

$$\sum_{n=0}^{\infty} e^{-nTs} = \lim_{n \to \infty} \frac{a(1 - r^{n+1})}{1 - r} = \lim_{n \to \infty} \frac{1[1 - e^{-(n+1)Ts}]}{1 - e^{-Ts}} = \frac{1}{1 - e^{-Ts}} \tag{6.74}$$

By combining Eqs. (6.73) and (6.74) and taking into account that the time variable in the integral of Eq. (6.73) is arbitrary, the Laplace transform of the periodic function $f(t)$ becomes

$$\mathcal{L}[f(t)] = \frac{\int_0^T e^{-st_1} f(t_1)\,dt_1}{1 - e^{-sT}} = \frac{\int_0^T e^{-st} f(t)\,dt}{1 - e^{-sT}} \tag{6.75}$$

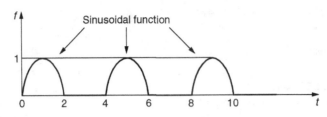

FIGURE 6.9

Train of Sinusoidal Pulses.

Example 6.9

Find the Laplace transform of the periodic function drawn in Figure 6.9. Consider that time is measured in seconds.

Solution

Figure 6.9 indicates the function's period is $T = 4$ s. The sinusoidal portions have an amplitude of 1; and because this maximum value occurs at $t = 1$ s for the first time, it follows that the sinusoidal portion can be expressed as $\sin(\omega t)$ with $\omega = \pi/2$. As a consequence, $f(t)$ is expressed as

$$f(t) = \begin{cases} \sin\left(\frac{\pi}{2}t\right), & nT \le t \le \left(n + \frac{1}{2}\right)T \\ 0, & \left(n + \frac{1}{2}\right)T \le t \le (n + 1)T \end{cases} = \begin{cases} \sin\left(\frac{\pi}{2}t\right), & 4n \le t \le 4\left(n + \frac{1}{2}\right) \\ 0, & \left(n + \frac{1}{2}\right)4 \le t \le 4(n + 1) \end{cases} \tag{6.76}$$

Application of Eq. (6.75) with consideration of the fact that the function $f(t)$ is zero for the second interval (namely, for t between 2 and 4 s) leads to

$$\mathcal{L}[f(t)] = \frac{\displaystyle\int_0^2 e^{-st} \sin\left(\frac{\pi}{2}t\right)dt}{1 - e^{-4s}} \tag{6.77}$$

The integral in the numerator of Eq. (6.77) can be solved by parts or by applying the symbolic calculation capability of MATLAB®; the final result is

$$\mathcal{L}[f(t)] = \frac{2\pi(1 + e^{-2s})}{(\pi^2 + 4s^2)(1 - e^{-4s})} \tag{6.78}$$

The Convolution Theorem

The *convolution theorem* offers an elegant alternative to finding the inverse Laplace transform of a function that can be written as the product of two functions, without using the simple fraction expansion process, which, at times, could be quite complex, as we see later in this chapter. The convolution theorem is based on the *convolution of two functions* $f(t)$ and $g(t)$. According to the definition, the convolution of $f(t)$ and $g(t)$ is

$$f(t) * g(t) = \int_0^t f(\tau)g(t - \tau)d\tau \qquad (6.79)$$

It is straightforward to demonstrate that the convolution of two functions is a commutative operation. If the change of variable is used, $\tau_1 = t - \tau$ in Eq. (6.79), this equation changes to

$$\int_0^t f(\tau)g(t - \tau)d\tau = -\int_t^0 f(t - \tau_1)g(\tau_1)d\tau_1$$

$$= \int_0^t g(\tau_1)f(t - \tau_1)d\tau_1 = g(t) * f(t) \qquad (6.80)$$

The convolution theorem (whose demonstration is given in the companion website Chapter 6) states that

$$\mathcal{L}[f(t) * g(t)] = F(s)G(s) \qquad (6.81)$$

which leads to

$$\mathcal{L}^{-1}[F(s)G(s)] = f(t) * g(t) = \int_0^t f(\tau)g(t - \tau)d\tau \qquad (6.82)$$

Equation (6.82) indicates that the inverse Laplace transform of the product of two functions in the s domain is equal to the convolution of the original (time-domain) functions.

Example 6.10

Determine the inverse Laplace transform of the function $X(s) = 1/[(s^2 + a^2)(s^2 + b^2)]$ using the convolution theorem.

Solution

The given function can be expressed in product form as

$$X(s) = \frac{1}{s^2 + a^2} \times \frac{1}{s^2 + b^2} = F(s)G(s) \qquad (6.83)$$

which indicates that

$$F(s) = \frac{1}{s^2 + a^2}; \quad G(s) = \frac{1}{s^2 + b^2} \qquad (6.84)$$

therefore, the time-domain counterparts of the functions in Eq. (6.84) are

$$f(t) = \frac{1}{a}\sin(at); \; g(t) = \frac{1}{b}\sin(bt) \tag{6.85}$$

By applying the convolution theorem and integral, the function $x(t)$ is calculated as

$$x(t) = f(t) * g(t) = \frac{1}{ab}\int_0^t \sin(a\tau)\sin[b(t-\tau)]d\tau \tag{6.86}$$

The sine product under the integral can be written as

$$\sin(a\tau)\sin[b(t-\tau)] = \frac{1}{2}\{\cos[(a+b)\tau - bt] - \cos[(a-b)\tau + bt]\} \tag{6.87}$$

Substituting Eq. (6.87) into Eq. (6.86) and solving the resulting two integrals yields

$$x(t) = \frac{1}{ab(a^2 - b^2)}[a\sin(bt) - b\sin(at)] \tag{6.88}$$

6.2 SOLVING DIFFERENTIAL EQUATIONS BY THE DIRECT AND INVERSE LAPLACE TRANSFORMS

The theorem that applies the direct Laplace transform to derivatives of functions is fundamental in solving linear differential equations and linear differential equation systems. While the Laplace transform excels in solving linear differential equations with constant coefficients, there are cases where the Laplace transform can be applied to solve linear differential equations with time-dependent coefficients. For constant-coefficient linear differential equations, application of the Laplace transform changes a time-domain differential equation into an algebraic one in the s domain, where the unknown (the Laplace transform of the time-dependent unknown function of the differential equation) can be determined easily. After finding the unknown function in the s domain, the process of partial- (or simple-) fraction expansion needs to be applied to simplify the usually complex functions, simplification of which enable migration back to the time domain using regular Laplace pairs and properties. To summarize, the steps that are needed to solve differential equations by means of the Laplace transforms are

- Formulate the mathematical model with differential equation(s) in the time domain.
- Apply the Laplace transform(s) to the time-dependent differential equation(s).
- Solve the resulting s-domain algebraic equation(s) for the Laplace transform(s) of the time-domain unknown function(s).

- Use partial-fraction expansion of the s-domain unknown function(s) to sim-plify them.
- Apply the inverse Laplace transform(s) to the simplified s-domain unknown function(s) and thus determine the original time-domain unknown functions.

6.2.1 Analytical and MATLAB® Partial-Fraction Expansion

Most often, rather complicated functions are produced in the s domain by application of the direct Laplace transform, functions that cannot be directly converted into the time domain using relatively simple transformation pairs and adequate properties. As a consequence, simplifications need to be performed before the inverse Laplace transformation can be utilized conveniently. Laplace-domain functions are obtained in fraction form:

$$F(s) = \frac{N(s)}{D(s)} \tag{6.89}$$

where $N(s)$ (N standing for numerator) and $D(s)$ (D meaning denominator) are poly-nomials in s. When the degree of $N(s)$ is larger than or equal to the degree of $D(s)$, it is necessary to divide $N(s)$ by $D(s)$ until a quotient is produced together with another polynomial fraction where the degree of the new numerator is smaller than the degree of the denominator.

Example 6.11

Determine the original function $f(t)$, knowing its Laplace transform is $F(s) = (s^4 + 3s^3 + 6s^2 + 6)/(s^2 + 2s + 5)$.

Solution

Division of the numerator to the denominator yields the following sum:

$$F(s) = s^2 + s - 1 - \frac{3s - 11}{s^2 + 2s + 5} \tag{6.90}$$

$F(s)$ can further be conditioned to the form

$$F(s) = s^2 + s - 1 - 3 \times \frac{s + 1}{(s + 1)^2 + 2^2} + 7 \times \frac{2}{(s + 1)^2 + 2^2} \tag{6.91}$$

which enables application of the inverse Laplace transform resulting in

$$f(t) = \frac{d^2 \delta(t)}{dt^2} + \frac{d\delta(t)}{dt} - \delta(t) - \left[3\cos(2t) - 7\sin(2t)\right]e^{-t} \tag{6.92}$$

Still, division of the numerator to the denominator of Eq. (6.89) may not be sufficient, because the resulting polynomial fraction is too complex (particularly due to the nature of its denominator), and the procedure known as *partial-fraction expansion* needs to be applied, as we discuss next. Partial fraction expansion can be performed analytically or using specialized MATLAB® commands.

Analytical Partial-Fraction Expansion

There are *four cases*, as determined by the following denominator roots:

- Real and simple (distinct).
- Complex (or imaginary) and simple (distinct).
- Real and multiple (repeated).
- Complex and multiple (repeated).

In each of these cases, different partial-fraction expansion rules apply, and several unknown coefficients in the numerators of these fractions need to be determined by employing three methods. For simple roots, the method of *fraction combination* (which is presented in the companion website Chapter 6) or the *cover-up method* can be used to determine the unknown fraction coefficients. For multiple roots, the method of the *s derivative* can be utilized in conjunction with either of the other two methods just mentioned. A couple of examples are studied next; the full treatment of the four cases mentioned previously can be found in the companion website Chapter 6.

Example 6.12

Calculate the inverse Laplace transform of the function $F(s) = (s + 3)/[(s^2 + 3s + 2)(s^2 + 1)]$ using partial-fraction expansion and the cover-up method.

Solution

The function $F(s)$ can be written as

$$F(s) = \frac{s + 3}{(s + 1)(s + 2)(s^2 + 1)} \tag{6.93}$$

and the particular fraction expansion of it is of the form

$$\frac{s + 3}{(s + 1)(s + 2)(s^2 + 1)} = \frac{a}{s + 1} + \frac{b}{s + 2} + \frac{cs + d}{s^2 + 1} \tag{6.94}$$

It can be seen that, in any of the simple fractions in the right-hand side of Eq. (6.94), the difference between the degree of the polynomial in the denominator and the degree of the polynomial in the numerator is 1.

The cover-up method determines the coefficients individually and sequentially, and it is advantageous to use this method in applications involving a large number of coefficients (fractions). To find the coefficient a, Eq. (6.94) is multiplied by $(s + 1)$ then s is made equal to -1, which results in

$$a = \left.\frac{s + 3}{(s + 2)(s^2 + 1)}\right|_{s=-1} = 1 \tag{6.95}$$

This process is equivalent to "covering" $(s + 1)$ on the left-hand side of Eq. (6.94) and evaluating the remainder of the fraction for the value of s that zeroes the denominator of the fraction that contains a on the right-hand side of Eq. (6.94); that is, $s = -1$. By using this procedure, b is found by covering $(s + 2)$ on the left-hand side of Eq. (6.94) and taking $s = -2$ for the remainder:

$$b = \left.\frac{s + 3}{(s + 1)(s^2 + 1)}\right|_{s=-2} = -\frac{1}{5} \tag{6.96}$$

The coefficients c and d are determined similarly, by first multiplying through Eq. (6.94) by $(s^2 + 1)$, then considering that $s^2 + 1 = 0$. The last operation can be done for either of the two complex roots of $s^2 + 1 = 0$; that is, $s = -j$ or $s = j$. For $s = j$, for instance, the following is obtained:

$$cs + d = \left.\frac{s + 3}{(s + 1)(s + 2)}\right|_{s=j} \tag{6.97}$$

which is equivalent to the following identity:

$$d - 3c + (3d + c)j = 3 + j \tag{6.98}$$

By equating the real parts and then the imaginary parts in the two sides of Eq. (6.98), b and c are obtained as $c = -4/5$ and $d = 3/5$. As a consequence, $F(s)$ can be written as

$$F(s) = \frac{1}{s + 1} - \frac{1}{5} \times \frac{1}{s + 2} - \frac{4}{5} \times \frac{s}{s^2 + 1} + \frac{3}{5} \times \frac{1}{s^2 + 1} \tag{6.99}$$

whose inverse Laplace transform is

$$f(t) = e^{-t} - \frac{1}{5}e^{-2t} - \frac{4}{5}\cos t + \frac{3}{5}\sin t \tag{6.100}$$

■

Example 6.13

Determine the inverse-Laplace transform of the function $F(s) = 1/[s(s + 1)^2]$ using partial-fraction expansion together with the cover-up and s-derivative methods.

Solution

The roots of the denominator are $s = 0$ and $s = -1$, which is real and of the order 2 of multiplicity. The partial-fraction expansion of $F(s)$ is

$$\frac{1}{s(s + 1)^2} = \frac{a}{s} + \frac{b}{(s + 1)^2} + \frac{c}{s + 1} \tag{6.101}$$

The coefficients a, b, and c can be determined with a method that combines the cover-up procedure with a derivation technique, as shown next. The coefficient a is calculated using the cover-up method as

$$a = \frac{1}{(s + 1)^2}\bigg|_{s=0} = 1 \tag{6.102}$$

The coefficient b, the one corresponding to largest degree denominator of the two fractions with $s + 1$ in the denominator, is directly found by means of the cover-up method as

$$b = \frac{1}{s}\bigg|_{s=-1} = -1 \tag{6.103}$$

To find coefficient c, Eq. (6.101) is multiplied by $(s + 1)^2$, which results in

$$\frac{1}{s} = \frac{a}{s}(s + 1)^2 + b + c(s + 1) \tag{6.104}$$

By applying the derivation with respect to s in Eq. (6.104), the following equation is obtained:

$$-\frac{1}{s^2} = \frac{2a(s + 1)}{s} - \frac{a(s + 1)^2}{s^2} + c \tag{6.105}$$

For $s = -1$ in Eq. (6.105), the value of c is found:

$$c = -\frac{1}{s^2}\bigg|_{s=-1} = -1 \tag{6.106}$$

As a consequence, the function $F(s)$ is

$$F(s) = \frac{1}{s} - \frac{1}{(s + 1)^2} - \frac{1}{s + 1} \tag{6.107}$$

whose inverse Laplace transformation yields the original function:

$$f(t) = 1 - (t + 1)e^{-t} \tag{6.108}$$

MATLAB® Partial-Fraction Expansion

Partial-fraction expansion of the function $F(s)$ given in Eq. (6.89) results in

$$F(s) = k(s) + \frac{r_1}{s - p_1} + \frac{r_2}{s - p_2} + \cdots + \frac{r_i}{s - p_i} + \frac{r_{i+1}}{(s - p_i)^2} + \cdots$$

$$+ \frac{r_{i+k-1}}{(s - p_i)^k} + \cdots + \frac{r_n}{s - p_n} \tag{6.109}$$

where $k(s)$ is the *direct-terms polynomial*, $r_1, r_2, \ldots, r_n$ are *residues*, and $p_1, p_2, \ldots, p_n$ are *poles*. Please note that the pole p_i is a pole of multiplicity k. The poles can be either real or complex. MATLAB® has a simple set of commands that are centered around the residue command, which enable partial-fraction expansion in a simple way, as shown in the next example.

Example 6.14

Use MATLAB® to determine the partial-fraction expansion of the Laplace-domain function:

$$F(s) = \frac{s^2 - 8s + 15}{s^4 + 7s^3 + 18s^2 + 20s + 8}$$

Solution

In MATLAB®, the coefficients of $N(s)$, which is the numerator of $F(s)$, and $D(s)$, which is the denominator of $F(s)$, need to be first defined, then the residue command can be entered:

```
» n = [1,-8,15];
» d = [1,7,18,20,8];
» [r,p,k] = residue(n,d)
```

and MATLAB® returns

```
r =
  -24.0000
  -23.0000
  -35.0000
   24.0000
p =
   -2.0000
   -2.0000
   -2.0000
   -1.0000
k =
   []
```

The interpretation of these MATLAB® results is that the partial-fraction expansion of $F(s)$ is

$$F(s) = -\frac{24}{s + 2} - \frac{23}{(s + 2)^2} - \frac{35}{(s + 2)^3} + \frac{24}{s + 1} \qquad (6.110)$$

6.2.2 Linear Differential Equations with Constant Coefficients

Let us assume the linear differential equation is of the general form

$$c_n \frac{d^n x(t)}{dt^n} + c_{n-1} \frac{d^{n-1} x(t)}{dt^{n-1}} + \cdots + c_1 \frac{dx(t)}{dt} + c_0 x(t) = f(t) \qquad (6.111)$$

where $f(t)$ is the input (or excitation) function, $x(t)$ is the output (unknown) function, and the coefficients c_0 to c_n are constant. Applying the direct Laplace transformation to Eq. (6.111) (and let us assume all the initial conditions are zero, to simplify this generic model), the following algebraic equation is obtained:

$$\left(c_n s^n + c_{n-1} s^{n-1} + \cdots + c_1 s + c_0 \right) X(s) = F(s) \qquad (6.112)$$

which provides the s-domain solution:

$$X(s) = \frac{F(s)}{c_n s^n + c_{n-1} s^{n-1} + \cdots + c_1 s + c_0} \qquad (6.113)$$

The time-domain solution is formally calculated as

$$x(t) = \mathcal{L}^{-1}[X(s)] \qquad (6.114)$$

Example 6.15

The MEMS device shown in Figure 6.10, consisting of a mass $m = 10^{-6}$ kg and two beam springs of total stiffness $k = 2 \times 10^{-6}$ N/m, is acted upon by a an external force f. The microsystem motion is opposed by a viscous damping force whose damping coefficient is $c = 3 \times 10^{-6}$ N-s/m. Calculate the system response (displacement) for the following forcing functions:
a. $f(t) = 2 \ \mu N$.
b. $f(t) = 2t \ \mu N$.
Consider that the initial conditions in both cases are

$$x(0) = 0; \ \left. \frac{dx(t)}{dt} \right|_{t=0} = 0$$

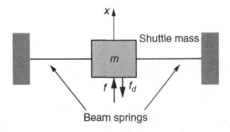

FIGURE 6.10

Micromechanism with Shuttle Mass and Two Beam Springs.

Solution

The Laplace transform is applied to the differential equation of this problem (which, as shown in Chapter 2, is $m\ddot{x} + c\dot{x} + kx = t$) taking into account the parameter values and given initial conditions; this leads to

$$X(s) = \frac{F(s)}{ms^2 + cs + k} = \frac{F(s)\,10^{-6}}{(s^2 + 3s + 2)10^{-6}} = \frac{F(s)}{(s + 1)(s + 2)} \tag{6.115}$$

where $X(s)$ is the Laplace transform of $x(t)$ and $F(s)$ is the Laplace transform of $f(t)$. International System of units (N, m, s, kg) have been used in Eq. (6.115).

a. In the case where $f(t) = 2$, its Laplace transform is $F(s) = 2/s$ so that Eq. (6.115) transforms into

$$X(s) = \frac{2}{s(s + 1)(s + 2)} \tag{6.116}$$

All roots of the denominator are real and distinct; therefore, $X(s)$ is of the form

$$X(s) = \frac{a}{s} + \frac{b}{s + 1} + \frac{c}{s + 2} \tag{6.117}$$

The coefficients are easily found by either of the two methods mentioned, and it is quite straightforward to check that $a = c = 1$ and $b = -2$. As a consequence, $X(s)$ becomes

$$X(s) = \frac{1}{s} - \frac{2}{s + 1} + \frac{1}{s + 2} \tag{6.118}$$

The inverse Laplace transform is now applied to $X(s)$, which results in the solution $x(t)$:

$$x(t) = \mathcal{L}^{-1}[X(s)] = 1 - 2e^{-t} + e^{-2t} \tag{6.119}$$

b. The Laplace transform of $f(t) = 2t$ is $F(s) = 2/s^2$; and therefore,

$$X(s) = \frac{2}{s^2(s + 1)(s + 2)} \tag{6.120}$$

The partial fraction expansion of $X(s)$ is

$$\frac{2}{s^2(s + 1)(s + 2)} = \frac{a}{s^2} + \frac{b}{s} + \frac{c}{s + 1} + \frac{d}{s + 2} \tag{6.121}$$

The coefficients c and d are found as follows:

$$\begin{cases} c = \dfrac{2}{s^2(s + 2)} \bigg|_{s=-1} = 2 \\[4mm] d = \dfrac{2}{s^2(s + 1)} \bigg|_{s=-2} = -\dfrac{1}{2} \end{cases} \tag{6.122}$$

To find the coefficient a, Eq. (6.121) is multiplied by s^2, which yields

$$\frac{2}{(s+1)(s+2)} = a + bs + \frac{cs^2}{s+1} + \frac{ds^2}{s+2} \tag{6.123}$$

By taking $s = 0$ in Eq. (6.123), it is found that $a = 1$. The s derivative is now applied to Eq. (6.123), which results in

$$b + \frac{cs^2 + 2cs}{(s+1)^2} + \frac{ds^2 + 4ds}{(s+2)^2} = -\frac{2(2s+3)}{(s+1)^2(s+2)^2} \tag{6.124}$$

Equation (6.124) has to be valid for all values of s and therefore for $s = 0$; it follows that $b = -3/2$.

As a consequence,

$$X(s) = \frac{1}{s^2} - \frac{3}{2} \times \frac{1}{s} + \frac{2}{s+1} - \frac{1}{2} \times \frac{1}{s+2} \tag{6.125}$$

Application of the inverse Laplace transform to $X(s)$ of Eq. (6.125) results in

$$x(t) = t - \frac{3}{2} + 2e^{-t} - \frac{1}{2}e^{-2t} \tag{6.126}$$

6.2.3 Use of MATLAB® to Calculate Direct and Inverse Laplace Transforms

In many instances, the s-domain unknowns are quite complex, and applying the inverse Laplace transform to retrieve their time-dependent counterparts is no easy task. MATLAB® has some strong capabilities allowing us to calculate both the direct and inverse Laplace transforms. MATLAB® Symbolic Math Toolbox™ enables calculation of the direct and inverse Laplace transforms of functions by using the `laplace (f(t))` and `ilaplace(F(s))` commands, as shown in the following example.

Example 6.16

Use analytic calculation and also MATLAB® to solve the differential equation of Example 6.15 for $f(t) = 2 \sin(t) \, \mu N$.

Solution

The Laplace transform of $f(t)$ in this case is

$$F(s) = \frac{2}{s^2 + 1} \tag{6.127}$$

which can also be determined using the following MATLAB® code:

```
» syms t
» f = 2*sin(t);
» pretty(laplace(f));
```

By means of $F(s)$ of Eq. (6.127), the Laplace-domain unknown becomes

$$X(s) = \frac{2}{(s^2 + 1)(s + 1)(s + 2)} \tag{6.128}$$

The simple fraction expansion of the right-hand side of Eq. (6.128) is of the form

$$X(s) = \frac{as + b}{s^2 + 1} + \frac{c}{s + 1} + \frac{d}{s + 2} \tag{6.129}$$

Using the cover-up method, the coefficients are found to be a = −3/5, b = 1/5, c = 1, d = −2/5. Substitution of these values into $X(s)$ of Eq. (6.129) and application of the inverse Laplace transform to the resulting numerical expression yields the following time-dependent solution:

$$x(t) = \frac{1}{5}[\sin(t) - 3\cos(t)] + e^{-t} - \frac{2}{5}e^{-2t} \tag{6.130}$$

In MATLAB®, the command `ilaplace`, as mentioned here, allows obtaining the time-domain original function $x(t)$ from $X(s)$, which needs to be defined symbolically. The MATLAB® command lines are

```
» syms s
» X = 2/((s^2+1)*(s+1)*(s+2));
» pretty(ilaplace(X))
```

which returns

```
 - 2/5 exp(-2 t) + exp(-t) - 3/5 cos(t) + 1/5 sin(t)
```

6.2.4 Linear Differential Equation Systems with Constant Coefficients

The method of Laplace transform can also be applied to multiple-DOF systems, and this results in several differential equations, depending generally on a number of time-dependent functions that is equal to the number of available equations. They can be transformed into algebraic equations that form an algebraic equations system whose unknowns are the Laplace transforms of the unknown time-dependent functions. These Laplace-transformed unknowns are solved for in the s domain algebraically, and then inverse Laplace transforms are applied to those to determine the original time-dependent unknown functions.

Example 6.17

For the electrical circuit of Figure 6.11, find and plot the currents through the inductor and the capacitor by deriving the mathematical model of the circuit and using the inverse Laplace transform capabilities of MATLAB®. Consider that all the initial conditions are zero and $R = 200 \ \Omega$, $L = 20$ H, $C = 10 \ \mu$F, and $v = 100$ V.

Solution

According to Kirchhoff's current law applied at one of the nodes highlighted in Figure 6.11, the three currents are connected as

$$i_R(t) = i_L(t) + i_C(t) \tag{6.131}$$

For the two meshes of the circuit, Kirchhoff's second law results in

$$\begin{cases} Ri_R(t) + L\dfrac{di_L(t)}{dt} = v \\ \dfrac{1}{C}\displaystyle\int i_C(t)\,dt - L\dfrac{di_L(t)}{dt} = 0 \end{cases} \tag{6.132}$$

The current i_R is substituted from Eq. (6.131) into Eq. (6.132), and after applying the Laplace transform to the two resulting equations, the following algebraic equations system is obtained:

$$\begin{cases} (R + Ls)I_L(s) + RI_C(s) = \dfrac{v}{s} \\ -LsI_L(s) + \dfrac{1}{Cs}I_C(s) = 0 \end{cases} \tag{6.133}$$

which can be written in vector-matrix form as

$$\begin{bmatrix} R + Ls & R \\ -Ls & \dfrac{1}{Cs} \end{bmatrix} \begin{Bmatrix} I_L(s) \\ I_C(s) \end{Bmatrix} = \begin{Bmatrix} \dfrac{v}{s} \\ 0 \end{Bmatrix} \tag{6.134}$$

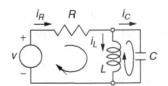

FIGURE 6.11

Electrical Circuit with Step Voltage Input.

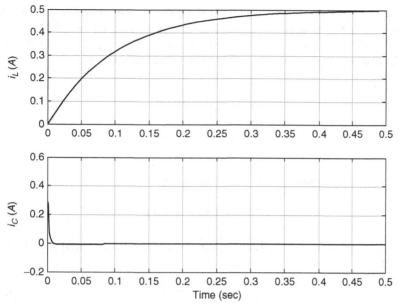

FIGURE 6.12

Time Variation of Currents in Inductor and Capacitor.

The unknown vector can be calculated as

$$
\begin{Bmatrix} I_L(s) \\ I_C(s) \end{Bmatrix} = \begin{bmatrix} R + Ls & R \\ -Ls & \dfrac{1}{Cs} \end{bmatrix}^{-1} \begin{Bmatrix} \dfrac{v}{s} \\ 0 \end{Bmatrix} = \begin{Bmatrix} \dfrac{v}{s(RLCs^2 + Ls + R)} \\ \dfrac{LCvs}{RLCs^2 + Ls + R} \end{Bmatrix} \tag{6.135}
$$

By using MATLAB® symbolic calculation capability, the unknowns are found to be

$$
\begin{cases} i_L(t) = 0.5 - [0.5\cosh(239.8t) + 0.52\sinh(239.8t)]e^{-250t} \\ i_C(t) = [0.5\cosh(239.8t) - 0.52\sinh(239.8t)]e^{-250t} \end{cases} \tag{6.136}
$$

These currents are plotted against time in Figure 6.12.   ■

6.2.5 Laplace Transformation of Vector-Matrix Differential Equations

We have seen that the Laplace transform can be applied individually to differential equations making up a system. Differential equations describing the dynamic response of multiple-DOF systems can be formulated in vector- matrix form, and the question arises whether the Laplace transform (direct and inverse) can operate directly on this form. It is actually quite straightforward to extend the Laplace transform from functions to collections of functions (which are vectors). Let us look at a vector $\{x(t)\}$ consisting of two components, $x_1(t)$ and $x_2(t)$. The Laplace transforms of the two components (assuming they exist) are determined as

$$\begin{cases} X_1(s) = \int_0^\infty x_1(t)\, e^{-st}\, dt \\[2mm] X_2(s) = \int_0^\infty x_2(t)\, e^{-st}\, dt \end{cases} \tag{6.137}$$

and they can be assembled into the vector $X(s)$ as

$$\{X(s)\} = \{X_1(s)\ \ X_2(s)\}^t \tag{6.138}$$

We can therefore define the Laplace transform of the vector $\{x(t)\}$ into the vector $\{X(s)\}$ as

$$\mathcal{L}[\{x(t)\}] = \{X(s)\} \tag{6.139}$$

In conclusion, application of the Laplace transform to a vector results in another vector whose components are the Laplace transforms of the original (time-dependent) vector.

Extensions can be made, and it can simply be shown that, given the vectors $\{x_1(t)\}$ and $\{x_2(t)\}$, as well as the matrices $[A_1]$ and $[A_2]$ (made up of constants only), the following linear relationship holds:

$$\mathcal{L}\big[[A_1]\{x_1(t)\} + [A_2]\{x_2(t)\}\big] = [A_1]\{X_1(s)\} + [A_2]\{X_2(s)\} \tag{6.140}$$

We thus can conclude that the Laplace pairs and properties pertaining to scalar functions are also valid for vector functions, which could be particularly helpful in solving differential equations written in vector-matrix form. For instance, taking the first and second time derivatives of a vector $\{x(t)\}$ would formally lead to

$$\begin{cases} \mathcal{L}\left[\dfrac{d}{dt}\{x(t)\}\right] = s\{X(s)\} - \{x(0)\} \\[3mm] \mathcal{L}\left[\dfrac{d^2}{dt^2}\{x(t)\}\right] = s^2\{X(s)\} - s\{x(0)\} - \{\dot{x}(0)\} \end{cases} \tag{6.141}$$

where the vectors having the argument zero indicate initial conditions.

Consider a multiple-DOF undamped mechanical system whose vector-matrix equation modeling the forced response behavior is

$$[M]\{\ddot{x}(t)\} + [K]\{x(t)\} = \{f(t)\} \tag{6.142}$$

Substituting Eqs. (6.141) into Eq. (6.142) results in

$$[M]\left(s^2\{X(s)\} - s\{x(0)\} - \{\dot{x}(0)\}\right) + [K]\{X(s)\} = \{F(s)\} \qquad (6.143)$$

which can also be written as

$$\left(s^2[M] + [K]\right)\{X(s)\} = [M]\left(\{\dot{x}(0)\} + s\{x(0)\}\right) + \{F(s)\} \qquad (6.144)$$

The solution to Eq. (6.144) is therefore

$$\{X(s)\} = \left(s^2[M] + [K]\right)^{-1}\left([M]\left(\{\dot{x}(0)\} + s\{x(0)\}\right) + \{F(s)\}\right) \qquad (6.145)$$

and the time-domain solution vector is simply found by inverse-Laplace transforming $\{X(s)\}$.

Example 6.18

The MEMS system of Figure 6.13(a), which consists of two shuttle masses and two serpentine springs, is acted upon by a force f; the lumped-parameter model of this mechanical microsystem is sketched in Figure 6.13(b). Find the forced response of the mechanical microsystem by using the Laplace transform method applied to the vector-matrix form of the mathematical model. Consider that the initial conditions are zero, $f = 10^{-6}\,\delta(t)$ N, where $\delta(t)$ is the unit impulse, and $m_1 = m_2 = m = 10^{-6}$ kg, $k_1 = k_2 = k = 10^{-6}$ N/m.

Solution

For zero initial conditions, Eq. (6.145) reduces to

$$\{X(s)\} = \left(s^2[M] + [K]\right)^{-1}\{F(s)\} \qquad (6.146)$$

The equations of motion for the two-DOF mechanical system are

$$\begin{cases} m_1\ddot{x}_1 + (k_1 + k_2)x_1 - k_2x_2 = 0 \\ m_2\ddot{x}_2 - k_2x_1 + k_2x_2 = f \end{cases} \qquad (6.147)$$

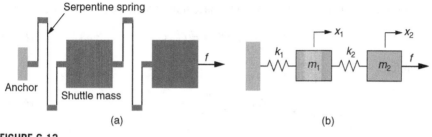

(a)   (b)

FIGURE 6.13

Two-DOF Translatory Mechanical System: (a) Physical Model; (b) Lumped-Parameter Model.

which can be written in the vector-matrix form of Eq. (6.142) with

$$[M] = \begin{bmatrix} m & 0 \\ 0 & m \end{bmatrix}; [K] = \begin{bmatrix} 2k & -k \\ -k & k \end{bmatrix}; \{f(t)\} = 10^{-6} \begin{Bmatrix} 0 \\ \delta(t) \end{Bmatrix} \qquad (6.148)$$

The two components of $\{X(s)\}$ are determined by using MATLAB® as

$$\begin{cases} X_1(s) = \dfrac{k}{m^2 s^4 + 3mks^2 + k^2} \\ X_2(s) = \dfrac{ms^2 + 2k}{m^2 s^4 + 3mks^2 + k^2} \end{cases} \qquad (6.149)$$

and using the inverse-Laplace transform with the numerical parameters of this example, the time response is

$$\begin{cases} x_1(t) = 0.72 \sin (0.62t) - 0.28 \sin(1.62t) \\ x_2(t) = 1.17 \sin (0.62t) + 0.17 \sin(1.62t) \end{cases} \qquad (6.150)$$

where the two displacements are measured in micrometers. If symbolic MATLAB® calculation is used, [r,p,k] commands are first needed to obtain the $X_1(s)$ and $X_2(s)$ of Eq. (6.149).

6.2.6 Solving Integral and Integral-Differential Equations by the Convolution Theorem

The convolution theorem can be used to solve *integral equations*, namely, equations that contain terms where the time-domain unknown function is under the integral operator. Let us assume the mathematical model of a system consists of the following integral equation:

$$x(t) + \int_0^t x(\tau)f(t - \tau)d\tau = g(t) \qquad (6.151)$$

where the functions f and g are known time-dependent functions and $x(t)$ is the unknown function. Application of the Laplace transform to Eq. (6.151) results in

$$X(s) + X(s)F(s) = G(s) \qquad (6.152)$$

Therefore,

$$X(s) = \frac{G(s)}{1 + F(s)} \qquad (6.153)$$

which yields

$$x(t) = \mathcal{L}^{-1}[X(s)] = \mathcal{L}^{-1}\left[\frac{G(s)}{1 + F(s)}\right] \qquad (6.154)$$

Example 6.19

Solve the following integral equation knowing that $x(0) = 0$:

$$x(t) + \int_0^t x(t - \tau)e^{-t}d\tau = \sin(t)$$

Solution

In this case the functions $f(t)$ and $g(t)$ are

$$f(t) = e^{-t}; \ g(t) = \sin(t) \tag{6.155}$$

Therefore,

$$F(s) = \frac{1}{s + 1}; \ G(s) = \frac{1}{s^2 + 1} \tag{6.156}$$

so that, in accordance with Eq. (6.153), $X(s)$ is

$$X(s) = \frac{s + 1}{(s + 2)(s^2 + 1)} \tag{6.157}$$

Simple fraction expansion of $X(s)$ results in

$$\frac{s + 1}{(s + 2)(s^2 + 1)} = \frac{a}{s + 2} + \frac{bs + c}{s^2 + 1} \tag{6.158}$$

It can be shown by one of the known methods that $a = -1/5$, $b = 1/5$ and $c = 3/5$. As a consequence,

$$x(t) = \mathcal{L}^{-1}[X(s)] = \frac{1}{5}(3\sin t + \cos t - e^{-2t}) \tag{6.159}$$

There are situations where the mathematical model of a dynamical system is formed of one or several differential equations that also include the unknown(s) under an integral; such equations are named *integral-differential* and can be solved under certain conditions using the convolution theorem. ∎

Example 6.20

Solve the following integral-differential equation:

$$\dot{x}(t) + \int_0^t x(\tau)(t - \tau)d\tau = 4$$

with $x(0) = 0$.

Solution

Application of the Laplace transform to the integral-differential equation of the problem yields

$$sX(s) + \frac{X(s)}{s^2} = \frac{4}{s} \tag{6.160}$$

which results in

$$X(s) = \frac{4s}{s^3 + 1} = \frac{4s}{(s + 1)(s^2 - s + 1)} = \frac{a}{s + 1} + \frac{bs + c}{s^2 - s + 1} \tag{6.161}$$

The constants are found to be $a = -4/3$, $b = c = 4/3$. Now applying the inverse Laplace transforms to the two fractions on the right-hand side of Eq. (6.161), the solution is obtained as

$$x(t) = \frac{4}{3} \left\{ \left[\cos\left(\frac{\sqrt{3}}{2}t\right) + \sqrt{3} \sin\left(\frac{\sqrt{3}}{2}t\right) \right] e^{0.5t} - e^{-t} \right\} \tag{6.162}$$

6.2.7 Linear Differential Equations with Time-Dependent Coefficients

To demonstrate the applicability of the Laplace method to linear differential equations with time-dependent coefficients, a few more properties and theorems of the Laplace transforms are presented (and demonstrated) as solved examples in the companion website Chapter 6. One of these proprieties,

$$\mathcal{L}[tf(t)] = -\frac{dF(s)}{ds} \tag{6.163}$$

is used in the next example.

Example 6.21

Solve the differential equation $d^2x(t)/dt^2 = t\sinh(t)$ knowing that $x(0) = 0$ and

$$\frac{dx(t)}{dt}\bigg|_{t=0} = 0.$$

Solution

The direct Laplace transform is applied to both sides of the given differential equation based on Eq. (6.163), which results in

$$s^2 X(s) = -\frac{d}{ds}\left(\mathcal{L}[\sinh(t)]\right) \tag{6.164}$$

Completion of the calculations on the right-hand side of Eq. (6.164), which takes into account that $\mathcal{L}[\sinh t] = 1/(s^2 - 1)$, results in

$$s^2 X(s) = \frac{2s}{(s^2 - 1)^2} \tag{6.165}$$

Equation (6.165) gives $X(s)$:

$$X(s) = \frac{2}{s(s^2 - 1)^2} \tag{6.166}$$

The inverse Laplace transform of $X(s)$ is

$$x(t) = 2[1 - \cosh(t)] + t \sinh(t) \tag{6.167}$$

■

Example 6.22

Determine $x(t)$, the solution to the differential equation

$$t \frac{d^2 x(t)}{dt^2} + \frac{dx(t)}{dt} = 1$$

for $x(0) = 0$; $\dot{x}(0) = 0$. Use direct and inverse Laplace transforms.

Solution

Because only the derivatives of the unknown function are involved here, the following substitution can be used:

$$\frac{dx(t)}{dt} = y(t) \tag{6.168}$$

As a consequence, the original differential equation changes to

$$t \frac{dy(t)}{dt} + y(t) = 1 \tag{6.169}$$

The Laplace transform is applied to this differential equation, which yields

$$dY(s) = -\frac{ds}{s^2} \tag{6.170}$$

Integration of Eq. (6.170) results in

$$Y(s) = \frac{1}{s} + C_1 \tag{6.171}$$

where C_1 is an arbitrary constant of integration. The inverse Laplace transform is applied to Eq. (6.171), which yields

$$y(t) = 1 + C_1 \delta(t) \tag{6.172}$$

where $\delta(t)$ is the delta Dirac function. Knowing $y(t)$ enables us to determine $x(t)$ by performing integration in Eq. (6.168). Let us integrate between the limits of zero and infinity (this is consistent with the definition of any Laplace-operable function), which results in

$$x(t) = t + C_1 \int_0^\infty \delta(t)\,dt \tag{6.173}$$

A basic property of the delta Dirac function is that

$$\int_0^\infty \delta(t)\,dt = 0 \tag{6.174}$$

therefore, Eq. (6.173) gives the solution

$$x(t) = t \tag{6.175}$$

It can simply be checked that $x = t$ is indeed the solution to the differential equation of the problem.

6.3 TIME-DOMAIN SYSTEM IDENTIFICATION FROM LAPLACE-DOMAIN INFORMATION

A time-domain dynamic system can be identified by analyzing it in the Laplace domain. The topic is particularly important when the frequency-response of a system is known—this can be achieved through experimental means, forinstance. Chapter 9 studies the frequency response approach in more detail. In this section, we aim to identify a single-DOF time-defined dynamic system from a given function to the Laplace domain, as shown next.

Let us assume a single-DOF dynamic system is defined by the generic differential equation:

$$a\ddot{x}(t) + b\dot{x}(t) + cx(t) = f(t) \tag{6.176}$$

with a, b, and c being constants and the initial conditions being nonzero. Application of the Laplace transform to Eq. (6.176) enables expressing the Laplace transform of the unknown $x(t)$ as

$$X(s) = \frac{ax(0)s + a\dot{x}(0) + bx(0) + F(s)}{as^2 + bs + c} \tag{6.177}$$

where $F(s)$ is the Laplace transform of the forcing function $f(t)$. Equation (6.177) can be written in the generic form

$$X(s) = \frac{ds + e + F(s)}{as^2 + bs + c} \tag{6.178}$$

Equations (6.177) and (6.178) are general, as they encompass all possible elements, forcing, and initial conditions for a single-DOF system. Particular cases can be derived from them by annulling various term(s) in either the numerator or denominator, terms that correspond to missing elements, forcing functions, or initial conditions.

Example 6.23

Identify a single-DOF system whose Laplace-domain transform is

$$X(s) = \frac{s(5s^2 - s + 5)}{(s^2 + 1)(2s^2 + s + 1)}.$$

Solution

The expression of $X(s)$ can be transformed as follows:

$$X(s) = \frac{5s^3 - s^2 + 5s - 1 + 1}{(s^2 + 1)(2s^2 + s + 1)} = \frac{(5s - 1)(s^2 + 1) + 1}{(s^2 + 1)(2s^2 + s + 1)}$$

$$= \frac{(5s - 1) + \dfrac{1}{s^2 + 1}}{(2s^2 + s + 1)} \tag{6.179}$$

Comparison of Eqs. (6.177) and (6.179) shows that

$$F(s) = \frac{1}{s^2 + 1} \tag{6.180}$$

which yields

$$f(t) = \sin(t) \tag{6.181}$$

The same comparison also indicates that $a = 2$, $b = 1$, $c = 1$, and

$$\begin{cases} ax(0) = 5 \\ a\dot{x}(0) + bx(0) = -1 \end{cases} \tag{6.182}$$

The equation system (6.182) is solved for the two initial conditions, which are

$$\begin{cases} x(0) = \dfrac{5}{2} \\ \dot{x}(0) = -\dfrac{7}{4} \end{cases} \tag{6.183}$$

As a consequence, the time-domain differential equation that corresponds to $X(s)$ of this example is

$$2\ddot{x} + \dot{x} + x = \sin(t) \tag{6.184}$$

and the initial conditions are given in Eqs. (6.183).

When Eq. (6.184) is the mathematical model of a mechanical system, one possibility is that the system is a translatory one, consisting of a mass $m = 2$ kg, a damper with a damping coefficient of $c = 1$ N-s/m, a spring of stiffness $k = 1$ N/m, and a sinusoidal force $f = \sin(t)$ N acting on the mass in the motion direction. The initial displacement of this mechanical system is equal to 5/2 m and the initial velocity is equal to $-7/4$ m/s. The differential equation can also represent an electrical system, having its elements (resistor, capacitor, and inductor) disposed in series in the case where the source is a voltage one; alternatively, the same elements and a current source can be connected in parallel.

SUMMARY

This chapter introduces the direct and inverse Laplace transforms as tools for solving differential equations that are mathematical models of mechanical, electrical, or fluid or thermal systems. By using Laplace transform pairs as well as properties of the Laplace transforms, differential equations are transformed into algebraic equations in the Laplace (s) domain, which allow algebraic solution for the unknown Laplace transforms, followed by identification of the original time-domain unknowns. Using the technique of partial-fraction expansion, the direct and inverse Laplace transforms can be used to solve linear differential equations with constant and time-varying coefficients. MATLAB® symbolic calculation of partial-fraction expansion and Laplace transforms also is presented through examples. The Laplace transformation is used extensively in subsequent chapters, which are dedicated to transfer functions, state space modeling, and frequency-domain analysis.

PROBLEMS

6.1   Find the Laplace transforms of the following functions and check the results with MATLAB®:

(a) $f(t) = \sin(t)\cos(3t)$.

(b) $f(t) = \cos(2t)\cos(4t)$.

(c) $f(t) = \sin(2t)\sin(3t)$.

6.2 Same question as in Problem 6.1 for the functions

(a) $f(t) = \cosh(4t)\,\sinh(3t)$.

(b) $f(t) = \cosh(t)\,\cosh(5t)$.

(c) $f(t) = \sinh(2t)\,\sinh(3t)$.

6.3 Find the Laplace transforms of $\sinh(\omega t)$ and $\cosh(\omega t)$, the hyperbolic sine and cosine functions, considering that known are the Laplace transforms of $\sin(\omega t)$ and $\cos(\omega t)$.

6.4 Use the theorem that expresses the Laplace transform of the first time derivative to calculate the Laplace transforms of the functions $f_1(t) = t\,\sin(\omega t)$ and $f_2(t) = t\,\cos(\omega t)$, when known are the Laplace transforms of $\sin(\omega t)$ and $\cos(\omega t)$.

6.5 Same question as in Problem 6.4 for the functions $f_1(t) = te^{-at}, f_2(t) = t^2 e^{-at}$, when known is the Laplace transform of e^{-at} (a being a positive real constant).

6.6 Calculate the Laplace transforms of the following functions both analytically and with MATLAB®:

(a) $f(t) = te^{-at}\,\sinh(\omega t)$.

(b) $f(t) = te^{-at}\,\cosh(\omega t)$.

6.7 Calculate the Laplace transform of the function $f(t) = \cosh(\omega t)\,\sin(\omega t) - \sinh(\omega t)\,\cos(\omega t)$ analytically and with MATLAB®.

6.8 Demonstrate that $\mathcal{L}[t^2 f(t)] = d^2 F(s)/ds^2$; use the mathematical induction to calculate $\mathcal{L}[t^n f(t)]$.

6.9 Demonstrate that

$$\mathcal{L}\left[\frac{d^n f(t)}{dt^n}\right] = s^n F(s) - s^{n-1} f(0) - s^{n-2}\left.\frac{df(t)}{dt}\right|_{t=0} - \cdots - \left.\frac{df^{n-1}(t)}{dt^{n-1}}\right|_{t=0}$$

when the function $f(t)$ and its first $(n-1)$ derivatives are all of exponential order.

6.10 By using partial-fraction expansion and the inverse Laplace transform, calculate the original functions $f(t)$ for the following s-domain functions. Check the results with MATLAB® for both partial-fraction expansion and inverse Laplace transform calculations:

(a) $F(s) = m/[(s+m)(s+n)]$.

(b) $F(s) = (2s+1)/[(s^2+3s+2)(s-3)]$.

(c) $F(s) = (2s+1)/[s(s^2+s+1)]$.

6.11 Same question as in Problem 6.10 for the following functions:

(a) $F(s) = (s + 1)/[(s^2 + 7s + 12)(s^2 + 2)]$.

(b) $F(s) = 2/[s^2(s + 3)]$.

(c) $F(s) = 4s^3/(s^4 - \omega^4)$.

6.12 Same question as in Problem 6.10 for the following functions:

(a) $F(s) = (3s^4 + 2s^2 - 1)/[(s^2 + 1)(s^2 + s - 2)]$.

(b) $F(s) = (s^2 - 3s + 1)/[s^2(s + 2)^2]$.

(c) $F(s) = (2s + 1)/[s(s^2 + 2)^2]$.

6.13 Find $f(t)$ for

$$F(s) = \frac{s + 3}{(s + 1)^2(s + 2)^2}e^{-4s}$$

Verify the result by using MATLAB®.

6.14 Using just basic Laplace pairs and properties, find the inverse Laplace transform of the function $F(s) = 4e^{-3s}/[s(s^2 + 25)]$. Verify the obtained result with MATLAB®.

6.15 The rotary mechanical system of Figure 6.14 is formed of a cylinder with a mass moment of inertia J, a damper with a damping coefficient c, and a spring of stiffness k. Calculate the final value of the cylinder rotation angle when a unit impulse torque is applied to it. Use symbolic calculation and MATLAB® to confirm the result.

6.16 An unknown constant voltage is applied to a series circuit formed of a resistor $R = 40 \, \Omega$ and an inductor L, as shown in Figure 6.15. It is determined that the steady-state current in the circuit is $i(\infty) = 0.1$ A. Calculate the source voltage.

6.17 Find the Laplace transform of the function shown in Figure 6.16.

6.18 Calculate the Laplace transform of the function sketched in Figure 6.17.

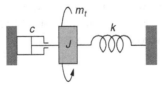

FIGURE 6.14

Rotary Mechanical System with Inertia, Damping, and Stiffness Properties under Unit Impulse Input.

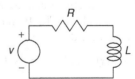

FIGURE 6.15

Electrical Circuit with Constant Voltage Input.

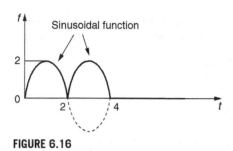

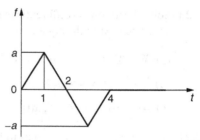

FIGURE 6.16

Truncated Sinusoidal-Segment Function.

FIGURE 6.17

Truncated Multisegment Function.

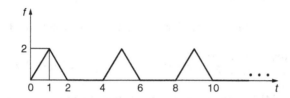

FIGURE 6.18

Periodic Multisegment Function.

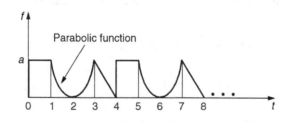

FIGURE 6.19

Periodic Multiline Function.

6.19 Calculate the Laplace transform of the periodic function sketched in Figure 6.18.

6.20 Calculate the Laplace transform of the periodic function shown in Figure 6.19.

6.21 Calculate the inverse Laplace transforms of the following functions without applying partial-fraction expansion and verify the results by MATLAB®:

   (a) $X(s) = 6s/(s^2 + 9)^2$.

   (b) $X(s) = 1/(s^3 + 4s^2)$.

6.22 Use the convolution theorem to determine the inverse Laplace transforms of the following functions:

   (a) $X(s) = 1/(s^2 - \omega^2)^2$.

   (b) $X(s) = 1/[s^2(s^2 + a)]$.

6.23 Solve the following differential equations subject to the initial conditions that are specified next to the equations using direct and inverse Laplace transforms:

(a) $\ddot{x} - 3\dot{x} + x = 0$; $x(0) = 1$; $\dot{x}(0) = 0$.

(b) $\ddot{x} + 2\dot{x} + x = \cos(2t)$; $x(0) = 0$; $\dot{x}(0) = 0$.

(c) $4\ddot{x} + 6\dot{x} + x = e^{-2t}$; $x(0) = 0$; $\dot{x}(0) = 2$.

6.24 Answer the same question as in Problem 6.23 for the following system:

$$\begin{cases} 3\dot{x}_1 - x_1 - 2\dot{x}_2 + x_2 = 0 \\ \dot{x}_1 + 2\dot{x}_2 + x_1 = \sin t \end{cases}; \quad x_1(0) = 0; x_2(0) = 0.$$

6.25 Solve the following system of differential equations and initial conditions:

$$\begin{cases} \ddot{x}_1 + 200x_1 - 80x_2 = 0 \\ \ddot{x}_2 - 80x_1 + 80x_2 = \delta(t) \end{cases}; \quad x_1(0) = 0; x_2(0) = 0; \dot{x}_1(0) = 0; \dot{x}_2(0) = 0.$$

Propose a MEMS mechanical subsystem that can be modeled by means of these equations.

6.26 Solve the following system of integral-differential equations:

$$\begin{cases} \dot{x}_1 + 3x_1 + 10\int_0^t x_1 dt - 10\int_0^t x_2 dt = 50 \\ \dot{x}_2 + 10\int_0^t x_2 dt - 10\int_0^t x_1 dt = 0 \end{cases}; \quad \begin{array}{l} x_1(0) = 0; x_2(0) = 0; \\ \dot{x}_1(0) = 0; \dot{x}_2(0) = 0. \end{array}$$

Propose an electrical system that can be modeled by means of these equations.

6.27 Use the convolution theorem to solve the integral equation

$$2x(t) + \int_0^t x(\tau)e^{-(t-\tau)}\cos[3(t-\tau)]d\tau = e^{-t}$$

knowing that $x(0) = 0$.

6.28 Solve the following differential equation (after Sneddon 1972):

$$t\frac{d^2x(t)}{dt^2} - \frac{dx(t)}{dt} - tx(t) = 0$$

knowing that $x(0) = 0$.

6.29 Identify a single-DOF mechanical micro system whose Laplace-domain transform coordinate is defined as $X(s) = 3/[s(s^2 + 3s + 1)]$.

6.30 A single-DOF electrical system has its time-domain coordinate transformed into the Laplace domain as $X(s) = (3s^2 + 7s + 3)/[s^2(s^2 + 2s + 0.006)]$. Determine the system components, forcing function, and initial conditions.

Suggested Reading

T. F. Bogart, Jr., *Laplace Transforms and Control Systems Theory for Technology*, John Wiley & Sons, New York, 1982.

I. H. Sneddon, *The Use of Integral Transforms*, McGraw-Hill, New York, 1972.

W. R. LePage, *Complex Variables and the Laplace Transform for Engineers*, McGraw-Hill, New York, 1961.

C. J. Savant, Jr., *Fundamentals of the Laplace Transformation*, McGraw-Hill, New York, 1962.

W. T. Thomson, *Laplace Transformation*, 2nd Ed. New York, Prentice-Hall, 1960.

R. V. Churchill, *Operational Mathematics*, McGraw-Hill, New York, 1958.

E. J. Watson, *Laplace Transforms and Applications*, Van Nostrand, New York, 1981.

L. Debnath, *Integral Transforms and Their Applications*, CRC Press, Boca Raton, FL, 1995.

Transfer Function Approach

7

OBJECTIVES

Based on the mathematical models developed by the methods presented in Chapters 2, 3, 4, and 5 for mechanical, electrical, fluid, and thermal systems and using the Laplace transform of Chapter 6, this chapter introduces you to the transfer function approach by studying the following topics:

- The transfer function as a connector between input and output into the Laplace domain for single-input, single-output and multiple-input, multiple-output mechanical, electrical, fluid, and thermal systems.
- Calculating the transfer function from the mathematical model of a dynamic system.
- Using the complex impedance as a tool for modeling mechanical, electrical, fluid, and thermal systems and deriving the transfer function directly in the Laplace domain.
- Application of the transfer function in obtaining the forced response and the free response with nonzero initial conditions.
- Utilization of MATLAB® and Simulink® in transfer function applications.

INTRODUCTION

Using the Laplace transform technique, this chapter introduces the transfer function as a Laplace-domain operator characterizing the properties of a given dynamic system and connecting the input to the output. We see that the transfer function for single-input, single-output (SISO) systems and transfer function matrix for multiple-input, multiple-output (MIMO) systems can be used to determine the forced time response and the free time response with nonzero initial conditions. The complex impedance is introduced as a transfer function at the element level, which enables deriving the transfer function of dynamic systems directly in the Laplace domain. MATLAB® and Simulink® examples illustrating the use of transfer function by means of built-in capabilities are incorporated alongside many solved examples that use the

DOI: 10.1016/B978-0-240-81128-4.00007-6

standard transfer function formulation and solution. The transfer function concept is also important in studying system dynamics by means of the state space modeling approach (the topic of Chapter 8) and frequency-domain analysis (the focus of Chapter 9).

7.1 THE TRANSFER FUNCTION CONCEPT

A major objective in system dynamics, as seen in all previous chapters, is finding the system response (or *output*) that corresponds to a specific forcing (or *input*), which means solving the differential equations defining the behavior of a given dynamic system. Using direct and inverse Laplace operators, differential equations are converted into algebraic ones whose s-domain solution is transformed into the time-domain solution. This procedure has to be utilized every time a different input is applied to the same dynamic system. Consider a first-order system whose input $u_1(t)$ is connected to the output $y_1(t)$ by the following differential equation:

$$a_1 \dot{y}_1(t) + a_0 y(t) = u_1(t) \tag{7.1}$$

Laplace transforming the left- and right-hand sides of Eq. (7.1) with zero initial conditions results in

$$Y_1(s) = \frac{1}{a_1 s + a_0} U_1(s) \tag{7.2}$$

Another input $u_2(t)$ applied to the same dynamic system generates the Laplace-domain solution:

$$Y_2(s) = \frac{1}{a_1 s + a_0} U_2(s) \tag{7.3}$$

By this process, finding the Laplace-domain solution technically requires three operations: calculating the Laplace transform of the left-hand side of Eq. (7.1), calculating the Laplace transform of the input, and determining the Laplace-domain solution. For n inputs, $3 \times n$ operations are needed. If $n = 10$, $3 \times 10 = 30$ operations are necessary.

This process can be simplified by noticing the following ratio (the transfer function) is constant for a given dynamic system:

$$G(s) = \frac{Y_1(s)}{U_1(s)} = \frac{Y_2(s)}{U_2(s)} = \frac{1}{a_1 s + a_0} \tag{7.4}$$

For n inputs, evaluating the n outputs means calculating the function $G(s)$ once, plus evaluating the Laplace transform of the input, followed by determining the Laplace transform of the output (by multiplying $G(s)$ by the Laplace-transform of the input), so two operations per each input are needed. As a consequence, the total number of operations is $2 \times n + 1$. For $n = 10$, this latter process (the *transfer function approach*, as we see shortly) needs only $2 \times 10 + 1 = 21$ operations compared to the 30 operations necessary in the straightforward Laplace-transformation procedure.

The transfer function approach simplifies greatly the task of obtaining a system's response (output) both conceptually and through the use of built-in MATLAB® commands. Let us first consider an example that will help highlight the content of this chapter.

Example 7.1

The translatory mechanical system of Figure 7.1 is formed of a body of mass $m = 1$ kg that is connected to a damper with a damping coefficient $c = 1$ N-s/m and a spring of stiffness $k = 1$ N/m. A force u (the input) acts on the body in a direction parallel to the damper spring, which generates a rectilinear motion y (the output) of the body. Determine the ratio of the Laplace transforms of the output $Y(s)$ to the input $U(s)$ for zero initial conditions. For the particular case where the input is a unit impulse, calculate the system time response $y(t)$, as well as the steady-state response $y(\infty)$.

Solution

Newton's second law of motion is applied to the body of mass m, which yields

$$m\ddot{y}(t) + c\dot{y}(t) + ky(t) = u(t) \tag{7.5}$$

Applying the Laplace transform to Eq. (7.5) with *zero initial conditions* results in

$$G(s) = \frac{Y(s)}{U(s)} = \frac{1}{ms^2 + cs + k} \tag{7.6}$$

The function denoted by $G(s)$ in Eq. (7.6) is the *transfer function* of the mechanical system sketched in Figure 7.1, and it will be the focus of this chapter. Note that the transfer function is defined by only the system parameters m, c, and k. However, for a given input $u(t)$, such as $\delta(t)$, the unit impulse, whose Laplace transform is $U(s)$, the Laplace transform of the output can be evaluated from Eq. (7.6) as

$$Y(s) = G(s)U(s) \tag{7.7}$$

With $G(s)$ of Eq. (7.6) and $U(s) = 1$, Eq. (7.7) becomes

$$Y(s) = \frac{1}{ms^2 + cs + k} = \frac{1}{s^2 + s + 1} = \frac{2}{\sqrt{3}} \times \frac{\dfrac{\sqrt{3}}{2}}{\left(s + \dfrac{1}{2}\right)^2 + \left(\dfrac{\sqrt{3}}{2}\right)^2} \tag{7.8}$$

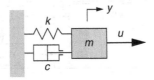

FIGURE 7.1

Forced Mass–Damper–Spring Translatory Mechanical System.

The time-domain response $y(t)$ is obtained by inverse Laplace transforming $Y(s)$ of Eq. (7.8), which is:

$$y(t) = \frac{2}{\sqrt{3}} e^{-0.5t} \sin\left(\frac{\sqrt{3}}{2} t\right)$$

(7.9)

The value of y when $t \to \infty$ is 0; this steady-state value can also be obtained by using the final-value theorem of Chapter 6, according to which: $y(\infty) = \lim_{t \to \infty} y(t) = \lim_{s \to 0} s Y(s) = 0$. ■

This example emphasizes the main role of the transfer function (which is defined more systematically shortly), namely, obtaining the Laplace-domain output as the product between the transfer function and the Laplace transform of the input, followed by calculation of the time-domain output from its counterpart in the s domain. The transfer function can also be utilized to

- Calculate the natural frequencies (eigenfrequencies) by solving the characteristic equation (the transfer function denominator made zero); more details are given in the Chapter 9.
- Model and solve control problems, as discussed in the website Chapter 11.

In many applications, the transfer function is determined as in Example 7.1 by applying the Laplace transform to the differential equation representing the time-domain mathematical model. Another modality of establishing the transfer function, as is shown in this chapter, is by direct operation in the Laplace domain through complex impedances (denoted by $Z(s)$, they are defined in a subsequent section). Figure 7.2 shows schematically these interactions, which are based on the transfer function of a SISO system but are also valid for MIMO systems. As is shown in this chapter, the actual forcing (or input) $u(t)$ can be combined with the nonzero initial conditions (denoted by IC in Figure 7.2) to generate an equivalent forcing term, $U_e(s)$, in the Laplace domain. With $G(s)$ and $U_e(s)$ determined, the time-domain response is calculated by inverse Laplace transforming $Y(s) = G(s)U_e(s)$.

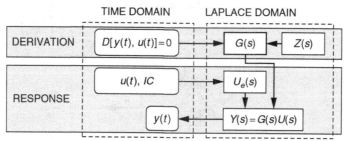

FIGURE 7.2

Derivation of Transfer Function Model and Time-Domain Response Calculation.

7.2 TRANSFER FUNCTION MODEL FORMULATION

As indicated in Figure 7.2, an important step in evaluating the time-domain response of a dynamic system is calculation of the transfer function. Analytically, this can be achieved either by starting from the time-domain differential equations (the mathematical model) or by working directly in the Laplace domain with complex impedances, which, as shown a bit later, are s-domain counterparts of the parameters defining the mechanical, electrical, fluid, and thermal system elements. A transfer function model can also be obtained from an existing zero-pole-gain (zpk) model using MATLAB®. These methods are discussed in the following.

 ## 7.2.1 Analytical Approach

Using time-domain mathematical models or the complex impedance approach to obtain transfer function models are procedures discussed in this section.

Transfer Function from the Time-Domain Mathematical Model

This section defines the transfer function more systematically and analyzes the modalities of calculating the transfer function for SISO systems and the transfer function matrix for MIMO systems using time-domain mathematical models of dynamic systems.

SISO Systems

A SISO system, as introduced in Chapter 1, has one input and one output that are part of a differential equation, as suggested in Figure 7.3(a), where the operator D indicates the time-domain equation is a differential one, including derivatives of the input $u(t)$ and the output $y(t)$.

Transfer function definition: For a SISO system, the *transfer function G(s)* is the *ratio* of the *Laplace transform of the output Y(s)* to the *Laplace transform of the input U(s)* for *zero initial conditions*, which is mathematically expressed as

$$G(s) = \left. \frac{Y(s)}{U(s)} \right|_{IC=0} \tag{7.10}$$

Equation (7.10) is quite powerful as it allows evaluating the dynamic system response into the Laplace domain.

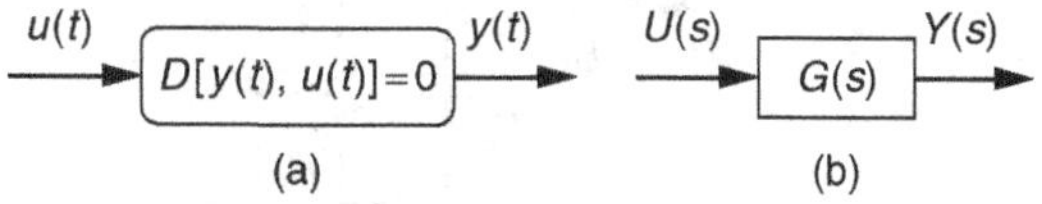

$$u(t) \longrightarrow \boxed{D[y(t), u(t)]=0} \xrightarrow{y(t)} \qquad U(s) \longrightarrow \boxed{G(s)} \xrightarrow{Y(s)}$$

(a)   (b)

FIGURE 7.3

SISO System with Input and Output Connected in the (a) Time Domain by a Differential Equation; (b) Laplace Domain by the Transfer Function.

Example 7.2

Derive the mathematical model of the electrical system of Figure 7.4 by relating the output voltage $v_o(t)$ to the input voltage $v_i(t)$. Use the model to calculate the transfer function $V_o(s)/V_i(s)$.

Solution

The following equations can be written for the electrical system of Figure 7.4 by applying Kirchhoff's voltage law (Chapter 4):

$$\begin{vmatrix} \dfrac{1}{C_1}\int i_1(t)dt = R_1[i(t) - i_1(t)] \\[2mm] \dfrac{1}{C_2}\int i_2(t)dt = R_2[i(t) - i_2(t)] \\[2mm] v_i(t) = R_1[i(t) - i_1(t)] + R_2[i(t) - i_2(t)] \\[2mm] v_o(t) = R_2[i(t) - i_2(t)] \end{vmatrix} \qquad (7.11)$$

Laplace transforming the first two Eqs. (7.11) with zero initial conditions yields

$$\begin{vmatrix} I_1(s) = \dfrac{R_1 C_1 s}{R_1 C_1 s + 1} I(s) \\[3mm] I_2(s) = \dfrac{R_2 C_2 s}{R_2 C_2 s + 1} I(s) \end{vmatrix} \qquad (7.12)$$

Substitution of $I_1(s)$ and $I_2(s)$ of Eqs. (7.12) into the equations produced by Laplace transforming the last two Eqs. (7.11) result, after some algebra, in the following s-domain voltages:

$$\begin{vmatrix} V_i(s) = \dfrac{R_1 R_2(C_1 + C_2)s + R_1 + R_2}{(R_1 C_1 s + 1)(R_2 C_2 s + 1)} I(s) \\[3mm] V_o(s) = \dfrac{R_2}{R_2 C_2 s + 1} I(s) \end{vmatrix} \qquad (7.13)$$

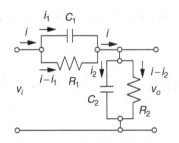

FIGURE 7.4

Electrical System with Resistors and Capacitors.

The required transfer function is obtained from Eqs. (7.13) as

$$G(s) = \frac{V_o(s)}{V_i(s)} = \frac{R_1 R_2 C_1 s + R_2}{R_1 R_2 (C_1 + C_2)s + R_1 + R_2} \tag{7.14}$$

Several remarks are needed about the preceding:

1. The transfer function approach is applicable only to dynamic systems described by linear differential equations with constant coefficients.
2. Although it is expressed as the ratio of the output to the input in the s-domain, the transfer function solely describes the system. In other words, for a specified system, with unique characteristics, the transfer function is also unique.
3. The order of the system (which is also the order of the differential equation describing the input-output relationship) is identical to the degree of the s polynomial in the denominator of the transfer function, which is named the *characteristic polynomial*.

MIMO Systems

Multiple-input, multiple-output systems have more than one component at either the input or the output; multiple-DOF systems, which have been studied in previous chapters, are MIMO systems because they possess more than one DOF and therefore have multiple outputs. Figure 7.5(a) illustrates a generic dynamic system defined by p differential equations, denoted by D_1 to D_p, and therefore by p outputs (the differential equations solution components y_1, y_1, ..., y_p); also m inputs, u_1, u_1, ..., u_m, are acting on the system.

As shown in the following example, the Laplace transforms of the input components, $U_1(s)$, $U_2(s)$, ..., $U_m(s)$, can be collected into an input vector $\{U(s)\}$, whereas the Laplace transforms of the output components are gathered, similarly, into another vector, the output vector $\{Y(s)\}$.

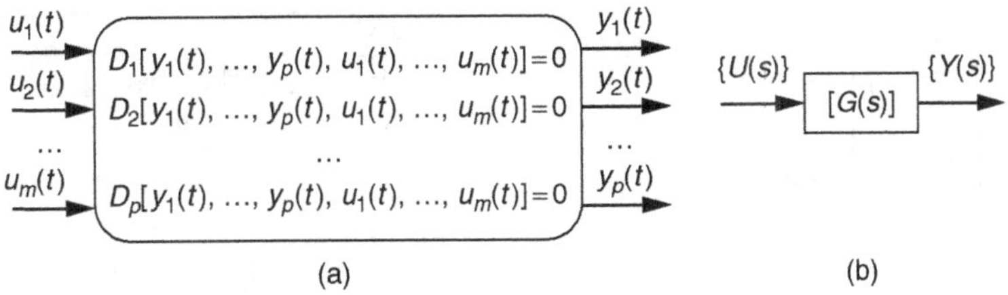

(a)   (b)

FIGURE 7.5

MIMO System with Input and Output Components Connected in the (a) Time Domain by a Differential Equations System; (b) Laplace Domain by the Transfer Function Matrix.

Transfer function matrix definition: For a MIMO system, the *transfer function matrix* $[G(s)]$ connects the *Laplace transform of the output vector*, $\{Y(s)\}$, to the *Laplace transform of the input vector*, $\{U(s)\}$ for zero initial conditions:

$$\{Y(s)\} = [G(s)]\{U(s)\} \tag{7.15}$$

Example 7.3

The MEMS actuator system of Figure 7.6(a) consists of two shuttle masses that can move horizontally and are coupled by means of a serpentine spring. The mass m_1 is also supported by two identical beam springs and is subjected to frontal damping on its left by the air squeezed between mass and an anchored wall. Electrostatic transverse actuation is applied to both masses such that opposite forces are generated on m_1 and m_2. Express the transfer function matrix using a lumped-parameter model of this MEMS device where the output vector is formed of the two masses' displacements.

Solution

The lumped-parameter mechanical model of the MEMS device of Figure 7.6(a) is shown in Figure 7.6(b). It has two inputs (the forces u_1 and u_2) and two outputs (the mass displacements y_1 and y_2); therefore, it is a MIMO system with $m = p = 2$. It should be mentioned that the system is also a two-DOF system, and the fact that the number of outputs is identical to the number of DOFs is not accidental; this feature can oftentimes be utilized to simplify the choice of DOFs once the output parameters have been determined.

Newton's second law of motion yields the mathematical model of this mechanical system:

$$\begin{cases} m_1\ddot{y}_1 = u_1 - c\dot{y}_1 - k_1 y_1 - k_2(y_1 - y_2) \\ m_2\ddot{y}_2 = -u_2 - k_2(y_2 - y_1) \end{cases} \tag{7.16}$$

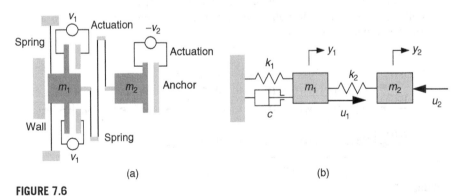

(a) (b)

FIGURE 7.6

Linear-Motion MEMS Device: (a) Physical Model; (b) Lumped-Parameter Mechanical Model with Two-Force Input and Two-Displacement Output.

Application of the Laplace transform to the two Eqs. (7.16) with zero initial conditions yields

$$\begin{cases} (m_1 s^2 + cs + k_1 + k_2) Y_1(s) - k_2 Y_2(s) = U_1(s) \\ -k_2 Y_1(s) + (m_2 s^2 + k_2) Y_2(s) = -U_2(s) \end{cases} \tag{7.17}$$

Equations (7.17) are expressed in vector-matrix form:

$$\begin{bmatrix} m_1 s^2 + cs + k_1 + k_2 & -k_2 \\ -k_2 & m_2 s^2 + k_2 \end{bmatrix} \begin{Bmatrix} Y_1(s) \\ Y_2(s) \end{Bmatrix} = \begin{Bmatrix} U_1(s) \\ -U_2(s) \end{Bmatrix} \tag{7.18}$$

which can be reformulated as

$$\begin{Bmatrix} Y_1(s) \\ Y_2(s) \end{Bmatrix} = \begin{bmatrix} m_1 s^2 + cs + k_1 + k_2 & -k_2 \\ -k_2 & m_2 s^2 + k_2 \end{bmatrix}^{-1} \begin{Bmatrix} U_1(s) \\ -U_2(s) \end{Bmatrix} \tag{7.19}$$

If $\{Y(s)\} = \{Y_1(s), Y_2(s)\}^t$ is the output vector and $\{U(s)\} = \{U_1(s), -U_2(s)\}^t$ is the input vector, Eq. (7.19) shows, by comparison to the definition Eq. (7.15), that the transfer function matrix of this example is the matrix connecting $\{Y(s)\}$ to $\{U(s)\}$ in Eq. (7.19).

A similar result is obtained if Eq. (7.16) is written directly into vector-matrix form after application of the Laplace transform:

$$\begin{bmatrix} m_1 & 0 \\ 0 & m_2 \end{bmatrix} s^2 \begin{Bmatrix} Y_1(s) \\ Y_2(s) \end{Bmatrix} + \begin{bmatrix} c & 0 \\ 0 & 0 \end{bmatrix} s \begin{Bmatrix} Y_1(s) \\ Y_2(s) \end{Bmatrix} + \begin{bmatrix} k_1 + k_2 & -k_2 \\ -k_2 & k_2 \end{bmatrix} \begin{Bmatrix} Y_1(s) \\ Y_2(s) \end{Bmatrix} = \begin{Bmatrix} U_1(s) \\ -U_2(s) \end{Bmatrix} \tag{7.20}$$

which reduces to Eq. (7.18) after right-factoring the vector $\{Y_1(s)\ Y_2(s)\}^t$ and adding up the corresponding matrix components in the left-hand side of Eq. (7.20). The transfer matrix $[G(s)]$ is determined as

$$[G(s)] = \begin{bmatrix} m_1 s^2 + cs + k_1 + k_2 & -k_2 \\ -k_2 & m_2 s^2 + k_2 \end{bmatrix}^{-1}$$

$$= \begin{bmatrix} \dfrac{m_2 s^2 + k_2}{D(s)} & \dfrac{k_2}{D(s)} \\ \dfrac{k_2}{D(s)} & \dfrac{m_1 s^2 + cs + k_1 + k_2}{D(s)} \end{bmatrix} = \begin{bmatrix} G_{11}(s) & G_{12}(s) \\ G_{21}(s) & G_{22}(s) \end{bmatrix} \tag{7.21}$$

where $G_{11}(s)$, $G_{12}(s) = G_{21}(s)$ and $G_{22}(s)$ are regular (scalar) transfer functions that connect specific input and output components; $G_{12}(s)$, for instance, relates $Y_1(s)$ to $U_2(s)$. The denominator $D(s)$ of Eq. (7.21) is

$$D(s) = m_1 m_2 s^4 + m_2 cs^3 + [m_1 k_2 + m_2(k_1 + k_2)] s^2 + ck_2 s + k_1 k_2 \tag{7.22}$$

It is also important to point out that all individual transfer functions have the same denominator, which is the *characteristic polynomial*. Its roots are the *eigenfrequencies*, which can be calculated by taking $s = j\omega \left(j = \sqrt{-1}\right)$ in Eq. (7.22) and solving the resulting *characteristic equation* $D(j\omega) = 0$. As mentioned before, Chapter 9 studies this topic.

Transfer Function from Complex Impedances

The notion of *complex impedance* is a tool derived by means of the transfer function approach, which allows modeling dynamic systems and deriving transfer functions directly in the Laplace domain. In electrical systems, for instance, the transfer function between current and voltage defines the impedance. The qualifier *complex* indicates that the impedance is a function of the variable s, which is a complex number, as learned in Chapter 6.

The complex impedance, denoted by $Z(s)$, Figure 7.7, relates an input amount $\text{In}(s)$ to the corresponding output amount $\text{Out}(s)$ through a structural relationship, depending on the type of system (mechanical, electrical, fluid, or thermal) and the type of element. Its defining equation is

$$Z(s) = \left.\frac{\text{Out}(s)}{\text{In}(s)}\right|_{IC=0} \tag{7.23}$$

Complex Impedances for Mechanical, Electrical, Fluid, and Thermal Systems

Table 7.1 synthesizes the complex impedances corresponding to the defining elements of mechanical, electrical, fluid, and thermal systems. Full details on the derivation of all impedances given in Table 7.1 are given in the companion website Chapter 7.

Complex Impedance System Modeling and Analysis

Several examples are studied illustrating the use of complex impedances in deriving transfer functions for SISO and MIMO electrical, thermal, fluid, and mechanical systems.

Electrical Systems One important consequence of using the complex impedance is that relationships based on Ohm's law can be used to apply other laws, such as Kirchhoff's first and second laws, by considering that all electrical components behave as resistors through their corresponding complex impedances. Because of

FIGURE 7.7

Complex Impedance $Z(s)$ as a Transfer Function for a Generic Dynamic System Component.

Table 7.1 Complex Impedances of Mechanical, Electrical, Fluid, and Thermal Systems

System and Element	Time-Domain Relationship	In(s)	Out(s)	Z(s)
Mechanical				
Translation				
Mass	$f(t) = md^2x(t)/dt^2$	Displacement, $X(s)$	Force, $F(s)$	ms^2
Damping	$f(t) = cdx(t)/dt$			cs
Stiffness	$f(t) = kx(t)$			k
Rotation				
Mass moment of inertia	$m(t) = Jd^2\theta(t)/dt^2$	Rotation, $\Theta(s)$	Moment, $M(s)$	Js^2
Damping	$m(t) = cd\theta(t)/dt$			cs
Stiffness	$m(t) = k\theta(t)$			k
Electrical				
Inductance	$v(t) = Ldi(t)/dt$	Current, $I(s)$	Voltage, $V(s)$	Ls
Resistance	$v(t) = Ri(t)$			R
Capacitance	$i(t) = Cdv(t)/dt$			$1/(Cs)$
Fluid				
Pressure-defined				
Inductance	$p(t) = I_{f,p}dq_f(t)/dt$	Fluid flow rate, $Q_f(s)$	Pressure $P(s)$	$I_{f,p}s$
Resistance	$p(t) = R_{f,p}q_f(t)$			$R_{f,p}$
Capacitance	$q_f(t) = C_{f,p}dp(t)/dt$			$1/(C_{f,p}s)$
Head-defined				
Inductance	$h(t) = I_{f,h}dp_f(t)/dt$	Fluid flow rate, $Q_f(s)$	Head, $H(s)$	$I_{f,h}s$
Resistance	$h(t) = R_{f,h}q_f(t)$			$R_{f,h}$
Capacitance	$q_f(t) = C_{f,h}dh/dt$			$1/(C_{f,h}s)$
Thermal				
Resistance	$\theta_{th}(t) = R_{th}q_{th}(t)$	Thermal flow rate, $Q_{th}(s)$	Temperature, $\Theta_{th}(s)$	R_{th}
Capacitance	$q_{th}(t) = C_{th}d\theta_{th}(t)/dt$			$1/(C_{th}s)$

that, electrical impedances are connected in series, parallel, and in mixed formations exactly like resistances. According to Figure 7.8, the series and parallel equivalent impedances of two electrical impedances Z_1 and Z_2 are

$$Z_s(s) = Z_1(s) + Z_2(s); \quad \frac{1}{Z_p(s)} = \frac{1}{Z_1(s)} + \frac{1}{Z_2(s)} \tag{7.24}$$

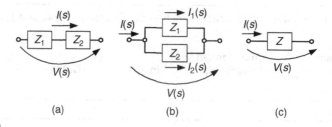

FIGURE 7.8

Connection of Electrical Impedances: (a) Series; (b) Parallel; (c) Equivalent Impedance.

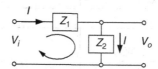

FIGURE 7.9

Single-Stage Electrical System with Two Generic Complex Impedances.

Example 7.4

Determine the transfer function $V_o(s)/V_i(s)$ of the two-component electrical system represented in Figure 7.9, and based on it, evaluate the transfer function of the electrical system shown in Figure 7.4 of Example 7.2.

Solution

The transfer function of this circuit is determined by using Kirchhoff's second law for the two loops shown in Figure 7.9:

$$\begin{cases} V_i(s) = Z_1 I(s) + Z_2 I(s) \\ V_o(s) = Z_2 I(s) \end{cases} \tag{7.25}$$

The transfer function is therefore

$$G(s) = \frac{V_o(s)}{V_i(s)} = \frac{Z_2}{Z_1 + Z_2} \tag{7.26}$$

The electrical system of Figure 7.4 is of the form sketched in the generic Figure 7.9, where the two impedances are

$$\begin{cases} Z_1(s) = \dfrac{R_1}{R_1 C_1 s + 1} \\ Z_2(s) = \dfrac{R_2}{R_2 C_2 s + 1} \end{cases} \tag{7.27}$$

each being calculated as a parallel-connection impedance. By substituting Eqs. (7.27) into the transfer function of Eq. (7.26), the latter one yields the transfer function of Eq. (7.14).

Example 7.5

Calculate the transfer function $V_o(s)/V_i(s)$ of the two-component operational amplifier shown in Figure 7.10(a) and then determine the transfer function of the two-stage operational amplifier circuit of Figure 7.10(b).

Solution

The negative feedback in the system of Figure 7.10(a) ensures that the positive and negative input ports have the same zero voltage because the positive input port is connected to the ground. At the same time, due to the high input impedance of the operational amplifier, the current I that passes through Z_1 is identical to the one passing through Z_2:

$$\frac{V_i(s) - 0}{Z_1} = \frac{0 - V_o(s)}{Z_2} \tag{7.28}$$

Equation (7.28) can be rewritten to highlight the circuit transfer function as

$$G(s) = \frac{V_o(s)}{V_i(s)} = -\frac{Z_2}{Z_1} \tag{7.29}$$

The minus sign in Eq. (7.29) indicates the inverting effect in the s domain, which is similar to the time-domain inverting effect, demonstrated in Chapter 4, for the two resistors R_1 and R_2.

The electrical system of Figure 7.10(b) is formed of two individual stages serially connected such that the output of the first stage is the input to the second stage, as illustrated in Figure 7.11; the total transfer function can be expressed as the product of the two stages' transfer functions:

$$G^*(s) = G_1(s)G_2(s) = \left(-\frac{Z_2}{Z_1}\right) \times \left(-\frac{Z_4}{Z_3}\right) = \left(-\frac{R_2}{R_1}\right) \times \left(-\frac{1}{R_3Cs}\right)$$

$$= \frac{R_2}{R_1R_3Cs} \tag{7.30}$$

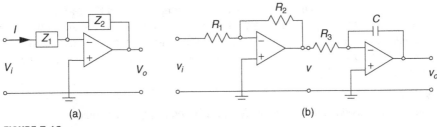

(a)   (b)

FIGURE 7.10

(a) Two-Component, One-Stage Inverting Operational Amplifier Electrical System;
(b) Two-Stage Operational Amplifier System.

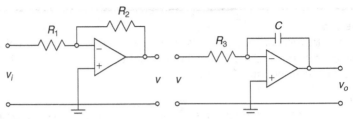

FIGURE 7.11

Two-Stage Operational Amplifier System with Separate Stages.

When considering the whole electrical system of Figure 7.10(b), because the current through the four components is the same, the following equations can be written in the Laplace domain:

$$\begin{cases} I(s) = \dfrac{V_i(s) - 0}{R_1} = \dfrac{0 - V(s)}{R_2} \\[2mm] I(s) = \dfrac{V(s) - 0}{R_3} = \dfrac{0 - V_o(s)}{1/(Cs)} \end{cases} \qquad (7.31)$$

By combining the two Eqs. (7.31), the following transfer function is obtained:

$$G(s) = \frac{V_o(s)}{V_i(s)} = \frac{R_2}{R_1 R_3 Cs} \qquad (7.32)$$

It can be seen that the actual transfer function of Eq. (7.32) is identical to the product of the two partial transfer functions as in Eq. (7.30), which consider that the electrical system can be split into two serially connected subsystems. ■

Thermal Systems A thermal system example is studied next using the complex impedance approach and a node-analysis procedure (similar to the one introduced in Chapter 4 for electrical systems).

Example 7.6

Find the transfer function matrix of the one-room thermal system of Figure 7.12 using the complex impedance approach. Consider that the input components are the thermal flow rate q_i and the outdoors temperature θ_2, whereas the output is the indoor temperature θ_1. The enclosed space has a thermal capacity C_{th} and the four identical walls are defined by a thermal resistance R_{th}.

Solution

The thermal impedances provided in Table 7.1 become, for this example,

$$Z_{R_{th}} = \frac{\Theta_1(s) - \Theta_2(s)}{Q_o(s)} = R_{th}; \; Z_{C_{th}} = \frac{\Theta_1(s)}{Q_i(s) - Q_o(s)} = \frac{1}{C_{th}s} \qquad (7.33)$$

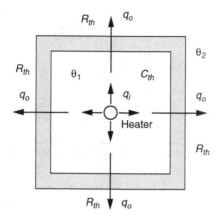

FIGURE 7.12

Four-Wall, One-Room Thermal System with Internal Heating.

By expressing $Q_o(s)$ of the first Eq. (7.33) and substituting it in the second Eq. (7.33), the following equation is obtained:

$$(R_{th}C_{th}s + 1)\Theta_1(s) = \Theta_2(s) + R_{th}Q_i(s) \tag{7.34}$$

which can be reformulated in vector form as

$$\Theta_1(s) = \left\{ \frac{1}{R_{th}C_{th}s + 1} \quad \frac{R_{th}}{R_{th}C_{th}s + 1} \right\} \left\{ \begin{matrix} \Theta_2(s) \\ Q_i(s) \end{matrix} \right\} \tag{7.35}$$

The transfer function connecting the output $\Theta_1(s)$ to the input vector $\{\Theta_2(s)\ Q_i(s)\}^t$ is the row vector of Eq. (7.35):

$$[G(s)] = \left\{ \frac{1}{R_{th}C_{th}s + 1} \quad \frac{R_{th}}{R_{th}C_{th}s + 1} \right\} \tag{7.36}$$

Let the impedance-based circuit of Figure 7.13 be the model for the actual thermal system of Figure 7.12.

Treating heat flow rates as electrical currents and the temperatures as voltages, the *node-analysis method* can be applied to the thermal circuit of Figure 7.13. The source flow rate at node 1 is the sum of flow rates passing through the impedances $Z_{C_{th}}$ and $Z_{R_{th}}$, which is expressed as

$$Q_i(s) = \frac{\Theta_1(s)}{Z_{C_{th}}} + \frac{\Theta_1(s) - \Theta_2(s)}{Z_{R_{th}}}; \ Q_i(s) = C_{th}s\Theta_1(s) + \frac{\Theta_1(s) - \Theta_2(s)}{R_{th}} \tag{7.37}$$

The second Eq. (7.37) changes to Eq. (7.34), and therefore the complex impedance model of Figure 7.13 correctly represents the original thermal system.

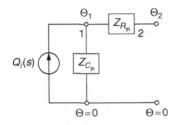

FIGURE 7.13

Impedance-Based Thermal Circuit as a Model Candidate for the Actual System of Figure 7.12.

Fluid Systems As the case was with electrical and thermal systems, impedances can be utilized to model fluid systems directly into the Laplace domain, as illustrated in the following example.

Example 7.7

Use an impedance-based mathematical model of the liquid system shown in Figure 7.14 and determine the transfer function matrix considering that the input to the system consists of the volume flow rates q_{i1} and q_{i2}, whereas the liquid heads h_1 and h_2 are the system's output. The system is formed of two tanks of capacitances C_{l1} and C_{l2} and two valves of resistances R_{l1} and R_{l2}.

Solution

Using the fluid impedance definitions of Table 7.1, the following equations can be written for the fluid resistances and capacitances indicated in Figure 7.14:

$$Z_{C_{l1}} = \frac{H_1(s)}{Q_{i1}(s) - Q(s)}; \quad Z_{R_{l1}} = \frac{H_1(s) - H_2(s)}{Q(s)};$$

$$Z_{C_{l2}} = \frac{H_2(s)}{Q(s) + Q_{i2}(s) - Q_o(s)}; \quad Z_{R_{l2}} = \frac{H_2(s)}{Q_o(s)} \tag{7.38}$$

Referring to Eqs. (7.38), the flow rate $Q(s)$ is substituted from the second equation in the first one, and the flow rate $Q_o(s)$ is substituted from the fourth equation in the third one; the following equations are obtained:

$$\left[\left(1 + \frac{Z_{C_{l1}}}{Z_{R_{l1}}} \right) H_1(s) - \frac{Z_{C_{l1}}}{Z_{R_{l1}}} H_2(s) = Z_{C_{l1}} Q_{i1}(s) \right.$$

$$\left. -\frac{Z_{C_{l2}}}{Z_{R_{l1}}} H_1(s) + \left(1 + \frac{Z_{C_{l2}}}{Z_{R_{l1}}} + \frac{Z_{C_{l2}}}{Z_{R_{l2}}} \right) H_2(s) = Z_{C_{l2}} Q_{i2}(s) \right. \tag{7.39}$$

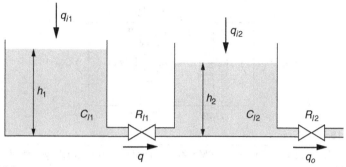

FIGURE 7.14

Two-Tank Liquid-Level System.

The impedances of Eqs. (7.39) are

$$Z_{C_{l1}} = \frac{1}{C_{l1}s}; Z_{C_{l2}} = \frac{1}{C_{l2}s}; Z_{R_{l1}} = R_{l1}; Z_{R_{l2}} = R_{l2} \tag{7.40}$$

By substituting them in Eqs. (7.39), the latter ones become

$$\begin{bmatrix} \dfrac{1}{R_{l1}} + C_{l1}s & -\dfrac{1}{R_{l1}} \\ -\dfrac{1}{R_{l1}} & \dfrac{1}{R_{l1}} + \dfrac{1}{R_{l2}} + C_{l2}s \end{bmatrix} \begin{Bmatrix} H_1(s) \\ H_2(s) \end{Bmatrix} = \begin{Bmatrix} Q_{i1}(s) \\ Q_{i2}(s) \end{Bmatrix} \tag{7.41}$$

Because the head vector of the left-hand side of Eq. (7.41) is the output and the volume flow rate of the right-hand side of Eq. (7.41) is the input, the transfer function matrix is

$$[G(s)] = \begin{bmatrix} \dfrac{1}{R_{l1}} + C_{l1}s & -\dfrac{1}{R_{l1}} \\ -\dfrac{1}{R_{l1}} & \dfrac{1}{R_{l1}} + \dfrac{1}{R_{l2}} + C_{l2}s \end{bmatrix}^{-1}$$

$$= \begin{bmatrix} \dfrac{R_{l1}R_{l2}C_{l2}s + R_{l1} + R_{l2}}{D(s)} & \dfrac{R_{l2}}{D(s)} \\ \dfrac{R_{l2}}{D(s)} & \dfrac{R_{l1}R_{l2}C_{l1}s + R_{l2}}{D(s)} \end{bmatrix} \tag{7.42}$$

the denominator polynomial being $D(s) = R_{l1}R_{l2}C_{l1}C_{l2}s^2 + [R_{l2}(C_{l1} + C_{l2}) + R_{l1}C_{l1}]s + 1$.

Let us check whether a relationship identical to that of Eq. (7.41) can be set for the liquid circuit shown in Figure 7.15, which is proposed as an impedance model of the physical (actual) system.

Considering that liquid flow rates and heads in liquid systems play the roles of currents and voltages in electrical systems, respectively, the node analysis method (introduced

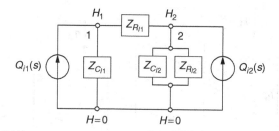

FIGURE 7.15

Impedance-Based Liquid Circuit as a Model Candidate for the Actual System of Figure 7.14.

for electrical systems in Chapter 4) can also be applied to liquid (and fluid, in general) systems by using complex impedances. Assuming the heads at the lower points on the circuit of Figure 7.15 are zero, the flow rates connecting at nodes 1 and 2 can be expressed, by virtue of conservation, as

$$
\begin{cases}
Q_{i1}(s) = \dfrac{H_1(s)}{Z_{C_{l1}}} + \dfrac{H_1(s) - H_2(s)}{Z_{R_{l1}}} \\[3mm]
Q_{i2}(s) = \dfrac{H_2(s) - H_1(s)}{Z_{R_{l1}}} + \dfrac{H_2(s)}{Z_{C_{l2}}} + \dfrac{H_2(s)}{Z_{R_{l2}}}
\end{cases}
\tag{7.43}
$$

Substituting Eqs. (7.40) into Eqs. (7.43) and using matrix notation, the latter equations become Eq. (7.41). As a consequence, the impedance-based circuit of Figure 7.15 models correctly the actual liquid system of Figure 7.14.   ■

Mechanical Systems Mechanical impedances can be combined in series and in parallel; therefore, equivalent impedances can be calculated for either case. Figure 7.16 illustrates the serial and the parallel connection of translatory-motion spring-type and damper-type impedances.

The companion website Chapter 7 derives the two equivalent mechanical impedances, which are

$$
\frac{1}{Z_s} = \frac{1}{Z_1} + \frac{1}{Z_2}; \; Z_p = Z_1 + Z_2
\tag{7.44}
$$

As shown in Table 7.1, the input to a mechanical impedance is a displacement (or rotation angle) and the output is a force (or moment). By taking into account that parallel-connected mechanical impedances have the same displacement, while series-connected mechanical impedances transmit the same force, *mechanical impedance circuits* can be designed for mechanical systems. This process is also aided by the *Newton's second law of motion*, which can be expressed in terms of impedances, as shown in the following example.

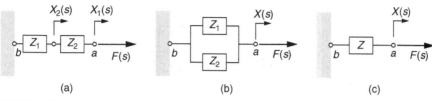

FIGURE 7.16

Connections of Elastic and Damping Mechanical Impedances: (a) Series; (b) Parallel; (c) Equivalent.

Example 7.8

a. Use the single DOF mechanical system of Figure 7.1 to demonstrate *Newton's second law of motion applied to mechanical impedances*, according to which the sum of impedance forces (resulting from inertia, damping, or stiffness) is equal to the sum of externally applied Laplace-transformed forces at any degree of freedom; draw a corresponding mechanical impedance circuit; and express the system's transfer function $Y(s)/U(s)$.

b. Use the complex impedance approach and an appropriate mechanical impedance circuit to determine the transfer function matrix of the MEMS sketched in Figure 7.6 of Example 7.3.

Solution

a. The dynamic equation of motion for the mass-damper-spring system of Figure 7.1 is given in Eq. (7.1). Its Laplace transform with zero initial conditions leads to

$$(ms^2 + cs + k)Y(s) = U(s) \tag{7.45}$$

which can also be written as

$$Z_m Y(s) + Z_d Y(s) + Z_e Y(s) = U(s) \tag{7.46}$$

where Z_m, Z_d, and Z_e are the mass, damping, and elastic mechanical impedances, respectively, as given in Table 7.1. Equation (7.45) illustrates the statement of Newton's second law of motion in terms of impedances. Let us analyze the circuit of Figure 7.17, which has mechanical impedances as elements and a force source. This circuit indicates that all mechanical impedances have the same displacement $Y(s)$. By considering that forces play the role of voltages (both represent forcing) and displacements are similar to currents, Kirchhoff's voltage law applied to the circuit of Figure 7.17 results in Eq. (7.45). Equation (7.45) enables expressing the transfer function that has also been obtained in Eq. (7.6).

b. Analyzing the lumped-parameter model of Figure 7.6(b), which corresponds to the MEMS of Figure 7.6(a), it can be seen that the mass m_1, the damper c, and the spring k_1 undergo the same displacement y_1; as a consequence, the three impedances are connected in parallel. The middle spring k_2 incurs a deformation that is the

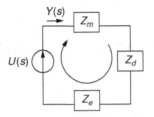

FIGURE 7.17

Mechanical Impedance Circuit for the Mechanical System of Figure 7.1.

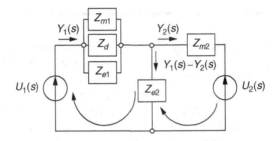

FIGURE 7.18

Mechanical Impedance Circuit for the MEMS Lumped-Parameter Model of Figure 7.6(b).

difference between its end points displacements, $y_1 - y_2$, whose Laplace transform is the input to this spring's impedance. On the other hand, the mass m_2 undergoes a displacement y_2. The impedance-form of Newton's second law of motion can be written for each of the two DOFs based on the definition that has just been demonstrated at point (a):

$$\begin{cases} (Z_{m1} + Z_d + Z_{e1}) Y_1(s) + Z_{e2}(Y_1(s) - Y_2(s)) = U_1(s) \\ Z_{m2} Y_2(s) + Z_{e2}(Y_2(s) - Y_1(s)) = -U_2(s) \end{cases} \tag{7.47}$$

Equations (7.47) took into account the law of action and reaction, in terms of the force generated by the middle spring k_2. By using the impedance definitions of Table 7.1, Eqs. (7.47) can be written in a form identical to Eqs. (7.17), which are derived in Example 7.3; therefore, the same transfer function derived from the MEMS mathematical model is obtained using the complex impedance approach.

Let us analyze the mechanical impedance circuit of Figure 7.18, which contains two meshes and where the external forces $U_1(s)$ and $U_2(s)$ have been represented as sources.

The impedance Newton's second law that has just been introduced can be applied for each of the two meshes, and this actually results in Eqs. (7.47) that have been obtained by applying Newton's second law in the impedance variant for each of the two DOF. It therefore means that the circuit of Fig. 7.18 is a valid representation of the lumped-parameter model of Fig. 7.6(b). ∎

7.2.2 MATLAB® Approach

The transfer function is formulated in MATLAB® by the `tf` command, where the coefficients of the numerator and denominator have to be defined in descending order of powers. For the transfer function $G(s) = (6s + 3)/(5s^2 + s + 4)$, for instance, the numerator and denominator are `num = [6, 3]`, `den = [5, 1, 4]`, and the transfer function is defined as `sys = tf (num, den)`. The sequence just discussed can also be written directly as `sys = tf ([6, 3], [5, 1, 4])`. Note: Any other function name or symbol can be used instead of `sys`.

The transfer function (tf) model is actually one of the four *linear time invariant (LTI) models* (or *objects*) available in the MATLAB® Control System Toolbox™; more details on LTIs are offered in Appendix C. Whenever the `tf` command is issued, MATLAB® generates and stores the corresponding transfer function model. Another MATLAB® LTI object is *zero-pole-gain* (zpk), which can build a model that stores the zeroes (*z*), poles (*p*), and constant gain (*k*) of a specific function (such as a transfer function). According to this model, a transfer function can be written in the form

$$G(s) = \frac{k(s - z_1)(s - z_2)...(s - z_m)}{(s - p_1)(s - p_2)...(s - p_n)} \tag{7.48}$$

Whenever the zeroes, poles, and gain are known, the MATLAB® command `zpk(z, p, k)` generates a zero-pole-gain LTI object. MATLAB® allows conversion between tf and zpk models, as illustrated in the following example.

Example 7.9

An *s*-domain function is defined by the zeroes 1 and 2; the poles −1, −2, and −3; and a gain of 5. Build the MATLAB® zero-pole-gain object corresponding to these data then convert this object into a transfer function model. Reconvert the obtained transfer function model into a zero-pole-gain model and confirm that the original zpk model is retrieved.

Solution

The following MATLAB® command sequence is used to generate the zpk model from the given data, convert that model into a tf model, and reconvert the tf object into the original zpk model:

```
>> f = zpk([1,2],[-1,-2,-3],5)
Zero/pole/gain:
5 (s-1) (s-2)
------------------
(s+1) (s+2) (s+3)
>> g = tf(f)
Transfer function:
5 s^2 - 15 s + 10
----------------------
```

```
s^3 + 6 s^2 + 11 s + 6
>> zpk(g)
Zero/pole/gain:
5 (s-2) (s-1)
-----------------
(s+3) (s+2) (s+1)
```

7.3 TRANSFER FUNCTION AND THE TIME RESPONSE

Employing the transfer functions derived by one of the two methods presented in section 7.2, we determine the time response of SISO and MIMO systems both analytically and by means of MATLAB®. The generic formulation considers that both forcing and nonzero initial conditions can be applied.

7.3.1 SISO Systems

The transfer function approach to determining the time response of SISO systems is discussed in this section.

Analytical Approach

The general case of a second-order system subjected to both forcing and nonzero initial conditions is studied here, out of which result the particular cases of the forced response with zero initial conditions and the free response with nonzero initial conditions.

For such a dynamic system, the differential equation is

$$a_2\ddot{y}(t) + a_1\dot{y}(t) + a_0 y(t) = u(t) \tag{7.49}$$

with the initial conditions: $\dot{y}(0) \neq 0$; $y(0) \neq 0$. Laplace transforming Eq. (7.49) yields

$$Y(s) = \frac{1}{a_2 s^2 + a_1 s + a_0}[U(s) + a_2 y(0)s + a_2 \dot{y}(0) + a_1 y(0)] = G(s)U_e(s) \tag{7.50}$$

where $G(s)$ is the transfer function of the second-order system and $U_e(s)$ is the *equivalent input (forcing) function* that combines the effects of the actual input function $U(s)$ and the initial conditions, as introduced in Figure 7.2; these functions are

$$G(s) = \frac{1}{a_2 s^2 + a_1 s + a_0}; \ U_e(s) = U(s) + a_2 y(0)s + a_2 \dot{y}(0) + a_1 y(0) \tag{7.51}$$

The time-domain counterpart of $U_e(s)$ is

$$u_e(t) = \mathcal{L}^{-1}[U(s)] + a_2 y(0)\frac{d}{dt}\delta(t) + [a_2 \dot{y}(0) + a_1 y(0)]\delta(t) \tag{7.52}$$

For *zero initial conditions*, Eq. (7.52) reduces to

$$u_e(t) = \mathcal{L}^{-1}[U(s)] \tag{7.53}$$

which represents equivalent input. Conversely, when $U(s) = 0$, which means $u(t) = 0$, Eq. (7.52) simplifies to

$$u_e(t) = a_2 y(0)\frac{d}{dt}\delta(t) + [a_2\dot{y}(0) + a_1 y(0)]\delta(t) \tag{7.54}$$

which describes equivalent input of a second-order system with nonzero initial conditions.

It can simply be shown that the transfer-function model of a *first-order system* is derived from the model of the second-order system, previously formulated, by taking $a_2 = 0$ in Eqs. (7.51) and (7.52).

According to the *initial-value theorem*, the initial value of the output is calculated as

$$y(0) = \lim_{t \to 0} y(t) = \lim_{s \to \infty} sY(s) = \lim_{s \to \infty} sG(s)U_e(s) \tag{7.55}$$

Similarly, the final value of the time-domain response can be determined by using the *final-value theorem*:

$$y(\infty) = \lim_{t \to \infty} y(t) = \lim_{s \to 0} sY(s) = \lim_{s \to 0} sG(s)U_e(s) \tag{7.56}$$

Example 7.10
The pneumatic system of Figure 7.19(a) is formed of a container with a capacity $C_g = 1.2 \times 10^{-6}$ m-s², which is supplied with a gas whose pressure depends on time as $p_i = (1 + 4e^{-0.2t}) \times 10^5$ N/m². The gas passes through a constriction whose resistance is $R_g = 30,000$ m⁻¹s⁻¹. Use complex impedances to determine the pneumatic circuit corresponding to this system as well as its transfer function $P_o(s)/P_i(s)$; based on it,

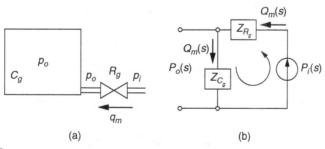

(a) (b)

FIGURE 7.19

Pneumatic System with Valve and Container: (a) Physical Model; (b) Pneumatic Impedance Circuit.

calculate the output (container) pressure $p_o(t)$ and plot it against time for zero initial conditions. Also determine the final (steady-state) value of $p_o(t)$.

Solution
Based on Figure 7.19(a) and on Table 7.1, the gas resistance and capacitance imped-ances are

$$Z_{R_g}(s) = \frac{P_i(s) - P_o(s)}{Q_m(s)}; \ Z_{C_g}(s) = \frac{P_o(s)}{Q_m(s)} \tag{7.57}$$

Eliminating $Q_m(s)$ between the two Eqs. (7.57) and taking into account that

$$Z_{R_g}(s) = R_g; Z_{C_g}(s) = \frac{1}{C_g s} \tag{7.58}$$

the complex transfer function is

$$G(s) = \frac{P_o(s)}{P_i(s)} = \frac{1}{R_g C_g s + 1} \tag{7.59}$$

Consider that the pneumatic impedance circuit of Figure 7.19(b) represents the physical system of Figure 7.19(a). Based on this circuit and applying an equivalent of the Kirchhoff's voltage law where pressure is used instead of voltage and volume flow rate for current, the following s-domain equations can be formulated:

$$\begin{cases} P_i(s) = Z_{R_g} Q_m(s) + Z_{C_g} Q_m(s) \\ P_o(s) = Z_{C_g} Q_m(s) \end{cases} \tag{7.60}$$

The pressure-defined impedances corresponding to fluid (gas) resistance and capaci-tance of Table 7.1 are substituted in Eqs. (7.60); by evaluating the $P_o(s)/P_i(s)$ ratio, the transfer function of Eq. (7.59) is obtained.

The Laplace transform of the output is

$$P_o(s) = P_i(s) G(s) = \frac{10^5}{1 + R_g C_g s} \left(\frac{1}{s} + \frac{4}{s + 0.2} \right) \tag{7.61}$$

The steady-state value of the container pressure is calculated as

$$p_o(\infty) = \lim_{t \to \infty} p_o(t) = \lim_{s \to 0} s P_o(s) = 10^5 \text{N/m}^2 \tag{7.62}$$

The container pressure is calculated by inverse Laplace transforming $P_o(s)$ of Eq. (7.61) as

$$p_o(t) \simeq \left(1 + 4e^{-0.2t} - 5e^{-28t} \right) \times 10^5 \text{N/m}^2 \tag{7.63}$$

which is plotted in Figure 7.20. It can be seen that the steady-state pressure value of 10^5 N/m² is reached after approximately 25 seconds.

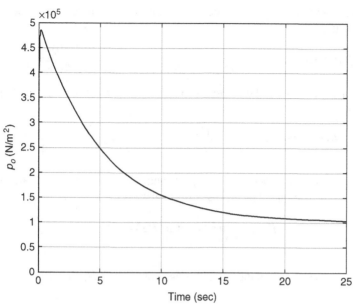

FIGURE 7.20

Container Pressure as a Function of Time.

Example 7.11

A single-mesh electrical system is formed of a resistor R, a capacitor C, an inductor L, and a voltage source providing a ramp voltage $v(t) = Vt$. Find the current $i(t)$ in the circuit when $4L = R^2 C$, knowing $i(0) = 0.1$ A and $q(0) = 0$ C. Plot $i(t)$ for $V = 100$ V, $R = 200$ Ω, and $L = 3$ H; also find the final value of the current.

Solution

The differential equation of the series electrical system is

$$Ri(t) + L\frac{d}{dt}i(t) + \frac{1}{C}\int i(t)\,dt = v(t) \tag{7.64}$$

which can be written in terms of charge as

$$L\ddot{q}(t) + R\dot{q}(t) + \frac{1}{C}q(t) = v(t) \tag{7.65}$$

According to Eqs. (7.49) and (7.51), the transfer function and the equivalent Laplace-transformed input are

$$G(s) = \frac{C}{LCs^2 + RCs + 1}; \; U_e(s) = \frac{V}{s^2} + Li(0) \tag{7.66}$$

As a consequence and for the particular condition of this example, the Laplace transform of the output is

$$Q(s) = G(s)U_e(s) = \frac{CV}{s^2(LCs^2 + RCs + 1)} + \frac{LCi(0)}{LCs^2 + RCs + 1} \qquad (7.67)$$

The Laplace transform of the current is expressed in terms of the Laplace transform of the charge as

$$I(s) = sQ(s) = \frac{CV}{s(LCs^2 + RCs + 1)} + \frac{LCi(0)s}{LCs^2 + RCs + 1} \qquad (7.68)$$

where this example's relationship among L, R, and C is used. The steady-state value of the current can be determined by means of the final value theorem as

$$i(\infty) = \lim_{t \to \infty} i(t) = \lim_{s \to 0} sI(s) = CV \qquad (7.69)$$

For the numerical values of this example, the final current is $i(\infty) = 0.03$ A.

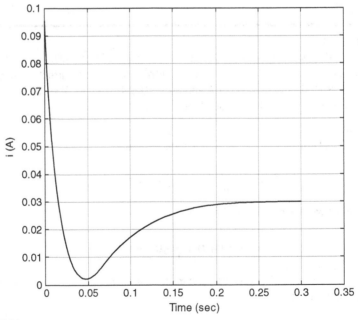

FIGURE 7.21

Current as a Function of Time.

By inverse Laplace transforming $I(s)$ of Eq. (7.68), the following time-dependent current is obtained:

$$i(t) = \frac{4LV}{R^2} - \frac{4LV + R^2 i(0)}{2RL} te^{-\frac{R}{2L}t} - \frac{4LV - R^2 i(0)}{R^2} e^{-\frac{R}{2L}t} \qquad (7.70)$$

The current $i(t)$ is plotted in Figure 7.21 for the numerical values of this example, which shows that, indeed, the final value (steady-state) value is 0.03 A. ◼

MATLAB® Approach

The Control System Toolbox™ of MATLAB® allows calculation and plotting of the forced response of dynamic systems by means of the transfer function. Once the mathematical model and the corresponding transfer function are obtained through modeling, MATLAB® has the capability of obtaining the system response (and plotting it) for various inputs, such as unit step, unit impulse, or an arbitrary form. As shown in a previous section of this chapter, the transfer function is determined in MATLAB® by using the `tf` command.

Response to Unit Step Input, the `Step` Function
The following sequence of MATLAB® code:

```
>> g = tf ([6, 3], [5, 1, 4]);
>> step(g)
```

produces the plot shown in Figure 7.22 as the response to a *unit step input* applied to the transfer function $G(s) = (6s + 3)/(5s^2 + s + 4)$.

It can be seen that the time range, time units, as well as title and axes labels have been introduced automatically. The MATLAB® command `ltiview` enables operating modifications in the obtained plot, such as changing the time range. The following sequence of commands produces the plot of Figure 7.23, which is the response based on the same transfer function over a time interval of 20 seconds:

```
>> g = tf ([6, 3], [5, 1, 4]);
>> ltiview
```

Appendix C gives a more detailed description of the `ltiview` command options; but in a nutshell, the plot of Figure 7.23 is obtained by selecting `File`, `Import`, and the transfer function that has just been defined (either in the Workspace or in a MAT-file). By right clicking the plot space, the time range can be changed to 20 s from the default 60 s by selecting `Properties` and `Limits`. By right clicking the plot area again, it is possible to highlight system features such as the `Peak Response`, the `Settling Time`, the `Rise Time`, or the `Steady State` (all properties of second-order systems under step input). By choosing `Rise Time`, for instance, the point of Figure 7.23 is highlighted; and by clicking

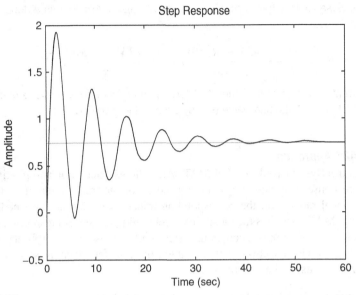

FIGURE 7.22

MATLAB® Plot with Basic Usage of the Step Function.

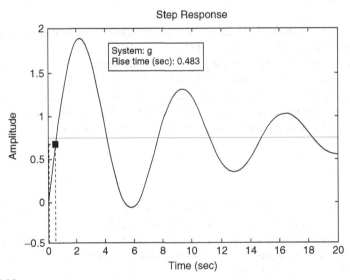

FIGURE 7.23

MATLAB® Plot with Unit Step Input and Modified Time Specifications by the ltiview Option.

it, the accompanying box is shown. In this system configuration, the rise time is the time necessary for the response to get from 10% to 90% of the final (steady-state) response. The ltiview command enables changing the type of input from unit step, as also described in Appendix C.

The unit step input can be used to find the free response of a system subjected to nonzero initial conditions as well. We saw in this section that the free response of a system under nonzero initial conditions is equivalent to that of the same system under equivalent forcing and zero initial conditions. It can be demonstrated that the free system can be converted to a forced system under unit step forcing, as shown in the next example.

Example 7.12

Prove that the standard step function can be used to determine the free response of a second-order system subjected to nonzero initial conditions. Using the MATLAB® tf and step commands, apply the methodology to a series electrical circuit that is formed of a capacitor $C = 50$ μF, a resistor $R = 300$ Ω, and an inductor $L = 40$ mH to plot the charge in the circuit as a function of time. An initial charge $q_0 = 0.1$ C is applied to the capacitor.

Solution

Let us consider a free second-order system whose differential equation is of the form

$$a\ddot{y}(t) + b\dot{y}(t) + cy(t) = 0 \qquad (7.71)$$

With nonzero initial conditions, the Laplace transform of Eq. (7.71) yields an equation that can be solved for $X(s)$, the Laplace transform of $x(t)$:

$$Y(s) = \frac{ay(0)s + a\dot{y}(0) + by(0)}{as^2 + bs + c} \qquad (7.72)$$

When considering a dummy forcing function $u(t)$ acts on the otherwise free dynamic system, it can simply be shown that the transfer function of that system is

$$G(s) = \frac{Y(s)}{U(s)} = \frac{1}{as^2 + bs + c} \qquad (7.73)$$

Equation (7.72) can be reformulated as

$$Y(s) = \frac{ay(0)s^2 + [a\dot{y}(0) + by(0)]s}{as^2 + bs + c} \times \frac{1}{s} = G'(s)\mathcal{L}[1(t)] \qquad (7.74)$$

which indicates that the original system with its transfer function $G(s)$ has been equivalently transformed into another system defined by the transfer function

$$G'(s) = \frac{ay(0)s^2 + [a\dot{y}(0) + by(0)]s}{as^2 + bs + c}$$

$$= \{ay(0)s^2 + [a\dot{y}(0) + by(0)]s\} G(s) \tag{7.75}$$

and acted upon by a unit step function $1(t)$.

The differential equation describing the free response of the electrical circuit is

$$L\ddot{q} + R\dot{q} + \frac{1}{C}q = 0 \tag{7.76}$$

and its original transfer function is

$$G(s) = \frac{1}{Ls^2 + Rs + 1/C} = \frac{1}{0.04s^2 + 300s + 20{,}000} \tag{7.77}$$

The modified transfer function, as in Eq. (7.75), is therefore

$$G'(s) = \frac{0.004s^2 + 30s}{0.04s^2 + 300s + 20{,}000} \tag{7.78}$$

The plot that gives the variation of $q(t)$ is shown in Figure 7.24.

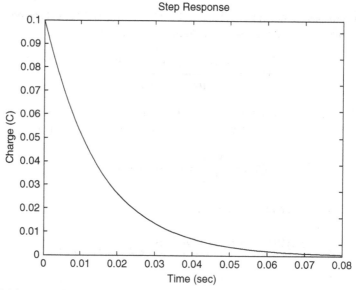

FIGURE 7.24

Charge Variation as a Function of Time.

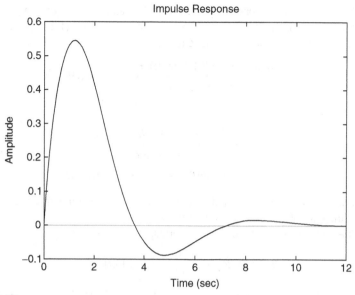

FIGURE 7.25

MATLAB® Plot with Basic Usage of the Impulse Function.

Response to Unit Impulse Input, the impulse Function

Another built-in MATLAB® input function is the *unit impulse*, which is considered to be applied under zero initial conditions. Consider the transfer function $G(s) = 1/(s^2 + s + 1)$. Usage of the following sequence of commands:

```
>> sys = tf (1, [1, 1, 1]);
>> impulse(sys)
```

generates the plot of Figure 7.25 for unit impulse forcing.

All the other features discussed for the step function as an input are also valid for the impulse function. Let us check, for instance, the following feature.

Example 7.13

Demonstrate that the regular impulse function can be used to model the response of a second-order system subjected to both nonunity impulse input and nonzero initial conditions. Apply the algorithm to the electric system of the previous Example 7.12 with a initial charge of that example. In addition, consider that the circuit includes a voltage source with $v = 40$ V, which opens and closes very fast thereafter.

Solution

The dynamic equation that describes the system is

$$a\ddot{y}(t) + b\dot{y}(t) + cy(t) = d\delta(t) \tag{7.79}$$

where the constant d multiplies the Dirac delta function $\delta(t)$, the unit impulse, which models the fast opening and closing of the voltage source. Applying the Laplace transform to Eq. (7.79) for nonzero initial conditions results in

$$Y(s) = \frac{ay(0)s + a\dot{y}(0) + by(0) + d}{as^2 + bs + c} \times \mathcal{L}[\delta(t)] \tag{7.80}$$

which indicates that

$$G'(s) = \frac{ay(0)s + a\dot{y}(0) + by(0) + d}{as^2 + bs + c}$$

$$= G(s)[ay(0)s + a\dot{y}(0) + by(0) + d] \tag{7.81}$$

is the transfer function of a modified system under forcing by a unit impulse. This demonstrates the equivalence stated in the example.

For the particular case of the electrical circuit, its equation is

$$L\ddot{q}(t) + R\dot{q}(t) + \frac{1}{C}q(t) = 40\delta(t) \tag{7.82}$$

and therefore the constants of the general model are $a = L$, $b = R$, $c = 1/C$, and $d = 40$. With the numerical values given in Example 7.12, the modified transfer function becomes

$$G'(s) = \frac{0.004s + 70}{0.04s^2 + 300s + 20,000} \tag{7.83}$$

The charge variation is shown in Figure 7.26.

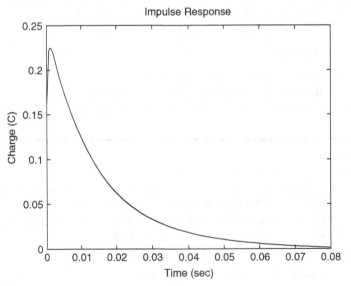

FIGURE 7.26

Charge in an RLC Series Circuit with Impulse Voltage and Initial Charge.

It can be seen that there is a spike in the charge response corresponding to the application of the impulsive voltage of 40 V; the charge decreases to zero thereafter. ■

Response to Arbitrary Input, the lsim Function

Arbitrary forcing can be generated in MATLAB® by the lsim function under zero initial conditions. The arbitrary function has to be defined before launching the calculating and plotting lsim command. Let us consider a dynamic system whose transfer function is $G(s) = 1/(s^2 + s + 1)$ and is acted upon by the forcing function $f(t) = 1/(t + 2)$. The following MATLAB® code

```
>> sys = tf (1, [1, 1, 1]);
>> t = [0:0.01:20];
>> f = 1./(t + 2);
>> lsim(sys,f,t)
```

generates the plot of Figure 7.27, where both the input (the curve starting from a value of 0.5) and the output are shown.

It is also possible to simulate a system that is under combined arbitrary forcing and nonzero initial conditions, as demonstrated in the following example. Caution has to be applied as only forcing of polynomial form can be used for this procedure.

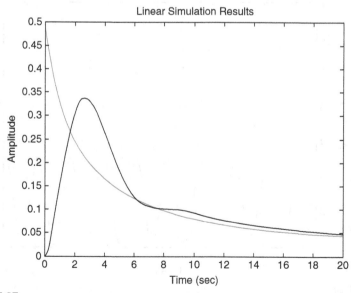

FIGURE 7.27

Second-Order System Response to Arbitrary Forcing with Zero Initial Conditions.

Example 7.14

Prove that the 1 s i m command, which is normally employed for forcing under zero initial conditions, can also be used for a forced system with nonzero initial conditions. Apply the result to a mechanical system formed of a mass $m = 0.1$ kg, a viscous damper $c = 0.2$ N-s/m, and a spring with $k = 10$ N/m under the action of a force $f(t) = 6 + 5 \sin(10t)$ N when the initial displacement of the mass is $y(0) = 0.1$ m.

Solution

Consider first a generic second-order system under an arbitrary forcing function $f(t)$ with the nonzero initial conditions $\dot{y}(0) \neq 0$; $y(0) \neq 0$. The mathematical model of the particular mechanical system of this example is described by the differential equation

$$m\ddot{y}(t) + c\dot{y}(t) + ky(t) = f(t) \tag{7.84}$$

Application of the Laplace transform to Eq. (7.84) yields

$$Y(s) = \frac{my(0)s + m\dot{y}(0) + cy(0) + F(s)}{(ms^2 + cs + k)F(s)} \times F(s) \tag{7.85}$$

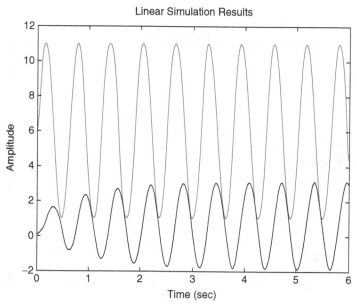

Linear Simulation Results

FIGURE 7.28

Second-Order System Response to Arbitrary Forcing with Nonzero Initial Conditions.

which indicates that

$$G'(s) = \frac{my(0)s + m\dot{y}(0) + cy(0) + F(s)}{(ms^2 + cs + k)F(s)}$$

$$= \frac{my(0)s + m\dot{y}(0) + cy(0) + F(s)}{F(s)} \times G(s) \qquad (7.86)$$

is the transfer function of a modified system that is acted upon by the given function f(t). By using the numerical values of this example, the transfer function of Eq. (7.86) is

$$G'(s) = \frac{s^4 + 2s^3 + 700s^2 + 5,200s + 60,000}{60s^4 + 620s^3 + 13,000s^2 + 62,000s + 600,000} \qquad (7.87)$$

Figure 7.28 shows the response (lower curve) and forcing functions.   ■

7.3.2 MIMO Systems

Similar to SISO systems, the forced response of MIMO systems is studied in this section using the analytical approach and MATLAB®.

Analytical Approach

The *forced response* of first-order and second-order MIMO systems are analyzed separately by considering the general situation where both *forcing* and *initial conditions* are applied to the system. The *forced response with zero initial conditions* and the *free response with nonzero initial conditions* result as particular cases of the general formulation.

First-Order Systems

First-order MIMO systems are described by the differential equation

$$[a_1]\{\dot{y}(t)\} + [a_0]\{y(t)\} = \{u(t)\} \qquad (7.88)$$

The number of inputs, m, is normally smaller than the number of outputs (which is assumed equal to the number of differential equations of the mathematical model), p; however, by adding $p - m$ zeroes to the input, the two vectors have the same dimension. Application of the Laplace transform to Eq. (7.88) with nonzero initial conditions leads to

$$[a_1](s\{Y(s)\} - \{y(0)\}) + [a_0]\{Y(s)\} = \{U(s)\} \qquad (7.89)$$

After factoring out $\{Y(s)\}$ and grouping the remaining terms, Eq. (7.89) can be written as

$$\{Y(s)\} = (s[a_1] + [a_0])^{-1}\{U(s)\} + (s[a_1] + [a_0])^{-1}[a_1]\{y(0)\} \qquad (7.90)$$

The transfer function matrix corresponding to zero initial conditions for the first-order system is obtained from Eq. (7.90) as

$$[G(s)] = (s[a_1] + [a_0])^{-1} \tag{7.91}$$

because $\{Y(s)\} = [G(s)] \{U(s)\}$. As a consequence, Eq. (7.90) becomes

$$\{Y(s)\} = [G(s)]\{U_e(s)\} \tag{7.92}$$

with

$$\{U_e(s)\} = \{U(s)\} + [a_1]\{y(0)\} \tag{7.93}$$

being the Laplace transform of an *equivalent input vector*. When $\{y(0)\} = 0$, the equivalent forcing vector reduces to the actual one, $\{U(s)\}$; therefore, the modified Eq. (7.92) enables calculating the forced response with zero initial conditions in the Laplace domain. Conversely, when $\{U(s)\} = 0$, the equivalent forcing vector of Eq. (7.93) simplifies to

$$\{U_e(s)\} = [a_1]\{y(0)\} \tag{7.94}$$

and Eq. (7.92) provides the Laplace transform of the free response with nonzero initial conditions.

The initial and final values of the output vector $\{y(t)\}$ can be determined without prior knowledge of the vector itself, by means of $\{Y(s)\}$ and applying the initial- and final-value theorems:

$$\begin{cases} \{y(0)\} = \lim_{t \to 0}\{y(t)\} = \lim_{s \to \infty} s\{Y(s)\} = \lim_{s \to \infty} s[G(s)]\{U_e(s)\} \\ \{y(\infty)\} = \lim_{t \to \infty}\{y(t)\} = \lim_{s \to 0} s\{Y(s)\} = \lim_{s \to 0} s[G(s)]\{U_e(s)\} \end{cases} \tag{7.95}$$

The time-domain equivalent forcing is found by inverse Laplace transforming Eq. (7.93) as:

$$\{u_e(t)\} = \{u(t)\} + \delta(t)[I][a_1]\{y(0)\} \tag{7.96}$$

where $\delta(t)$ is the delta Dirac function and $[I]$ is the identity matrix.

Example 7.15

The two-tank liquid system of Figure 7.14 analyzed in Example 7.7 is defined by $C_{l1} = 20$ m², $C_{l2} = 16$ m², $R_{l1} = 2$ s-m² , $R_{l2} = 1.2$ s-m². Knowing the initial heads are $h_1(0) = 0.1$ m, $h_2(0) = 0$, and that the input volume flow rates are $q_{i1} = q_{i2} = 0$, use the transfer function determined in Example 7.7 to find the time response of the hydraulic system. Also determine the steady-state head values.

Solution

The transfer function matrix $[G(s)]$ was derived in Eq. (7.42) and the Laplace-domain system response is found from Eqs. (7.92) and (7.93) as

$$\{Y(s)\} = \{H(s)\} = [G(s)][a_1]\{y(0)\} = [G(s)][a_1]\{h(0)\} \tag{7.97}$$

with

$$[a_1] = \begin{bmatrix} C_{11} & 0 \\ 0 & C_{12} \end{bmatrix}; \{h(0)\} = \begin{Bmatrix} h_1(0) \\ h_2(0) \end{Bmatrix} \tag{7.98}$$

By using the numerical values of the example, the following Laplace-domain heads are obtained:

$$H_1(s) = \frac{384s + 32}{3840s^2 + 416s + 5}; H_2(s) = \frac{12}{3840s^2 + 416s + 5} \tag{7.99}$$

The final (steady-state) values of the heads are

$$\begin{cases} h_1(\infty) = \lim_{t \to \infty} h_1(t) = \lim_{s \to 0} sH_1(s) = 0 \\ h_2(\infty) = \lim_{t \to \infty} h_2(t) = \lim_{s \to 0} sH_2(s) = 0 \end{cases} \tag{7.100}$$

The time-dependent heads are found by inverse Laplace transforming Eqs. (7.99) as

$$\begin{cases} h_1(t) = 0.001e^{-0.054t}[94\cosh(0.04t) + 68\sinh(0.04t)] \\ h_2(t) = 0.077e^{-0.054t}\sinh(0.04t) \end{cases} \tag{7.101}$$

and they are plotted against time in Figure 7.29.

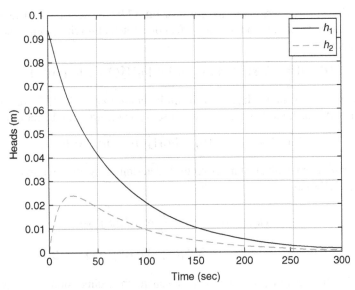

FIGURE 7.29

Time Variation of Hydraulic Heads.

Second-Order Systems

The vector-matrix differential equation of second-order MIMO systems is

$$[a_2]\{\ddot{y}(t)\} + [a_1]\{\dot{y}(t)\} + [a_0]\{y(t)\} = \{u(t)\} \tag{7.102}$$

where, again, the input and output vectors are assumed to have the same dimension. When the initial conditions are nonzero, that is, $\{\dot{y}(0)\} \neq \{0\}; \{y(0)\} \neq \{0\}$, the Laplace transform that is applied to Eq. (7.102) yields

$$[a_2](s^2\{Y(s)\} - s\{y(0)\} - \{\dot{y}(0)\}) + [a_1](s\{Y(s)\}$$
$$- \{y(0)\}) + [a_0]\{Y(s)\} = \{U(s)\} \tag{7.103}$$

The Laplace-transformed output vector $\{Y(s)\}$ is expressed from Eq. (7.103) as

$$\{Y(s)\} = (s^2[a_2] + s[a_1] + [a_0])^{-1}\{U(s)\}$$
$$+ (s^2[a_2] + s[a_1] + [a_0])^{-1}(s[a_2]\{y(0)\}$$
$$+ [a_2]\{\dot{y}(0)\} + [a_1]\{y(0)\}) \tag{7.104}$$

Taking into account that the regular (zero initial conditions) transfer function matrix $[G(s)]$ is the one connecting $\{Y(s)\}$ to $\{U(s)\}$,

$$[G(s)] = (s^2[a_2] + s[a_1] + [a_0])^{-1} \tag{7.105}$$

Eq. (7.104) can be written in the form of Eq. (7.92), where the equivalent forcing vector $U_e(s)$, which incorporates the effect of the actual forcing vector and the initial conditions, is

$$\{U_e(s)\} = \{U(s)\} + s[a_2]\{y(0)\} + [a_2]\{\dot{y}(0)\} + [a_1]\{y(0)\} \tag{7.106}$$

Equation (7.106) shows that, for zero initial conditions, $\{U_e(s)\} = \{U(s)\}$ and, for $\{U(s)\} = 0$, which models the free response with nonzero initial conditions:

$$\{U_e(s)\} = s[a_2]\{y(0)\} + [a_2]\{\dot{y}(0)\} + [a_1]\{y(0)\} \tag{7.107}$$

The time-domain equivalent input vector is determined by inverse Laplace transforming Eq. (7.106) as

$$\{u_e(t)\} = \{u(t)\} + \frac{d}{dt}\delta(t)[I][a_2]\{y(0)\}$$
$$+ \delta(t)[I]([a_2]\{\dot{y}(0)\} + [a_1]\{y(0)\}) \tag{7.108}$$

Note: The matrices $[a_2]$, $[a_1]$ and $[a_0]$ are square, their dimensions being equal to the dimensions of the output vector. In order to enable multiplications involving the input vector, its dimension needs to be equal to the output vector dimension. When this condition does not occur naturally, zeroes can be added to the vector of smaller dimension.

Example 7.16

The mechanical micro system sketched in Figure 7.30(a) consists of two shuttle masses, $m_1 = 2 \times 10^{-10}$ kg and $m_2 = 3 \times 10^{-10}$ kg, that can move horizontally and two identical serpentine springs. There is also viscous squeeze damping between the mass m_1 and the anchored wall with a damping coefficient $c = 0.01$ N-s/m. Determine the lumped-parameter model of this microdevice and consider that the initial displacement of m_1 is $y_1(0) = 2$ μm and the initial velocity of m_2 is $\dot{y}_2(0) = v_2(0) = 1$ μm/s. Determine the forcing function vector that renders the system equivalent to the forced response of the same system under zero initial conditions using the transfer function matrix and the specified nonzero initial conditions.

Solution

The lumped-parameter model of the physical model sketched in Figure 7.30(a) is shown in Figure 7.30(b), where the two serpentine springs have been replaced by identical springs of stiffness k and the interaction between mass m_1 and the wall of Figure 7.30(a) has been modeled by the damper c. This mechanical device has two outputs, the displacements y_1 and y_2; therefore, it is a multiple-output system and, as a consequence, a MIMO system. By using Newton's second law of motion (and dropping the time variable notation), the dynamic equations corresponding to the two masses are

$$\begin{cases} m_1 \ddot{y}_1 = -c\dot{y}_1 - ky_1 - k(y_1 - y_2) \\ m_2 \ddot{y}_2 = -k(y_2 - y_1) \end{cases} \tag{7.109}$$

which can be arranged in vector-matrix form as

$$\begin{bmatrix} m_1 & 0 \\ 0 & m_2 \end{bmatrix} \begin{Bmatrix} \ddot{y}_1 \\ \ddot{y}_2 \end{Bmatrix} + \begin{bmatrix} c & 0 \\ 0 & 0 \end{bmatrix} \begin{Bmatrix} \dot{y}_1 \\ \dot{y}_2 \end{Bmatrix} + \begin{bmatrix} 2k & -k \\ -k & k \end{bmatrix} \begin{Bmatrix} y_1 \\ y_2 \end{Bmatrix} = \begin{Bmatrix} 0 \\ 0 \end{Bmatrix} \tag{7.110}$$

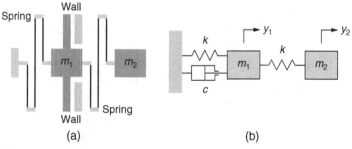

(a)   (b)

FIGURE 7.30

Translatory Mechanical Microsystem under Nonzero Initial Conditions: (a) Physical Model; (b) Lumped-Parameter Model.

Equation (7.110) shows that the two matrices of interest are

$$[a_2] = \begin{bmatrix} m_1 & 0 \\ 0 & m_2 \end{bmatrix}; [a_1] = \begin{bmatrix} c & 0 \\ 0 & 0 \end{bmatrix} \tag{7.111}$$

Now using Eq. (7.107) with $\{y(0)\} = \{y_1(0)\ 0\}^t$ and $\{v(0)\} = \{0\ v_2(0)\}^t$, the equivalent forcing vector becomes

$$\{f_e(t)\} = \left\{ m_1 y_1(0) \frac{d}{dt}\delta(t) + cy_1(0)\delta(t), \ m_2 \dot{y}_2(0)\delta(t) \right\}^t$$

$$= \left\{ 4 \times 10^{-16} \frac{d}{dt}\delta(t) + 2 \times 10^{-8}\delta(t), \ 3 \times 10^{-16}\delta(t) \right\}^t \tag{7.112}$$

Example 7.17

Use the node analysis method in conjunction with the complex impedance matrix approach to find the time response (consisting of the relevant voltages) of the electrical circuit of Figure 7.31(b). Known are $R_1 = 400\ \Omega$, $R_2 = 550\ \Omega$, $C = 250\ \mu F$, $L = 5$ H, $i = 20$ mA, and the initial conditions are zero.

Solution

Let us study the impedance-based circuit of Figure 7.31(a). Considering that the nodes c and d are grounded, it follows that $V_c = V_d = 0$. By applying Kirchhoff's first law at nodes a and b, the following equations are obtained:

$$\begin{cases} I(s) = \dfrac{V_a(s)}{Z_{R1}} + \dfrac{V_a(s) - V_b(s)}{Z_{R2}} \\ \dfrac{V_a(s) - V_b(s)}{Z_{R2}} = \dfrac{V_b(s)}{Z_L} + \dfrac{V_b(s)}{Z_C} \end{cases} \tag{7.113}$$

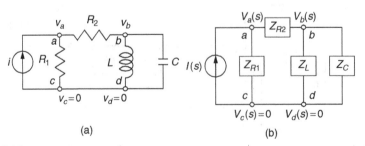

(a)   (b)

FIGURE 7.31

Electrical Circuit with Current Source: (a) Physical Model; (b) Complex Impedance Representation.

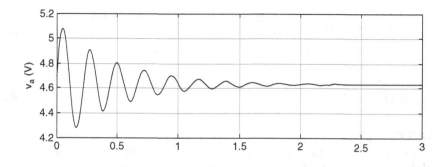

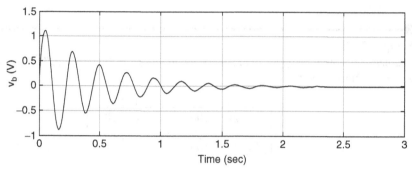

FIGURE 7.32

Node Voltages as Functions of Time.

which can be written as

$$
\begin{cases}
\left(\dfrac{1}{Z_{R1}} + \dfrac{1}{Z_{R2}}\right)V_a(s) - \dfrac{1}{Z_{R2}}V_b(s) = I(s) \\[2ex]
-\dfrac{1}{Z_{R2}}V_a(s) + \left(\dfrac{1}{Z_{R2}} + \dfrac{1}{Z_L} + \dfrac{1}{Z_C}\right)V_b(s) = 0
\end{cases}
\tag{7.114}
$$

The solution to Eqs. (7.114) is

$$
\begin{cases}
V_a(s) = \dfrac{R_1\left(R_2 CLs^2 + Ls + R_2\right)}{LC(R_1 + R_2)s^2 + Ls + R_1 + R_2} I(s) \\[2ex]
V_b(s) = \dfrac{R_1 Ls}{LC(R_1 + R_2)s^2 + Ls + R_1 + R_2} I(s)
\end{cases}
\tag{7.115}
$$

By applying the inverse Laplace transforms to $V_a(s)$ and $V_b(s)$ of Eqs. (7.115) with the numerical values of the parameters, the following time-domain voltages are obtained:

$$
\begin{cases}
v_a(t) = 4.63 + 0.5e^{-2.1t}\sin(28.2t) \\
v_b(t) = 1.2e^{-2.1t}\sin(28.2t)
\end{cases}
\tag{7.116}
$$

Figure 7.32 contains the plots of these voltages in terms of time. The initial and final values of the voltages v_a and v_b are calculated as

$$\begin{cases} v_a(0) = \lim_{s \to \infty} sV_a(s) = \dfrac{R_1 R_2 i}{R_1 + R_2}; \ v_a(\infty) = \lim_{s \to 0} sV_a(s) = \dfrac{R_1 R_2 i}{R_1 + R_2} \\ v_b(0) = \lim_{s \to \infty} sV_b(s) = 0; \ v_b(\infty) = \lim_{s \to 0} sV_b(s) = 0 \end{cases} \quad (7.117)$$

and their values are $v_a(0) = v_a(\infty) = 4.63$ V; $v_b(0) = v_b(\infty) = 0$ V, values confirmed on the plots of Figure 7.32. ■

MATLAB® Approach

By using the $\texttt{tf}$ function, it is possible to model and determine the time response of MIMO systems with MATLAB®. Let us consider the following example.

Example 7.18

Determine the system response produced by the input $u(t) = \{8 \sin(5t), 10/(t + 1)\}^t$ and the transfer function matrix

$$[G(s)] = \begin{bmatrix} \dfrac{2}{s^2 + 2s + 100} & \dfrac{4s + 20}{s^2 + 2s + 100} \\ \dfrac{-1}{s^2 + 2s + 100} & -\dfrac{2s}{s^2 + 2s + 100} \end{bmatrix}.$$

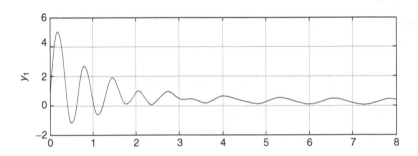

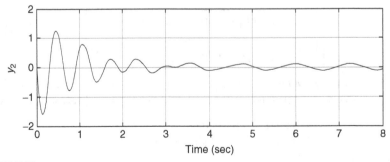

FIGURE 7.33

Response Curves for a Two-Input, Two-Output Dynamic System.

Solution

The transfer function matrix is of a 2 × 2 dimension, and because there are two input components, there also are two output components. Each element of the transfer function matrix is defined separately as g11, g12 (the elements of the first row) and g21, g22 (the elements of the second row). The first output, denoted by y1, is the sum of the response corresponding to the first input (denoted by y11) and the response corresponding to the second input (denoted by y12). A similar definition is given to the second output. The following MATLAB® sequence

```
>> g11 = tf([2],[1,2,100]);
>> g12 = tf([4,20],[1,2,100]);
>> g21 = tf([-1],[1,2,100]);
>> g22 = tf([-2,0],[1,2,100]);
>> t = 0:0.01:10;
>> u1 = 8*sin(5*t);
>> u2 = 10./(t+1);
>> lsim(g11,u1,t);
>> lsim(g12,u1,t);
>> lsim(g21,u2,t);
>> lsim(g22,u2,t);
>> [y11,t] = lsim(g11,u1,t);
>> [y12,t] = lsim(g12,u2,t);
>> [y21,t] = lsim(g21,u1,t);
>> [y22,t] = lsim(g22,u2,t);
>> y1 = y11+y12;
>> y2 = y21+y22;
>> subplot(2,1,1);
>> plot(t,y1), ylabel('y_1'), grid on, title('First Output')
>> subplot(2,1,2);
>> plot(t,y2), xlabel('Time (s)'), ylabel('y_2'), grid on,…
title('Second Output')
```

produces the plots of Figure 7.33.

7.4 USING SIMULINK® TO TRANSFER FUNCTION MODELING

The following examples illustrate the Simulink® environment application to create and simulate dynamic problems by means of transfer functions.

Example 7.19

Consider a series RLC electrical circuit, with $R = 3000$ Ω, $L = 1$ H, and $C = 0.002$ F, connected to a voltage source with $v = 80 \sin(10t)$ V. Considering that the resistor R has a nonlinearity of the saturation type with a saturation voltage of 50 V (both negative and positive), plot the resulting currents for the case where the resistance is purely linear (no saturation) and for the actual resistance (with saturation).

Solution

With the numerical data of this example, the transfer function that connects the input voltage to the output current

$$G(s) = \frac{I(s)}{V(s)} = \frac{Cs}{LCs^2 + RCs + 1} = \frac{0.002s}{0.002s^2 + 6s + 1} \tag{7.118}$$

A saturation-type nonlinearity takes into account that on a resistor the voltage-current relationship is linear (proportional) only up to the saturation voltage value; past that point, any increase in the input will keep the output constant. Figure 7.34 shows the blocks that are needed in this Simulink® model and their connections. Simulink® has the option of using saturation through the Saturation block under the Discontinuities category of the library—by double clicking it, its property window opens where you need to type 50 under the Upper limit and −50 under the Lower limit to indicate the saturation voltage values. The Sine Wave input was dragged from the Sources library; an Amplitude of 80 and a Frequency (rad/sec) of 10 need to be specified in its parameter window. The Transfer Fcn block is taken from the Continuous library and the numerator and denominator coefficients of the transfer functions have to be specified in the Function Block Parameters as:

```
Numerator coefficient:
[0.002,0]
Denominator coefficient:
[0.002,6,1]
```

A simulation time of 10 seconds is selected in the Stop time of the Configuration Parameters and then the model is run—the simulation result is shown in Figure 7.35, and it can be seen that the saturation has the effect of chopping the current peaks for both positive and negative values.

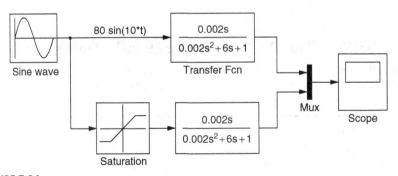

FIGURE 7.34

Simulink® Transfer Function Model of an Electrical System with Resistor Voltage Saturation.

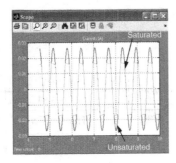

FIGURE 7.35

Simulink® Plot of the Saturated and Unsaturated Currents as Time Functions.

Example 7.20

A MIMO system is defined by the transfer function matrix

$$[G(s)] = \begin{bmatrix} \dfrac{1}{2s^2 + s + 1} & \dfrac{2}{2s^2 + s + 1} & \dfrac{1}{2s^2 + s + 1} \\ 0 & \dfrac{1}{2s^2 + s + 1} & \dfrac{3}{2s^2 + s + 1} \end{bmatrix}$$

Use Simulink® to determine and plot the time response of this system under the action of the input $\{u(t)\} = \{2,\ 1,\ sin(t)\}^t$.

Solution

The number of output components is two, as indicated by the number of rows of the transfer function, which connects the input to the output Laplace-transformed vectors according to $\{Y(s)\} = [G(s)]\{U(s)\}$. Simulink® has the capability of modeling MIMO systems and transfer function matrices by means of the same Transfer Fcn command introduced in the previous example. Specifically, the following specifications are needed in the Function Block Parameters:

```
Numerator coefficient:
[1,2,1;0,1,3]
Denominator coefficient:
[2,1,1]
```

Notice that while the numerator contains the numerator coefficients of the transfer function (and can be a matrix), the denominator can only be a vector collecting the coefficients of the denominator polynomial.

Because the transfer function block accepts formally one input and generates one output, the Merge block (taken from the Signal Routing library and shown in the Simulink® diagram of Figure 7.36) has to be used to model the three input components.

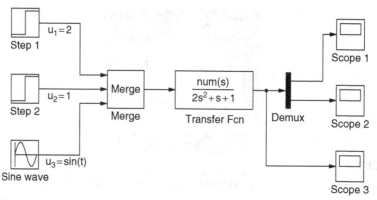

FIGURE 7.36

Simulink® Transfer Function Matrix Model of a MIMO System with Three Inputs and Two Outputs.

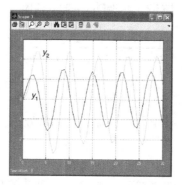

FIGURE 7.37

Simulink® Plot of the Two Outputs as Functions of Time.

This block, as the name suggests, merges several input signals into a single output signal; in our case three inputs need to be specified under `Parameters` in the `Function Block: Merge`. The two resulting time-domain outputs, y_1 and y_2, can be plotted separately through `Scope 1` and `Scope 2` using a Demux block (taken from the `Signal Routing` library). Alternately, the two output signals can be superimposed on a single plot (produced by the `Scope 3`) and this result is shown in Figure 7.37. ■

SUMMARY

This chapter introduces the transfer function, which enables connecting the input to the output of a dynamic system into the Laplace domain for single-input, single-output and multiple-input, multiple-output systems. It is shown how to derive the transfer function for SISO systems and the transfer function matrix for MIMO systems by starting from the system time-domain mathematical model or using complex impedances. The transfer function is further utilized to determine the free and forced time responses with nonzero initial conditions of mechanical, electrical, fluid, and thermal systems Examples that apply built-in MATLAB® commands and Simulink® to model and solve dynamic systems problems by means of the transfer function also are studied. The concept of transfer function is used in subsequent chapters to study the state space approach (Chapter 8) and the frequency-domain analysis (Chapter 9).

PROBLEMS

7.1   Evaluate the input function $u(t)$ of the following differential equation whose output is $y(t)$ and calculate the corresponding transfer functions:

(a) $\ddot{y}(t) + 0.1\dot{y}(t) + 75y(t) = \sin(10t) + \cos(10t)$

(b) $a_3\dddot{y}(t) + a_2\ddot{y}(t) + a_1\dot{y}(t) = e^{-2t} - te^{-2t}$

7.2   A two-input, two-output dynamic system is defined by the following differential equations system:

$$\begin{cases} \dddot{y}_1(t) + 2\ddot{y}_1(t) + 10\dot{y}_1(t) - 5\dot{y}_2(t) + 80y_1(t) - 60y_2(t) = u_1(t) \\ \ddot{y}_2(t) + 5\dot{y}_2(t) - 5\dot{y}_1(t) + 60y_2(t) - 60y_1(t) = u_2(t) \end{cases}.$$

Determine its transfer function matrix $[G(s)]$.

7.3   Determine the transfer function $\Theta_2(s)/M_a(s)$ for the shaft-gear mechanical system of Figure 7.38 using the mathematical model of this system assuming all physical parameters are known.

7.4   Apply the node analysis method in the time domain to determine the transfer function matrix between the input current and the relevant output voltages for the electrical circuit of Figure 7.39.

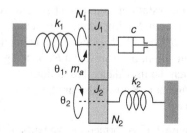

FIGURE 7.38

Rotary Mechanical Shaft-Gear System.

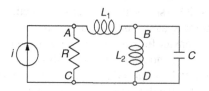

FIGURE 7.39

Electrical Circuit with Current Source.

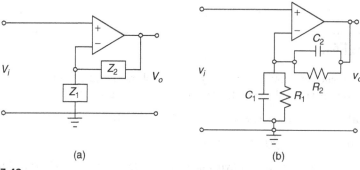

(a)   (b)

FIGURE 7.40

Noninverting Operational Amplifier: (a) Generic Impedance Model; (b) Actual Circuit with Resistors and Capacitors.

7.5 Using the complex impedance approach, find the transfer function $V_o(s)/V_i(s)$ of the operational amplifier system shown in Figure 7.40(a) and demonstrate that the system is a noninverting amplifier. Apply the result to calculate the transfer function of the system of Figure 7.40(b).

7.6 A thermometer defined by a thermal capacitance C_{th} and a thermal resistance R_{th} is placed in a bath whose temperature is $\theta_b(t)$. Use complex impedances to determine the transfer function $\Theta(s)/\Theta_b(s)$ where $\Theta(s)$ and $\Theta_b(s)$ are Laplace transforms of the thermometer and bath temperatures. Draw the impedance-based thermal system that generates the transfer function.

7.7 The tank-pipe liquid system of Figure 7.41 is defined by a resistance R_l and a capacitance C_l. Use complex impedances to determine the transfer functions $P_o(s)/P_i(s)$ and $H(s)/P_i(s)$ with $P_o(s)$ and $P_i(s)$, being the Laplace transforms of the output (tank) and input pressures, and $H(s)$, being the Laplace transform of the head $h(t)$. Sketch the impedance-based liquid system corresponding to the pressure transfer function. Known also are the gravitational acceleration g and liquid mass density ρ.

7.8 The MEMS of Figure 7.42 is formed of two pairs of beams and a shuttle mass that are placed on a massless platform. The platform is subjected to input displacement u and the shuttle mass displacement is measured capacitively. Use a

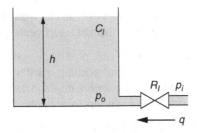

FIGURE 7.41

Liquid System with Tank and Pipe.

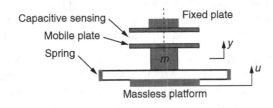

FIGURE 7.42

Physical Model of MEMS with Displacement Input for Translatory Motion.

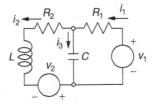

FIGURE 7.43

Two-Mesh Electrical System with Voltage Sources.

lumped-parameter model with known mass m, stiffness k, and damping coefficient c to formulate the mathematical model of the shuttle mass motion y and determine the transfer function $Y(s)/U(s)$. Also use complex impedances and a mechanical impedance circuit to calculate the system's transfer function.

7.9 Use the complex impedance approach to formulate the transfer function matrix for the electrical circuit of Figure 7.43, considering that the input components are the voltages v_1 and v_2 and the output is formed of the meaningful currents. Known are all the electrical components' parameters.

7.10 Use the complex impedance approach and an appropriate thermal circuit to determine the transfer function matrix of the two-room thermal system of Example 5.15 and sketched in Figure 5.23, considering that there is no cooler ($q_i = 0$). The output is formed by the two room temperatures θ_1 and θ_2, whereas the input is the outdoors temperature θ_o. Known are all the thermal components' parameters.

7.11 What type of system is the pump-tank-pipe liquid system of Figure 7.44 in terms of input and output? Find the corresponding transfer function (or transfer function matrix) of this system using complex impedances and a corresponding liquid circuit by using the system's parameters from Figure 7.44.

7.12 Consider the torsional mechanical microsystem shown in Figure 7.45, formed of three identical flexible bars and two identical rigid plates. Derive the time-domain mathematical model of this system and determine the transfer function matrix corresponding to the input m_t and the outputs θ_1 and θ_2. Confirm the obtained result by deriving the impedance-based transfer function matrix and the corresponding mechanical impedance circuit. Consider that known are the bars' torsional stiffness k and the plates mass moment of inertia J.

7.13 A dynamic system has 0 as its zero; −2 (order two of multiplicity), −3 as poles, and a gain of 2. Use MATLAB® to calculate the corresponding transfer function of this system then reconvert the obtained transfer function model into a zero-pole-gain model.

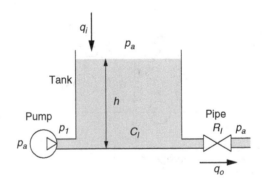

FIGURE 7.44

Liquid System with Pump, Tank, and Pipe.

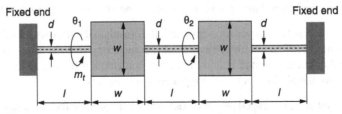

FIGURE 7.45

Torsional Mechanical Microsystem.

7.14 It was determined that the poles of a two-DOF system are -5 and -4. The zeroes and gains obtained when applying individual inputs u_1 and u_2 in correspondence with the individual outputs y_1 and y_2 (the two DOFs) follow:

	y_1	y_2
u_1	gain: 1, zeroes: none	gain: 2, zeroes: none
u_2	gain: 1, zeroes: $-1, -2$	gain: 3, zeroes: -2 (double)

Use one-line MATLAB® code to determine the corresponding transfer function matrix of this system.

7.15 A pneumatic system defined by a resistance R_g and capacitance C_g is subjected to a ramp input pressure $p_i(t) = P_i t$, where P_i is a constant. After a time $t_1 = 3$ s, it is determined that the output pressure is twice the input pressure constant P_i. Use the transfer function approach to determine the system's time constant. For a pressure $p_i = 50$ atm and zero initial conditions, plot the output pressure as a function of time.

7.16 Consider the heat transfer between the conditioned space of a room with four walls and the outdoors.

(a) Find the transfer function corresponding to the indoor and outdoor temperatures θ_1 and θ_2 assuming that $\theta_2 > \theta_1$.

(b) Determine θ_1 and plot it as a function of time when $\theta_2 = 40°C$. Known also are the following data for the wall: height $h = 4$ m, width $w = 10$ m, length $l = 0.2$ m, thermal conductivity $k = 0.04$ W/m-C; and for the air in the room space, mass density $\rho = 1.1$ kg/m³, constant-pressure specific heat $c_p = 1100$ J/kg-C and $\theta_1(0) = 20°C$.

7.17 Consider the mechanical system shown in Figure 7.46 with $c_1 = 0.5$ N-s/m, $c_2 = 0.8$ N-s/m, $c_3 = 0.3$ N-s/m, $k_1 = 100$ N/m, $k_2 = 120$ N/m, $k_3 = 150$ N/m. Use the complex impedance approach to find the equivalent complex impedance and the transfer function of this system. Plot the chain total displacement when a force $f = \sin(10t)$ N is applied at the free end.

7.18 A rotor system such as the one shown in Figure 7.47 consists of a disk with the mass moment of inertia $J = 0.1$ kg-m², a damper with viscous

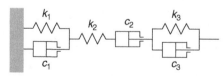

FIGURE 7.46

Spring-Damper Mechanical System.

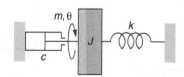

FIGURE 7.47

Elastic Shaft with Inertia and Damping.

damping coefficient $c = 6$ N-m-s/rad, and a rotary spring of unknown stiffness k. When applying a step input $m = 20$ N-m to the disk, it is determined experimentally that the steady-state response is $\theta_o(\infty) = 4°$. Evaluate the spring constant using the transfer function approach and plot the system response for a rotor initial velocity $\omega(0) = 40$ rad/s and the unit step input of MATLAB®.

7.19 Use the complex impedance approach to find the transfer function of the operational amplifier circuit of Figure 7.48 and use it to plot the output voltage v_o as a function of time for the following numerical values: $v_i = 30$ V, $R = 1$ kΩ, $C = 50$ mF, and $L = 200$ mF. Also calculate the initial and final values of the output voltage.

7.20 The electrical circuit of Figure 7.49 is designed for a variable capacity MEMS device; it contains a resistor with $R = 1000$ Ω, a voltage source with $v = 1$ V, and a transverse-motion variable capacitor with an initial gap $g_0 = 10$ μm, plate area $A = 10^{-7}$ m^2, and dielectric permittivity $\varepsilon_o = 8.8 \times 10^{-12}$ C^2-N^{-1}-m^{-2}. Use complex impedances and the transfer function approach to calculate the displacement of the mobile plate after a very short amount of time $t_1 = 10^{-9}$ s from the initial moment; at that time, the measured current in the circuit is $i_1 = 10$ μA.

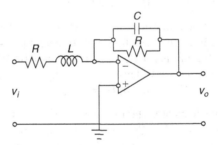

FIGURE 7.48

Operational Amplifier Circuit.

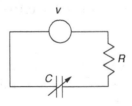

FIGURE 7.49

Electrical Circuit with Variable-Gap Capacitor and Bias Voltage Used as a Displacement Sensor.

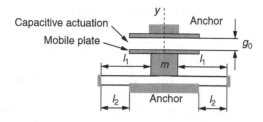

FIGURE 7.50

Physical Model of Spring-Mass Mechanical Microsystem with Capacitive Actuation for Translatory Motion.

7.21 The MEMS of Figure 7.50 is formed of two pairs of beams and a shuttle mass and is subjected to capacitive actuation. Use a lumped-parameter model to formulate the system's mathematical model for the motion about the y direction by also taking into account the damping between the fixed and mobile capacitor plates. All beams have a circular cross-section of diameter $d = 2$ μm. Known also are the shuttle mass $m = 9 \times 10^{-11}$ kg, $l_1 = 200$ μm, $l_2 = 100$ μm, Young's modulus $E = 165$ GPa, damping coefficient $c = 0.5$ N-s/m. Plot the system's response using its transfer function and MATLAB®'s impulse function when an impulsive electrostatic force $f = 1 \times 10^{-6} \delta(t)$ N is applied by the capacitive superimposed to an initial displacement $y(0) = 1.2$ μm. Ignore the mass contribution from the springs and verify whether the initial capacitive gap $g_0 = 8$ μm is sufficient for the mobile plate motion.

7.22 By using the transfer function $\Theta_1(s)/M_a(s)$, determined in Problem 7.3 for the rotary shaft-gear mechanical system shown in Figure 7.38, calculate and plot $\theta_1(t)$ using the step input command of MATLAB®. Known are $N_1 = 48$, $N_2 = 36$, $J_1 = 0.002$ kg-m², $J_2 = 0.0016$ kg-m², $c = 80$ N-m-s, $k_1 = 170$ N-m, $k_2 = 110$ N-m, $m_a = 15$ N-m, and $\theta_2(0) = 3°$.

7.23 The liquid system of Problem 7.7 and shown in Figure 7.41 carries water with $\rho = 1000$ kg/m³ and $\mu = 0.001$ N-s/m². The pipe diameter is $d_i = 0.07$ m and its total length is $l = 20$ m; the tank has a diameter $d = 3$m and an initial liquid head $h(0) = 1$ m. The input pressure is a function of time, $p_i = (1 + e^{-10t}) P$, where $P = 3 \times 10^5$ N/m². Use MATLAB® and its lsim input command to plot the tank pressure p_o in terms of time. Also calculate the steady-state value of the tank pressure.

7.24 Utilize the transfer function matrix connecting between the input current and the relevant output voltage of the electrical system of Problem 7.4 and shown in Figure 7.39 to plot the currents through the four electrical components by using MATLAB®. Also calculate the steady-state current values. Known are $R = 500$ Ω, $C = 200$ μF, $L_1 = 2$ H, $L_2 = 3$ H, and $i = 10$ mA.

7.25 Consider the following transfer function matrix:

$$[G(s)] = \begin{bmatrix} \dfrac{3s+2}{s^2+4s+25} & \dfrac{2}{s^2+4s+25} \\ \dfrac{1}{s^2+4s+25} & \dfrac{s+3}{s^2+4s+25} \end{bmatrix}.$$

Determine the system response under the following inputs: $u_1(t) = 1$ and $u_2(t) = e^{-t}$ using MATLAB® and its impulse command.

7.26 For the mechanical system of Figure 7.51, calculate the components of a forcing vector that will render this free system under the initial conditions $\{\dot{y}(0)\} = \{-0.2,\ 0.1\}'$m/s; $\{y(0)\} = \{0.03,\ 0.02\}'$m into a forced system with zero initial conditions; plot $y_1(t)$ and $y_2(t)$ for $m = 1$ kg, $c = 2$ N-s/m, and $k = 40$ N/m. Also determine the final values of the two displacements.

7.27 Use complex impedances and the transfer function approach for the circuit of Figure 7.52 to determine and plot the output voltage for $R_1 = 500\ \Omega$, $R_2 = 600\ \Omega$, $L = 400$ mH, $C = 0.004$ F, $v_1 = 50$ V, and $v_2 = 40$ V.

7.28 Plot the output voltage of the operational-amplifier circuit of Problem 7.5 and shown in Figure 7.40(b) for $C_1 = 3$ mF, $C_2 = 5$ mF, $R_1 = 100\ \Omega$, $R_2 = 85\ \Omega$, and $v_i = 110$ V using Simulink®. The resistors are nonlinear with their saturation limits being $+70$ V and -50 V.

7.29 Utilize the transfer function approach and Simulink® to plot the output flow rate of the liquid system shown in Figure 7.44 of Problem 7.11. Known are

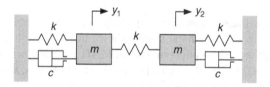

FIGURE 7.51

Free Mechanical System under Nonzero Initial Conditions.

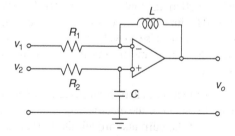

FIGURE 7.52

Operational Amplifier Circuit with Two Input Voltages.

$p_a = 10^5$ N/m², $p_1 = 2 \times 10^5$ N/m², $q_i = 0.02$ m³/s, $d = 1$ m (diameter of the tank), $\rho = 1000$ kg/m³, $d_i = 0.02$ m (diameter of the pipe), $l = 10$ m (length of pipe), $\mu = 0.001$ N-s/m².

7.30 Use Simulink® to plot the angles $\theta_1(t)$ and $\theta_2(t)$ of the torsional MEMS shown in Fig. 7.45 of Problem 7.12. The flexible connectors are of circular cross-section with a diameter $d = 200$ nm, a length $l = 10$ μm, and a shear modulus $G = 0.8 \times 10^{10}$ N/m². The rigid plates have a width $w = 100$ μm and thickness $t = 5$ μm. The material density is $\rho = 3600$ kg/m³. The input torque is $m_t = 10$ μN-nm, and the initial conditions are zero.

Suggested Reading

H. Klee, *Simulation of Dynamic Systems with MATLAB® and Simulink®*, CRC Press, Boca Raton, FL, 2007.

N. S. Nise, *Control System Engineering*, 5th Ed. John Wiley & Sons, New York, 2008.

D. G. Alciatore and M. B. Histand, *Introduction to Mechatronics and Measurement Systems*, 3rd Ed. McGraw-Hill, New York, 2007.

S. B. Hammond and D. K. Gehmlich, *Electrical Engineering*, McGraw-Hill, New York, 1970.

W. T. Thomson, *Laplace Transformation*, 2nd Ed. New York, Prentice-Hall, 1960.

I. Cochin, *Analysis and Design of Dynamic Systems*, Harper & Row Publishers, New York, 1980.

K. Ogata, *System Dynamics*, 4th Ed. Prentice Hall, Upper Saddle River, 2004.

W. J. Palm III, *System Dynamics*, 2nd Ed. McGraw-Hill, New York, 2009.

J. F. Lindsay and S. Katz, *Dynamics of Physical Circuits and Systems*, Matrix Publishers, Champaign, IL, 1978.

State Space Approach

OBJECTIVES

This chapter focuses on the state space approach to modeling dynamic systems in the time domain and determining the corresponding solution. The subject matter of the chapter is related to the modeling of Chapters 2, 3, 4, and 5, as well as with the Laplace transform of Chapter 6 and the transfer function of Chapter 7. The following topics are studied:

- The state space concept as a procedure for modeling the time-domain dynamics of SISO and MIMO systems and for obtaining solutions to the corresponding models.

- Deriving state space models directly in the time domain or via transfer functions in the Laplace domain in cases where the input includes or has no time derivatives in it.

- Analytical and MATLAB® methods of converting between transfer function or zero-pole-gain models and state space models.

- Nonlinear state space models and procedures for linearizing such models.

- The use of state space models to solve for the free response with nonzero initial conditions and for the forced response of dynamic systems.

- Specialized MATLAB® commands designed to model and solve state space problems.

- Application of Simulink® to graphically model and solve system dynamics problems by state space modeling.

INTRODUCTION

This chapter introduces the state space modeling method for SISO and MIMO dynamic systems into the time domain. The approach, which is a vector-matrix one, has the main advantage of being able to model systems with a large number of degrees of freedom as well as systems with nonlinearities. Similar to the transfer function approach, which was the subject of Chapter 7, the material presented here focuses on deriving state space models of dynamic systems and solving these models

DOI: 10.1016/B978-0-240-81128-4.00008-8

to determine the time response using analytical methods and MATLAB® custom commands. While the transfer function model belongs to the Laplace domain, a state space model operates in the time domain. The state space approach utilizes the same vector-matrix model for both SISO and MIMO dynamic systems. Converting between state space and transfer function or zero-pole-gain models is illustrated, so that a convenient modeling approach can be used for a specific problem. The capabilities of Simulink® to solve state space formulations, such as for systems with nonlinearities, are exemplified.

8.1 THE CONCEPT AND MODEL OF THE STATE SPACE APPROACH

The state space approach employs a unique vector-matrix formulation to model a dynamic system to determine the forced response of generally multiple-input, multiple-output (MIMO) systems, but single-input, single-output (SISO) systems can also be modeled using the same formulation and approach.

Mathematical models of MIMO systems having a large number of outputs (or DOFs) result in an equally large number of differential equations that need to be solved for these unknown outputs. Solutions can be determined by means of the Laplace transforms (as seen in Chapter 6) or using the transfer function matrix approach (as covered in Chapter 7). The question arises whether it is possible to employ a single mathematical model to capture the wide variety of particular dynamic systems in terms of number of inputs, number of outputs, and order of the differential equations. Consider, for instance, a MIMO dynamic system whose vector-matrix mathematical model is

$$[a_q]\{\overset{(q)}{y}(t)\} + [a_{q-1}]\{\overset{(q-1)}{y}(t)\} + \cdots + [a_1]\{\dot{y}(t)\} + [a_0]\{y(t)\} = [b]\{u(t)\} \qquad (8.1)$$

where $\{y(t)\}$ is the unknown output (response) vector and $\{u(t)\}$ is the specified input (forcing) vector—there might also be time derivatives of $\{u(t)\}$ on the right-hand side of Eq. (8.1). The maximum order of the differential equations in Eq. (8.1) is q, where

$$\{\overset{(q)}{y}(t)\}$$

indicates the q-order time derivative of $\{y(t)\}$.

The state space modeling approach utilizes the following variable transformation:

$$\{y(t)\} = [C]\{x(t)\} + [D]\{u(t)\} \qquad (8.2)$$

This uses the new unknown $\{x(t)\}$ instead of the original one $\{y(t)\}$ with the purpose of obtaining the following first-order differential equation instead of the original Eq. (8.1):

$$\{\dot{x}(t)\} - [A]\{x(t)\} = [B]\{u(t)\} \qquad (8.3)$$

which can also be written as

$$\{\dot{x}(t)\} = [A]\{x(t)\} + [B]\{u(t)\} \tag{8.4}$$

Equation (8.4) is known as the *state equation*; and by solving it, the *state vector* $\{x(t)\}$ is determined. The output vector $\{y(t)\}$ is subsequently calculated from the *output equation* (8.2). The *standard state space model* consists of the state equation, Eq. (8.4), and of the output equation, Eq. (8.2). The four matrices contain parameters defining the dynamic system being studied; and they are $[A]$, the *state matrix*, $[B]$, the *input matrix*, $[C]$, the *output matrix*, and $[D]$, the *direct transmission matrix*. Proper selection of the *state variables* (which are collected in *state vectors*, which, at their turn, form a *state space*) ensures that Eqs. (8.1) can always be applied, irrespective of the number of DOFs and of the system order. This topic is analyzed in the next section of this chapter.

Another form of the state space model, which is implemented in MATLAB®, generates the *descriptor state-space model*, whose state equation is

$$[E]\{\dot{x}(t)\} = [A']\{x(t)\} + [B']\{u(t)\} \tag{8.5}$$

where $[E]$, the *descriptor matrix*, is a square, nonsingular matrix, and $[A']$ and $[B']$ differ from the original $[A]$ and $[B]$, respectively.

Equations (8.1) and (8.2) define the forced response of a dynamic system by means of the state space model. When no input is acting on the dynamic system, the state space model simplifies to the one characterizing the free response in the homogeneous form:

$$\begin{aligned}\{\dot{x}(t)\} &= [A]\{x(t)\} \\ \{y(t)\} &= [C]\{x(t)\}\end{aligned} \tag{8.6}$$

or, in descriptor form, the state Eq. (8.5) is

$$[E]\{\dot{x}(t)\} = [A']\{x(t)\} \tag{8.7}$$

As Eqs. (8.2) and (8.4) suggest, the input and output vectors need not have the same dimensions. Assuming the input vector has the dimension m and the output vector has the dimension p, it can be determined what the dimensions of the other matrices are, based also on the dimension of the state vector, which is n. Equation (8.4) shows that $[A]$ has to be square with n rows and n columns; in other words $[A]$ is of $n \times n$ dimension. Similarly, the matrix $[B]$ is of $n \times m$ dimension, as it results from the same Eq. (8.4). From Eq. (8.2) it follows that $[C]$ has $p \times n$ dimension, whereas the matrix $[D]$ is of $p \times m$ dimension. A visual representation of the four matrices' dimensions is given in Figure 8.1.

Let us consider a few examples illustrating how to derive the state space models of dynamic systems from known mathematical models.

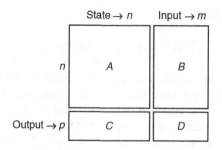

State → n   Input → m

n   A   B

Output → p   C   D

FIGURE 8.1

Dimension Connections of State Space Matrices.

Example 8.1

Derive a state-space model of the single-mesh electrical circuit sketched in Figure 8.2. Utilize the model to find the charge variation as a function of time for $R = 110\ \Omega$, $L = 3H$, $C = 1$ mF, $v = 60$ V, and zero initial conditions.

Solution

As shown in Chapter 4, the following differential equation is the mathematical model of the electrical system sketched in Figure 8.2:

$$L\ddot{q} + R\dot{q} + \frac{1}{C}q = v \tag{8.8}$$

where L is the inductance, R is the resistance, C is the capacitance, q is the charge, and v is the source voltage. Let us drop the time variable and select the variables x_1 and x_2 as

$$\begin{cases} x_1 = q \\ x_2 = \dot{q} \end{cases} \tag{8.9}$$

to be the state variables of the state vector

$$\{x\} = \{x_1 \ x_2\}^t \tag{8.10}$$

In Eqs. (8.8), (8.9), and (8.10), the variable t (time) has been dropped to simplify notation but all variables in this example (q, v, x_1, and x_2) are functions of time. The two Eqs. (8.9) indicate that

$$\dot{x}_1 = x_2 \tag{8.11}$$

On the other hand, Eq. (8.8) can be written in the form

$$\ddot{q} = -\frac{1}{LC}q - \frac{R}{L}\dot{q} + \frac{1}{L}v \tag{8.12}$$

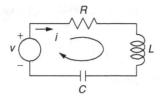

FIGURE 8.2

Single-Mesh Electrical System.

which can be formulated using the state variables as

$$\dot{x}_2 = -\frac{1}{LC}x_1 - \frac{R}{L}x_2 + \frac{1}{L}v \tag{8.13}$$

by virtue of Eqs. (8.9). The source voltage is the input to the system and therefore can be denoted by u, so $v = u$. Equations (8.11) and (8.13) are collected in vector-matrix form as

$$\begin{Bmatrix} \dot{x}_1 \\ \dot{x}_2 \end{Bmatrix} = \begin{bmatrix} 0 & 1 \\ -\dfrac{1}{LC} & -\dfrac{R}{L} \end{bmatrix} \begin{Bmatrix} x_1 \\ x_2 \end{Bmatrix} + \begin{Bmatrix} 0 \\ \dfrac{1}{L} \end{Bmatrix} u \tag{8.14}$$

Comparison of Eq. (8.14) to Eq. (8.4) indicates that Eq. (8.14) is the state equation with the state and input matrices being

$$[A] = \begin{bmatrix} 0 & 1 \\ -\dfrac{1}{LC} & -\dfrac{R}{L} \end{bmatrix}; [B] = \begin{Bmatrix} 0 \\ \dfrac{1}{L} \end{Bmatrix} \tag{8.15}$$

If the charge is the output, which can be denoted formally as $y = q$, then the first Eq. (8.9) becomes

$$y = x_1 \tag{8.16}$$

Equation (8.16) can be written in vector form as

$$y = \{1 \; 0\} \begin{Bmatrix} x_1 \\ x_2 \end{Bmatrix} + 0 \times u \tag{8.17}$$

which is of the form shown in Eq. (8.7). Comparison of this equation with Eq. (8.2) indicates that, indeed, Eq. (8.17) represents the output equation, and its defining matrices are

$$[C] = \{1 \; 0\}; [D] = 0 \tag{8.18}$$

The state Eq. (8.14) and the output Eq. (8.17) form the state space model of the electrical system of Figure 8.2.

The state equation can also be rendered in descriptor format by rewriting Eq. (8.8) as

$$L\ddot{q} = -\frac{1}{C}q - R\dot{q} + v \tag{8.19}$$

the case where Eq. (8.14) changes to

$$\begin{bmatrix} 1 & 0 \\ 0 & L \end{bmatrix}\begin{Bmatrix} \dot{x}_1 \\ \dot{x}_2 \end{Bmatrix} = \begin{bmatrix} 0 & 1 \\ -\dfrac{1}{C} & -R \end{bmatrix}\begin{Bmatrix} x_1 \\ x_2 \end{Bmatrix} + \begin{Bmatrix} 0 \\ 1 \end{Bmatrix}u \tag{8.20}$$

therefore,

$$[E] = \begin{bmatrix} 1 & 0 \\ 0 & L \end{bmatrix}; [A'] = \begin{bmatrix} 0 & 1 \\ -\dfrac{1}{C} & -R \end{bmatrix}; [B'] = \begin{Bmatrix} 0 \\ 1 \end{Bmatrix} \tag{8.21}$$

while the output equation and its matrices remain unchanged.

Note: In Chapter 4, the *potential electrical energy* stored by a capacitor was shown to be equal to $\frac{1}{2}(q^2/C)$, whereas the *kinetic electrical energy* related to an inductor was $\frac{1}{2}L\dot{q}^2$. Both energies describe a given system state, defined as a function of charge (through the potential energy) and of charge derivative, or current (through the kinetic energy). The particular choice of the state variables as the charge and the charge time derivative, Eq. (8.9), indicates the connection between parameters with physical significance (the energy terms) and state variables. This observation is also valid for other systems, for instance, mechanical ones, where the potential energy depends on displacement and the kinetic energy on velocity (the displacement's first derivative); for a second-order differential equation representing the mathematical model of a mechanical system, a common choice of state variables consists in displacement and velocity.

The state Eq. (8.4) is solved for $\{x(t)\}$ then the output Eq. (8.2) is solved for $\{y(t)\}$ by first calculating $\{X(s)\}$ and $\{Y(s)\}$. Laplace transforming Eqs. (8.4) and (8.2) for the particular initial conditions of this example and with zero initial conditions results in

$$\begin{cases} s\{X(s)\} = [A]\{X(s)\} + [B]\{U(s)\} \\ \{Y(s)\} = [C]\{X(s)\} + [D]\{U(s)\} \end{cases} \tag{8.22}$$

where $\{X(s)\}$, $\{U(s)\}$, and $\{Y(s)\}$ are the Laplace transforms of $\{x(t)\}$, $\{u(t)\}$, and $\{y(t)\}$, respectively. The first Eq. (8.22) yields $\{X(s)\}$ as

$$\{X(s)\} = (s[I] - [A])^{-1}[B]\{U(s)\} \tag{8.23}$$

which, substituted into the second Eq. (8.22), gives

$$\{Y(s)\} = [G(s)]\{U(s)\} \tag{8.24}$$

where

$$[G(s)] = [C](s[I] - [A])^{-1}[B] + [D] \tag{8.25}$$

is the *transfer function matrix*, introduced in Chapter 7. In this example, the dimension of $[G(s)]$ is 1×1; therefore, we actually have a scalar $G(s)$ transfer function connecting the Laplace transforms of the input voltage $v(t)$ and the charge $q(t)$. Using the numerical data of this example and MATLAB®, the transfer function is found to be

$$[G(s)] = \frac{1}{Ls^2 + Rs + 1/C} = \frac{1}{3s^2 + 110s + 1000} \tag{8.26}$$

The Laplace-transformed output variable is

$$Q(s) = Y(s) = G(s)U(s) = G(s)V(s) = \frac{60}{s(3s^2 + 110s + 1000)} \tag{8.27}$$

whose time-domain counterpart (charge) is

$$q(t) = 0.06 + 0.3e^{-20t} - 0.36e^{-\frac{50}{3}t} \tag{8.28}$$

■

Unlike the transfer function model, a state space model is not unique to any given dynamic system—this feature is demonstrated in the companion website Chapter 8—as illustrated in the following example.

■───

Example 8.2
Find another state space model for the electrical circuit of Figure 8.2 using the current i and the voltage across the capacitor, v_c, as state variables.

Solution
For the electrical system of Figure 8.2, Kirchhoff's voltage law can be written in the form

$$v_R + v_L + v_C = v \tag{8.29}$$

where v_R, v_L and v_C are the voltages across the resistor, inductor, and capacitor, respectively. To simplify notation, we will discontinue mentioning time as the independent variable in both physical and state space model variables; therefore, x will be used with the understanding that it is actually $x(t)$. By taking into account known current-voltage relationships, Eq. (8.29) can is expressed as

$$Ri + L\frac{di}{dt} + v_C = v \tag{8.30}$$

or as

$$\frac{di}{dt} = -\frac{R}{L}i - \frac{1}{L}v_C + \frac{1}{L}u \tag{8.31}$$

where it has been considered that v is the input to the system; that is, $v = u$. Another linear independent relationship between i and v_C, with v_C appearing as a derivative is

$$\frac{dv_C}{dt} = \frac{1}{C}i \qquad (8.32)$$

If we now select the state variables to be i and v_C, namely,

$$\begin{cases} \bar{x}_1 = i \\ \bar{x}_2 = v_C \end{cases} \qquad (8.33)$$

Eqs. (8.31) and (8.32) can be formulated in vector-matrix form as a state equation:

$$\begin{Bmatrix} \dot{\bar{x}}_1 \\ \dot{\bar{x}}_2 \end{Bmatrix} = \begin{bmatrix} -\dfrac{R}{L} & -\dfrac{1}{L} \\ \dfrac{1}{C} & 0 \end{bmatrix} \begin{Bmatrix} \bar{x}_1 \\ \bar{x}_2 \end{Bmatrix} + \begin{Bmatrix} \dfrac{1}{L} \\ 0 \end{Bmatrix} u \qquad (8.34)$$

where the [A] and [B] matrices are highlighted. By considering the output vector is defined by the components

$$\begin{cases} \bar{y}_1 = i \\ \bar{y}_2 = v_C \end{cases} \qquad (8.35)$$

the output equation can be written as

$$\begin{Bmatrix} \bar{y}_1 \\ \bar{y}_2 \end{Bmatrix} = \begin{bmatrix} 1 & 0 \\ 0 & 1 \end{bmatrix} \begin{Bmatrix} \bar{x}_1 \\ \bar{x}_2 \end{Bmatrix} + \begin{Bmatrix} 0 \\ 0 \end{Bmatrix} u \qquad (8.36)$$

where the matrices [C] and [D] are shown. Let us compare the state vectors of Examples 8.1 and 8.2, which are

$$\begin{aligned} \{x\} &= \{x_1 \ x_2\}' = \{q \ \dot{q}\}'; \\ \{\bar{x}\} &= \{\bar{x}_1 \ \bar{x}_2\}' = \{i \ v_c\}' \end{aligned} \qquad (8.37)$$

By considering known relationships between the electrical variables that enter the two vectors, the following transformation can be written:

$$\begin{Bmatrix} q \\ \dot{q} \end{Bmatrix} = \begin{bmatrix} 0 & C \\ 1 & 0 \end{bmatrix} \begin{Bmatrix} i \\ v_C \end{Bmatrix} \qquad (8.38)$$

which indicates the two state vectors are linearly related.　　　　　　　　■

Similar to the transfer function approach of Chapter 7, the material of this chapter is structured mainly in two segments: One is dedicated to deriving state space models and the other one to calculating the forced response from a state space model. Figure 8.3 depicts the two phases, their components, and the component relationships (to simplify representation, symbols for vectors and matrices have been dropped).

As suggested in Figure 8.3, a state space model can be obtained directly in the time domain from an existing mathematical model (such as the differential equations shown symbolically as $D[y(t), u(t)] = 0$) or by migrating from Laplace-domain

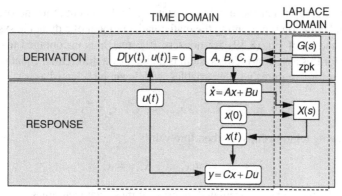

TIME DOMAIN

LAPLACE DOMAIN

DERIVATION $D[y(t), u(t)]=0$ — A, B, C, D ← $G(s)$ / zpk

RESPONSE $u(t)$ $\dot{x}=Ax+Bu$ $x(0)$ → $X(s)$ $x(t)$ ← $y=Cx+Du$

FIGURE 8.3

Derivation of State Space Models and Time-Domain Response Calculation.

transfer function or zero-pole-gain models. Calculating the time response $y(t)$ from a derived state space model is performed by first calculating the Laplace transform of the state vector, followed by inverse Laplace transformation and derivation of the time-domain output $y(t)$. An initial state vector $x(0)$ is also taken into consideration.

8.2 STATE SPACE MODEL FORMULATION

State space models can be derived either analytically or by means of MATLAB®, as discussed in the following sections.

 ### 8.2.1 Analytical Approach

The analytical procedures for state space model derivation presented here apply to systems without input time derivatives, systems with input time derivatives, systems that have already been defined by transfer functions, and nonlinear systems.

Dynamic Systems without Input Time Derivative

When the differential equations that constitute the mathematical model of a dynamic system contain input variables and no input variable time derivatives, the procedure of selecting the state variables is rather straightforward as will be discussed in the following. The cases of SISO and MIMO systems are analyzed separately.

SISO Systems

Single-input, single-output dynamic systems that do not include time derivatives of their input are described by a single differential equation of the form

$$a_n \overset{(n)}{y}(t) + a_{n-1} \overset{(n-1)}{y}(t) + \cdots + a_1 \dot{y}(t) + a_0 y(t) = u(t) \tag{8.39}$$

A rule of thumb is that the number of state variables is equal to the order of the differential equation representing the mathematical model of a dynamic system, provided a_0 and a_n of Eq. (8.39) are nonzero (the reader is encouraged to check the rationale for this assertion). This system is defined by a differential equation of order n and, therefore, requires n state variables, which can be selected as

$$x_1 = y, \ x_2 = \dot{y}, \ ..., \ x_{n-1} = \overset{(n-2)}{y}, \ x_n = \overset{(n-1)}{y} \tag{8.40}$$

The following relationships are therefore valid:

$$\dot{x}_1 = x_2, \ \dot{x}_2 = x_3, \ ..., \ \dot{x}_{n-1} = x_n \tag{8.41}$$

Equation (8.39) can be reformulated in terms of these state variables as

$$\dot{x}_n = -\frac{a_0}{a_n}x_1 - \frac{a_1}{a_n}x_2 - \cdots - \frac{a_{n-2}}{a_n}x_{n-1} - \frac{a_{n-1}}{a_n}x_n + \frac{1}{a_n}u \tag{8.42}$$

Equations (8.41) and (8.42) are collected into the vector-matrix equation

$$\begin{Bmatrix} \dot{x}_1 \\ \dot{x}_2 \\ \cdots \\ \dot{x}_{n-1} \\ \dot{x}_n \end{Bmatrix} = \begin{bmatrix} 0 & 1 & 0 & \cdots & 0 \\ 0 & 0 & 1 & \cdots & 0 \\ \cdots & \cdots & \cdots & \cdots & \cdots \\ 0 & 0 & 0 & \cdots & 1 \\ -\frac{a_0}{a_n} & -\frac{a_1}{a_n} & -\frac{a_2}{a_n} & \cdots & -\frac{a_{n-1}}{a_n} \end{bmatrix} \begin{Bmatrix} x_1 \\ x_2 \\ \cdots \\ x_{n-1} \\ x_n \end{Bmatrix} + \begin{Bmatrix} 0 \\ 0 \\ \cdots \\ 0 \\ \frac{1}{a_n} \end{Bmatrix} u \tag{8.43}$$

which represents the state equation where $[A]$ is the square matrix multiplying the state vector and $[B]$ is actually a column vector multiplying the scalar u. The output equation is determined by taking into account that the differential Eq. (8.34) has one unknown, y, which can be considered the output function. As a consequence,

$$y = \{1 \ \ 0 \ \ ... \ \ 0 \ \ 0\} \{x_1 \ \ x_2 \ \ ... \ \ x_{n-1} \ \ x_n\}^t + 0 \times u \tag{8.44}$$

and therefore $[C]$ is the row vector and $[D]$ is the zero scalar.

Examples 8.1 and 8.2 actually derive state space models of dynamic systems without input time derivatives. Let us analyze another example from the same category.

Example 8.3

The microresonator sketched in Figure 8.4(a) is actuated electrostatically by a comb drive. The shuttle mass is supported by two elastic beams, and there is viscous damping between the shuttle mass and the substrate. Use a lumped-parameter model to determine the mathematical model of this mechanical system and derive a state space model for it. The damping coefficient is c, one beam's spring constant is k, the shuttle mass is m, and the electrostatic actuation force is f.

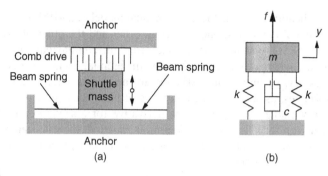

FIGURE 8.4

Microresonator Supported on Beam Springs, with Viscous Damping and Electrostatic Actuation: (a) Physical Model; (b) Lumped-Parameter Model.

Solution

The equation of motion for the lumped-parameter model of Figure 8.4(b) is

$$m\ddot{y} + c\dot{y} + 2ky = f \tag{8.45}$$

which is a second-order differential equation with the following coefficients and forcing: $a_2 = m$, $a_1 = c$, $a_0 = 2k$, and $u = f$. Based on these parameter assignments and according to Eqs. (8.43) and (8.44), the state space matrices are

$$[A] = \begin{bmatrix} 0 & 1 \\ -\dfrac{a_0}{a_2} & -\dfrac{a_1}{a_2} \end{bmatrix}; \ [B] = \begin{Bmatrix} 0 \\ \dfrac{1}{a_2} \end{Bmatrix}; \ [C] = \{1 \ 0\}; \ [D] = 0 \tag{8.46}$$

As a consequence, the generic state space Eq. (8.43) becomes

$$\begin{Bmatrix} \dot{x}_1 \\ \dot{x}_2 \end{Bmatrix} = \begin{bmatrix} 0 & 1 \\ -\dfrac{2k}{m} & -\dfrac{c}{m} \end{bmatrix} \begin{Bmatrix} x_1 \\ x_2 \end{Bmatrix} + \begin{Bmatrix} 0 \\ \dfrac{1}{m} \end{Bmatrix} u \tag{8.47}$$

and the output equation is

$$y = \{1 \ 0\} \begin{Bmatrix} x_1 \\ x_2 \end{Bmatrix} \tag{8.48}$$

Equations (8.47) and (8.48) form the state space model of the micromechanical system of Figure 8.4(a). ∎

MIMO Systems

The companion website Chapter 8 presents a formal procedure that can be used to derive state space models for MIMO systems with no input time derivative. A simplified version of that methodology is mentioned here, based on the following particular application. Assume the mathematical model of a MIMO system consists of two differential equations, one of second order and the other of first order. As discussed for SISO systems, the second-order differential equation needs two state variables, whereas for the first-order differential equation, one state variable is necessary, so three state variables are needed in total. A good choice of the state variables, as seen in the examples that have been studied thus far, is the time-domain variable and its first time derivative for a second-order differential equation, whereas for a first-order differential equation, the sole state variable can be the time-domain variable of that equation. Let us analyze another example and derive its state space model.

Example 8.4

Consider the mechanical system of Figure 8.5, where f_1 is the actuation force and f_3 is a friction force. Determine a state space model for this system when the output vector is formed of the displacements y_1, y_2, and y_3.

Solution

Newton's second law of motion applied to the three bodies of Figure 8.5 results in

$$\begin{cases} m_1\ddot{y}_1 = f_1 - c_1(\dot{y}_1 - \dot{y}_2) - k_1(y_1 - y_2) \\ m_2\ddot{y}_2 = -c_1(\dot{y}_2 - \dot{y}_1) - k_1(y_2 - y_1) - c_2(\dot{y}_2 - \dot{y}_3) - k_2(y_2 - y_3) \\ m_3\ddot{y}_3 = -f_3 - c_2(\dot{y}_3 - \dot{y}_2) - k_2(y_3 - y_2) \end{cases} \tag{8.49}$$

The mathematical model of this mechanical system consists of three second-order differential equations. As a consequence, we need $3 \times 2 = 6$ state variables, which are selected as

$$x_1 = y_1; \ x_2 = \dot{y}_1; \ x_3 = y_2; \ x_4 = \dot{y}_2; \ x_5 = y_3; \ x_6 = \dot{y}_3 \tag{8.50}$$

Equations (8.50) indicate the following state variable connections:

$$\dot{x}_1 = x_2; \ \dot{x}_3 = x_4; \ \dot{x}_5 = x_6 \tag{8.51}$$

Equations (8.49) can be written as

$$\begin{cases} \ddot{y}_1 = -\dfrac{k_1}{m_1}y_1 - \dfrac{c_1}{m_1}\dot{y}_1 + \dfrac{k_1}{m_1}y_2 + \dfrac{c_1}{m_1}\dot{y}_2 + \dfrac{1}{m_1}f_1 \\ \ddot{y}_2 = \dfrac{k_1}{m_2}y_1 + \dfrac{c_1}{m_2}\dot{y}_1 - \dfrac{k_1 + k_2}{m_2}y_2 - \dfrac{c_1 + c_2}{m_2}\dot{y}_2 + \dfrac{k_2}{m_2}y_3 + \dfrac{c_2}{m_2}\dot{y}_3 \\ \ddot{y}_3 = \dfrac{k_2}{m_3}y_2 + \dfrac{c_2}{m_3}\dot{y}_2 - \dfrac{k_2}{m_3}y_3 - \dfrac{c_2}{m_3}\dot{y}_3 - \dfrac{1}{m_3}f_3 \end{cases} \tag{8.52}$$

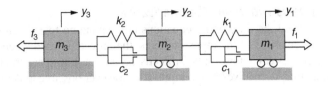

FIGURE 8.5

MIMO Translatory Mechanical System.

Collecting Eqs. (8.51) and (8.52) into a vector-matrix form results in the state equation

$$
\begin{Bmatrix} \dot{x}_1 \\ \dot{x}_2 \\ \dot{x}_3 \\ \dot{x}_4 \\ \dot{x}_5 \\ \dot{x}_6 \end{Bmatrix} =
\begin{bmatrix}
0 & 1 & 0 & 0 & 0 & 0 \\
-\dfrac{k_1}{m_1} & -\dfrac{c_1}{m_1} & \dfrac{k_1}{m_1} & \dfrac{c_1}{m_1} & 0 & 0 \\
0 & 0 & 0 & 1 & 0 & 0 \\
\dfrac{k_1}{m_2} & \dfrac{c_1}{m_2} & -\dfrac{k_1+k_2}{m_2} & -\dfrac{c_1+c_2}{m_2} & \dfrac{k_2}{m_2} & \dfrac{c_2}{m_2} \\
0 & 0 & 0 & 0 & 0 & 1 \\
0 & 0 & \dfrac{k_2}{m_3} & \dfrac{c_2}{m_3} & -\dfrac{k_2}{m_3} & -\dfrac{c_2}{m_3}
\end{bmatrix}
\begin{Bmatrix} x_1 \\ x_2 \\ x_3 \\ x_4 \\ x_5 \\ x_6 \end{Bmatrix}
$$

$$
+
\begin{bmatrix}
0 & 0 \\
\dfrac{1}{m_1} & 0 \\
0 & 0 \\
0 & 0 \\
0 & 0 \\
0 & -\dfrac{1}{m_3}
\end{bmatrix}
\begin{Bmatrix} u_1 \\ u_2 \end{Bmatrix}
\tag{8.53}
$$

where $u_1 = f_1$ and $u_2 = f_3$. The matrix multiplying the state vector in Eq. (8.53) is $[A]$ and the one multiplying the input vector is $[B]$. The output vector $\{y\} = \{y_1 \ y_2 \ y_3\}^t$ is connected to the state and input vectors as

$$
\begin{Bmatrix} y_1 \\ y_2 \\ y_3 \end{Bmatrix} =
\begin{bmatrix}
1 & 0 & 0 & 0 & 0 & 0 \\
0 & 0 & 1 & 0 & 0 & 0 \\
0 & 0 & 0 & 0 & 1 & 0
\end{bmatrix}
\begin{Bmatrix} x_1 \\ x_2 \\ x_3 \\ x_4 \\ x_5 \\ x_6 \end{Bmatrix}
+
\begin{bmatrix}
0 & 0 \\
0 & 0 \\
0 & 0
\end{bmatrix}
\begin{Bmatrix} u_1 \\ u_2 \end{Bmatrix}
\tag{8.54}
$$

The matrix multiplying the state vector in Eq. (8.54) is $[C]$ and the matrix multiplying the input vector in the same equation is $[D]$.

Dynamic Systems with Input Time Derivative

In situations where the differential equations corresponding to a dynamical model include time derivatives of the input variables, the procedure used when there were no input derivatives cannot be applied directly because a state space model does not accommodate time derivatives of the input. It is however possible to determine state space models using an additional function or vector, as shown in the following. Only coverage of SISO systems is provided here but the companion website Chapter 8 includes the study of MIMO systems.

Let us consider a SISO dynamic system whose mathematical model consists of the following differential equation:

$$a_n \overset{(n)}{y}(t) + a_{n-1} \overset{(n-1)}{y}(t) + \cdots + a_1 \dot{y}(t) + a_0 y(t) = b_q \overset{(q)}{u}(t) + b_{q-1} \overset{(q-1)}{u}(t) + \cdots$$
$$+ b_1 \dot{u}(t) + b_0 u(t) \qquad (8.55)$$

The state space variable choice made for SISO systems without input time derivatives cannot be applied here, simply because the standard state space model can have the term containing $u(t)$ only on the right-hand side of Eq. (8.55). Without loss of generality, let us assume that $n > q$. Application of the Laplace transform with zero initial conditions to Eq. (8.55) results in the following transfer function:

$$G(s) = \frac{Y(s)}{U(s)} = \frac{b_q s^q + b_{q-1} s^{q-1} + \cdots + b_1 s + b_0}{a_n s^n + a_{n-1} s^{n-1} + \cdots + a_1 s + a_0} \qquad (8.56)$$

A function $Z(s)$ is introduced in $G(s)$ of Eq. (8.56) as

$$G(s) = \frac{Z(s)}{U(s)} \times \frac{Y(s)}{Z(s)} = \frac{1}{a_n s^n + a_{n-1} s^{n-1} + \cdots + a_1 s + a_0}$$
$$\times \left(b_q s^q + b_{q-1} s^{q-1} + \cdots + b_1 s + b_0 \right) \qquad (8.57)$$

The following relationships can be written:

$$\begin{cases} \dfrac{Z(s)}{U(s)} = \dfrac{1}{a_n s^n + a_{n-1} s^{n-1} + \cdots + a_1 s + a_0} \\ \dfrac{Y(s)}{Z(s)} = b_q s^q + b_{q-1} s^{q-1} + \cdots + b_1 s + b_0 \end{cases} \qquad (8.58)$$

Cross-multiplication in Eqs. (8.58) leads to

$$\begin{cases} \left(a_n s^n + a_{n-1} s^{n-1} + \cdots + a_1 s + a_0 \right) Z(s) = U(s) \\ Y(s) = \left(b_q s^q + b_{q-1} s^{q-1} + \cdots + b_1 s + b_0 \right) Z(s) \end{cases} \qquad (8.59)$$

The inverse Laplace transform is applied to Eqs. (8.59), which results in

$$\begin{cases} a_n \overset{(n)}{z}(t) + a_{n-1} \overset{(n-1)}{z}(t) + \cdots + a_1 \dot{z}(t) + a_0 z(t) = u(t) \\ y(t) = b_q \overset{(q)}{z}(t) + b_{q-1} \overset{(q-1)}{z}(t) + \cdots + b_1 \dot{z}(t) + b_0 z(t) \end{cases} \tag{8.60}$$

The first of the two Eqs. (8.60) is now an equation in $z(t)$ that contains no time derivatives of the input $u(t)$ and, therefore, can be modeled using the standard procedure, which has been detailed in a previous section. This system needs n state variables, which can be defined as

$$x_1 = z, \quad x_2 = \dot{z}, \quad \ldots, \quad x_{n-1} = \overset{(n-2)}{z}, \quad x_n = \overset{(n-1)}{z} \tag{8.61}$$

Finding the state equation is done exactly as shown previously and is not detailed here. To determine the output equation, the second Eq. (8.60) is written as

$$y = b_0 x_1 + b_1 x_2 + \cdots + b_{q-1} x_q + b_q x_{q+1} \tag{8.62}$$

which can be reformulated in vector-matrix form as

$$y = \{b_0 \ b_1 \ \ldots \ b_q \ \ldots \ 0 \ 0\} \{x_1 \ x_2 \ \ldots \ x_m \ \ldots \ x_{n-1} \ x_n\}^t + 0 \times u \tag{8.63}$$

Note that, for a SISO system with a time derivative of the input, the numerator of the transfer function always contains terms in s; this problem is oftentimes referred to as *numerator dynamics*. Conversely, if a transfer function contains s terms in its numerator, the original input has derivatives in it.

Example 8.5

Find a state space representation for the electrical circuit of Figure 8.6 by using one state variable only.

Solution

Using the complex impedance approach, which was discussed in the Chapter 7, the transfer function is

$$\frac{V_o(s)}{V_i(s)} = \frac{Z_2}{Z_1 + Z_2} = \frac{R}{\dfrac{1}{Cs} + R} = \frac{RCs}{RCs + 1} \tag{8.64}$$

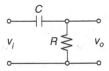

FIGURE 8.6

Single-Stage Capacitive/Resistive Electrical System.

As seen in Eq. (8.60), the transfer function numerator has an s term; therefore, the input voltage contains at least one time derivative. As a consequence, the procedure just outlined here needs to be applied. The ratio of Eq. (8.64) is written by means of the intermediate function $Z(s)$ as

$$\frac{V_o(s)}{V_i(s)} = \frac{V_o(s)}{Z(s)} \times \frac{Z(s)}{V_i(s)} \tag{8.65}$$

with

$$\frac{V_o(s)}{Z(s)} = RCs; \quad \frac{Z(s)}{V_i(s)} = \frac{1}{RCs + 1} \tag{8.66}$$

Cross-multiplication in the second Eq. (8.66), followed by application of the inverse Laplace transform, yields

$$\dot{z} = -\frac{1}{RC}z + \frac{1}{RC}v_i \tag{8.67}$$

If the state variable $x = z$ is chosen, Eq. (8.67) becomes the state equation

$$\dot{x} = -\frac{1}{RC}x + \frac{1}{RC}u \tag{8.68}$$

where the input voltage is considered to be the state space input, $u = v_i$.

Cross-multiplication followed by inverse Laplace transformation are applied now to the first Eq. (8.66), which, combined with Eq. (8.67), yields the output equation

$$y = -x + u \tag{8.69}$$

where $y = v_o$ is the output. Equations (8.68) and (8.69) form a state space model for the electrical system of Figure 8.6.

Conversions Between Transfer Function and State Space Models
Another modality of deriving state space models is by using an existing transfer function of a dynamic system. This section discusses the procedures that enable two-way conversions between transfer function and state space models.

Transformation of a Transfer Function Model into a State Space Model
The known transfer function of a SISO system (which is a scalar) can be converted into a state space model by following the steps indicated in the previous subsection, where the state space model of a SISO system with input derivatives is obtained using the transfer function concept and the intermediate function $Z(s)$. The companion website Chapter 8 covers the topic of conversion for MIMO systems.

Example 8.6

Determine a state space model by converting the transfer function model of the operational amplifier circuit sketched in Figure 7.10(b) of Example 7.5.

Solution

Equation (7.32) gives the system's transfer function, which can be written as

$$\frac{V_o}{V_i} = \frac{Z}{V_i} \times \frac{V_o}{Z} = \frac{1}{R_1 R_3 Cs} \times R_2 \tag{8.70}$$

with

$$\begin{cases} \dfrac{Z}{V_i} = \dfrac{1}{R_1 R_3 Cs} \\ \dfrac{V_o}{Z} = R_2 \end{cases} \tag{8.71}$$

Cross-multiplication in the first Eq. (8.71) and application of the inverse Laplace transformation to the resulting equation (with zero initial conditions) results in the following equation:

$$\dot{z} = \frac{1}{R_1 R_3 C} v_i \tag{8.72}$$

By selecting the state variable as $x = z$ and considering that the input is $u = v_i$ changes Eq. (8.72) to

$$\dot{x} = \frac{1}{R_1 R_3 C} u \tag{8.73}$$

which is the scalar form of the state equation with

$$A = 0; \; B = \frac{1}{R_1 R_3 C} \tag{8.74}$$

Cross-multiplication and inverse Laplace transformation with zero initial conditions are also applied to the second Eq. (8.71), which results in

$$y = R_2 x \tag{8.75}$$

where $y = v_o$. Equation (8.75) is the output equation with

$$C = R_2; \; D = 0 \tag{8.76}$$

Transformation of a State Space Model into a Transfer Function Model

This subsection studies the topic of transforming a state space model into the corresponding transfer function model. Example 8.1 demonstrates that $\{Y(s)\}$ and $\{U(s)\}$, the Laplace transforms of the output and input vectors, are connected by means of transfer function matrix $[G(s)]$, as shown in Eqs. (8.24) and (8.25). Since $[G(s)]$ is calculated based on the state space model matrices $[A]$, $[B]$, $[C]$, and $[D]$, it follows that the respective calculation procedure enables transformation of a state space model into a transfer function one for both SISO and MIMO systems. In Chapter 7, the matrix $[G(s)]$ is derived using a different formulation in the context of applying the transfer function concept to MIMO systems. Let us study the following example.

Example 8.7

Obtain the state space model for the mechanical system shown in Figure 8.7 and transform the resulting model into a transfer function model.

Solution

Figure 8.8 shows the free-body diagram corresponding to the massless point defined by the coordinate y.

Newton's second law of motion corresponding to the free-body diagram of Figure 8.8 results in

$$0 = -f_d - f_e \tag{8.77}$$

where f_d and f_e are the damping and elastic forces, defined as

$$\begin{cases} f_d = c(\dot{y} - \dot{u}) \\ f_e = ky \end{cases} \tag{8.78}$$

By combining Eqs. (8.77) and (8.78), the following equation is obtained:

$$c\dot{y} + ky = c\dot{u} \tag{8.79}$$

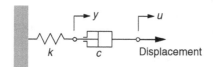

FIGURE 8.7

Translatory Mechanical System with Displacement Input.

FIGURE 8.8

Free-Body Diagram for the Mechanical System of Figure 8.7.

The mechanical system is SISO and its mathematical model consists of a first-order differential equation. As a result, a single state variable is needed. Because the input function appears as a time derivative, it is necessary to apply the Laplace transform to Eq. (8.79), which results in

$$\frac{Y(s)}{U(s)} = \frac{s}{s + \dfrac{k}{c}} \tag{8.80}$$

Using the intermediate function $Z(s)$ changes Eq. (8.80) to

$$\frac{Y(s)}{U(s)} = \frac{Z(s)}{U(s)} \times \frac{Y(s)}{Z(s)} = \frac{1}{s + \dfrac{k}{c}} \times s \tag{8.81}$$

which indicates the following selection needs to be made:

$$\begin{cases} \dfrac{Z(s)}{U(s)} = \dfrac{1}{s + \dfrac{k}{c}} \\[4mm] \dfrac{Y(s)}{Z(s)} = s \end{cases} \tag{8.82}$$

Cross-multiplication in the first Eq. (8.82) and application of the inverse Laplace transform to the resulting equation yields

$$c\dot{z} + kz = cu \tag{8.83}$$

Equation (8.83) no longer contains input derivatives; therefore, we can choose $x = z$ as the state variable. As a consequence, Eq. (8.83) becomes

$$\dot{x} = -\frac{k}{c}x + u \tag{8.84}$$

which is the state equation with $A = -k/c$ and $B = 1$. Cross-multiplication in the second Eq. (8.82) followed by inverse Laplace transformation results in

$$y = \dot{z} = \dot{x} \tag{8.85}$$

Combining Eqs. (8.84) and (8.85) yields

$$y = -\frac{k}{c}x + u \tag{8.86}$$

with $C = -k/c$ and $D = 1$. The transfer function corresponding to the state space model defined by the state Eq. (8.84) and output Eq. (8.86) is the one given in Eq. (8.80). This can readily be verified using the known equation $G(s) = C(sI - A)^{-1}B + D$.   ■

Nonlinear Systems

While it is almost always possible to derive a state space model for a nonlinear dynamic system, finding a solution to it can be problematic. In such cases, the *linearization technique* can be utilized to derive a linear state space model from the original, nonlinear one when the variables vary in small amounts around an *equilibrium* position. The companion website Chapter 8 analyzes the linearization of nonlinear nonhomogeneous state space models, that is, models that contain a nonzero input vector, but here we analyze only homogeneous nonlinear systems.

Consider a two-DOF dynamic system defined by one second-order differential equation and one first-order differential equations. As previously discussed in this chapter, one variant would use three state variables: two corresponding to the second-order differential equation and one for the first-order differential equation. Because the system has two DOFs, these variables can be considered as output variables. As a consequence, the following nonlinear state equations can be formulated:

$$\begin{cases} \dot{x}_1(t) = f_1(x_1, x_2, x_3) \\ \dot{x}_2(t) = f_2(x_1, x_2, x_3) \\ \dot{x}_3(t) = f_3(x_1, x_2, x_3) \end{cases} \tag{8.87}$$

where f_1, f_2, and f_3 are nonlinear functions of x_1, x_2, and x_3. Similarly, the output variables can be expressed as

$$\begin{cases} y_1(t) = h_1(x_1, x_2, x_3) \\ y_2(t) = h_2(x_1, x_2, x_3) \end{cases} \tag{8.88}$$

with h_1 and h_2 being nonlinear functions of x_1, x_2, and x_3.

If small (linear) variations of the time-dependent variables and functions of Eqs. (8.87) and (8.88) are considered (which is equivalent to using Taylor series expansions of the functions on the left-hand sides of Eqs. (8.87) and (8.88) about the equilibrium point), the following equations result from Eqs. (8.87):

$$\begin{cases} \delta\dot{x}_1(t) = \left(\dfrac{\partial f_1}{\partial x_1}\right)_e \delta x_1(t) + \left(\dfrac{\partial f_1}{\partial x_2}\right)_e \delta x_2(t) + \left(\dfrac{\partial f_1}{\partial x_3}\right)_e \delta x_3(t) \\[2ex] \delta\dot{x}_2(t) = \left(\dfrac{\partial f_2}{\partial x_1}\right)_e \delta x_1(t) + \left(\dfrac{\partial f_2}{\partial x_2}\right)_e \delta x_2(t) + \left(\dfrac{\partial f_2}{\partial x_3}\right)_e \delta x_3(t) \\[2ex] \delta\dot{x}_3(t) = \left(\dfrac{\partial f_3}{\partial x_1}\right)_e \delta x_1(t) + \left(\dfrac{\partial f_3}{\partial x_2}\right)_e \delta x_2(t) + \left(\dfrac{\partial f_3}{\partial x_3}\right)_e \delta x_3(t) \end{cases} \tag{8.89}$$

where the variables preceded by the symbol δ indicate small variations of those variables and the subscript e indicates the equilibrium point. Equations (8.89) can be arranged into the following vector matrix equation:

$$
\begin{Bmatrix} \delta\dot{x}_1(t) \\ \delta\dot{x}_2(t) \\ \delta\dot{x}_3(t) \end{Bmatrix} = \begin{bmatrix} \left(\dfrac{\partial f_1}{\partial x_1}\right)_e & \left(\dfrac{\partial f_1}{\partial x_2}\right)_e & \left(\dfrac{\partial f_1}{\partial x_3}\right)_e \\ \left(\dfrac{\partial f_2}{\partial x_1}\right)_e & \left(\dfrac{\partial f_2}{\partial x_2}\right)_e & \left(\dfrac{\partial f_2}{\partial x_3}\right)_e \\ \left(\dfrac{\partial f_3}{\partial x_1}\right)_e & \left(\dfrac{\partial f_3}{\partial x_2}\right)_e & \left(\dfrac{\partial f_3}{\partial x_3}\right)_e \end{bmatrix} \begin{bmatrix} \delta x_1(t) \\ \delta x_2(t) \\ \delta x_3(t) \end{bmatrix}
\tag{8.90}
$$

Equation (8.90) is the linearized version of the original, nonlinear state space equation, the first Eq. (8.87), and the matrix $[A]$ is formed of the partial derivatives of f_1, $f_2, \ldots, f_n$ in terms of $x_1, x_2, \ldots, x_n$, the evaluation being made at the equilibrium point. The state vector is formed of small variations of the original state vector's components.

If small variations are now considered for the nonlinear output Eqs. (8.88), the following equations are obtained:

$$
\begin{cases} \delta y_1(t) = \left(\dfrac{\partial h_1}{\partial x_1}\right)_e \delta x_1(t) + \left(\dfrac{\partial h_1}{\partial x_2}\right)_e \delta x_2(t) + \left(\dfrac{\partial h_1}{\partial x_3}\right)_e \delta x_3(t) \\[4mm] \delta y_2(t) = \left(\dfrac{\partial h_2}{\partial x_1}\right)_e \delta x_1(t) + \left(\dfrac{\partial h_2}{\partial x_2}\right)_e \delta x_2(t) + \left(\dfrac{\partial h_2}{\partial x_3}\right)_e \delta x_3(t) \end{cases}
\tag{8.91}
$$

which can be written in vector-matrix form as

$$
\begin{Bmatrix} \delta y_1(t) \\ \delta y_2(t) \end{Bmatrix} = \begin{bmatrix} \left(\dfrac{\partial h_1}{\partial x_1}\right)_e & \left(\dfrac{\partial h_1}{\partial x_2}\right)_e & \left(\dfrac{\partial h_1}{\partial x_3}\right)_e \\ \left(\dfrac{\partial h_2}{\partial x_1}\right)_e & \left(\dfrac{\partial h_2}{\partial x_2}\right)_e & \left(\dfrac{\partial h_2}{\partial x_3}\right)_e \end{bmatrix} \begin{bmatrix} \delta x_1(t) \\ \delta x_2(t) \\ \delta x_3(t) \end{bmatrix}
\tag{8.92}
$$

Equation (8.92) is the linearized form of the original, nonlinear output space Eq. (8.90). The output vector includes small variations of the original state vector's components, and the matrix $[C]$ connects the output and state vectors.

Example 8.8

The lumped-parameter model of a microcantilever that vibrates in a vacuum is the mass-spring system shown in Figure 8.9. Considering that the spring constant is nonlinear such that the elastic force is $f = 1/3y^3$, where y is the vertical displacement measured from the equilibrium position, find a state space representation of the system and derive the corresponding linearized state space model.

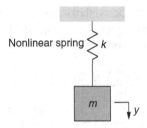

Nonlinear spring k

m

y

FIGURE 8.9

Lumped-Parameter Model of Microcantilever with a Nonlinear Spring.

Solution

At static equilibrium, the gravity force is equal to the elastic force; therefore,

$$mg = \frac{1}{3}y_e^3 \tag{8.93}$$

where y_e is the deformation of the spring at static equilibrium. If we now assume that the reference frame is moved to the position defined by y_e of Eq. (8.93), the equilibrium position, it can be shown that the equation of motion is

$$m\ddot{y} = -\frac{1}{3}y^3 \tag{8.94}$$

where y is the distance measured from the equilibrium position to an arbitrary position. The following state variables are selected:

$$x_1 = y; x_2 = \dot{y} \tag{8.95}$$

which enables formulating the equations

$$\begin{cases} \dot{x}_1 = x_2 \\ \dot{x}_2 = -\frac{1}{3m}x_1^3 \end{cases} \tag{8.96}$$

These are state equations, and the second one is nonlinear. The output equation is linear:

$$y = x_1 \tag{8.97}$$

Let us now linearize the state equation. If differentials in functions and in variables are considered, then the first Eq. (8.96), which is linear, can be written as

$$\delta\dot{x}_1 = \delta x_2 \tag{8.98}$$

The second Eq. (8.96) is reformulated based on the previous theory discussion as

$$\delta\dot{x}_2 = -\frac{1}{3m} \times \frac{\partial}{\partial x_1}(x_1^3)\bigg|_{x_1 = x_{1,e}} \delta x_1 = -\frac{x_{1,e}^2}{m} \delta x_1 \qquad (8.99)$$

Because $x_1 = y$ and y_e is provided in Eq. (8.93), Eq. (8.99) becomes

$$\delta\dot{x}_2 = -\sqrt[3]{\frac{9g^2}{m}} \delta x_1 \qquad (8.100)$$

Equations (8.98) and (8.100) can now be collected into the linearized state equation:

$$\begin{Bmatrix} \delta\dot{x}_1 \\ \delta\dot{x}_2 \end{Bmatrix} = \begin{bmatrix} 0 & 1 \\ -\sqrt[3]{\dfrac{9g^2}{m}} & 0 \end{bmatrix} \begin{Bmatrix} \delta x_1 \\ \delta x_2 \end{Bmatrix} \qquad (8.101)$$

where the state variables are δx_1 and δx_2. The output equation is simply obtained by differentiating Eq. (8.97) as

$$\delta y = \delta x_1 \qquad (8.102)$$

which is written in standard form as

$$\delta y = \{1 \quad 0\} \begin{Bmatrix} \delta x_1 \\ \delta x_2 \end{Bmatrix} \qquad (8.103)$$

The state Eq. (8.101) and output Eq. (8.103) form the linearized state space model of the mechanical system of Figure 8.9. ∎

8.2.2 MATLAB® Approach

This section illustrates the use of MATLAB® functions enabling conversions between transfer function or zero-pole-gain models and state space models.

Conversion between Transfer Function and State Space Models
A MATLAB® state space model or LTI (linear time invariant) object is defined by using a command of the type:

```
sys = ss (A, B, C, D)
```

where A, B, C, and D are the state space model defining matrices. If the command is issued as, for instance,

```
» sys = ss ([1, 3; 2, 5], [1; 3], [-1, 0;0, 5], [0; 1])
```

the following result will be generated:

```
a =                b =              c =                 d =

        x1   x2              u1               x1   x2               u1

   x1   1    3         x1    1        y1    -1    0         y1    0

   x2   2    5         x2    3        y2    0     5         y2    1
```

which is an explicit way to mention the vectors that connect to any of the system's matrices.

If the descriptor form of the state space is utilized instead of the regular form, then the MATLAB® command changes to

```
» sys = dss ( A', B', C, D, E)
```

where $[A'] \neq [A]$ and $[B'] \neq [B]$, as discussed in the introduction to this chapter. The MATLAB® functions that realize transformation of a transfer function model into a state space model, as well as the ones realizing the converse transformation, are examined in the following.

Transformation of a Transfer Function Model into a State Space Model

In instances where a transfer function or a transfer function matrix needs to be converted into a state space model, MATLAB® achieves this goal by means of two built-in functions. For a transfer function object, let us say sys1, which has previously been created, the function, tf2ss(sys1) or ss(sys1), generates a corresponding state space model.

Example 8.9

Use MATLAB® to convert the transfer function matrix

$$[G(s)] = \begin{bmatrix} \dfrac{3}{s^2 + 2s} & 1 \\ \dfrac{s}{s^2 + 2s} & \dfrac{1}{s^2 + 2s} \end{bmatrix}$$

into a space model.

Solution

The following MATLAB® sequence solves this example:

```
» s = tf ('s'); % command uses symbolic s in transfer function
» g = [3/(s^2 + 2*s), 1; s/(s^2 + 2*s), 1/(s^2 + 2*s)];
» f = ss(g);
```

To identify the four state space matrices, the following command can be added to this sequence:

```
» [a,b,c,d] = ssdata(f)
```

which returns

```
a =
```

-2	0	0	0
1	0	0	0
0	0	-2.0000	0
0	0	1.0000	0

```
b =
```

2	0
0	0
0	1
0	0

```
c =
```

0	1.5000	0	0
0.5000	0	0	1.0000

```
d =
```

0	1
0	0

As it can be seen from the result, the state space generated through conversion from the given transfer function uses five state variables (see the dimensions of [A]), two inputs, and two outputs (see the dimension of [D]), which is, obviously, just one possible solution from an infinite number of potential state space solutions. ◼

Transformation of a State Space Model into a Transfer Function Model

For a specified state-space model, identified as sys1, for instance, the command ss2tf(sys1) or tf(sys1) generates the corresponding transfer function model.

Example 8.10

The following matrices define a state space model:

$$A = \begin{bmatrix} 1 & 0 \\ 0 & 1 \end{bmatrix}; \ B = \begin{Bmatrix} 0 \\ 0 \end{Bmatrix}; \ C = \{2 \ 1\}; \ D = 2$$

Use MATLAB® to find the corresponding transfer function model.

Solution
The MATLAB® commands can be the following ones:

```
» a = eye (2);
» b = zeros (2, 1);
» c = [2, 1];
» sys1 = ss( a, b, c, 2);
» sys2 = tf (sys1)
```

The result of this sequence is 2. In other words, the transfer function is a gain equal to 2. If, instead of the last line of the sequence, we use

```
» [num, den] = ss2tf (a, b, c, d)
```

The result is

```
num   =
   2  -4    2
den   =
   1  -2    1
```

which indicates the transfer function is $G(s) = \dfrac{2x^2 - 4x + 2}{x^2 - 2x + 1} = 2$　　■

Conversion between Zero-pole-gain and State Space Models
Conversion between transfer function and zero-pole-gain models is presented in Chapter 7. Similarly, MATLAB® enables conversion between state space (ss) and zpk models, as well. Assume that a linear time invariant object sys1 has been defined as a zpk model; the MATLAB® command

```
» sys2=ss(sys1)
```

converts the zpk sys1 model into another ss model, labeled sys2. Conversely, when a ss model, named sys2, is available, it can be transformed into a zpk model, named sys1, by means of the MATLAB® command

```
» sys1=zpk(sys2)
```

Example 8.11
A dynamic system's zero-pole-gain model is defined by a gain of 2, its zeroes are 1 (double) and 3, and its poles are 0 and −1 (triple). Determine a state space model corresponding to the given zero-pole-gain model then confirm that the state space model can be converted back to the original zpk model.

Solution
The following MATLAB® sequence solves this example:

```
» sys1 = zpk([1,1,3],[-1,-1,-1,0],2)
Zero/pole/gain:
2 (s-1)^2 (s-3)
---------------
s (s+1)^3
» sys2 = ss(sys1)
```

a =

	x1	x2	x3	x4
x1	0	-1.414	-2	1
x2	0	-1	-2.828	1.414
x3	0	0	-1	2
x4	0	0	0	-1

b =

	u1
x1	0
x2	0
x3	0
x4	4

c =

	x1	x2	x3	x4
y1	-0.5	-0.7071	-1	0.5

d =

	u1
y1	0

```
Continuous-time model.
» sys3=zpk(sys2)
Zero/pole/gain:
2 (s-3) (s-1)^2
---------------
s (s+1)^3
```

It can be seen that the state space model rendered by MATLAB® (one of the many possible models) is the one of a SISO system (as indicated by the dimensions of the matrix [D], which uses four state variables (see the dimensions of matrix [A]). By applying the zpk command to the state space model, the original zpk object is retrieved. ∎

8.3 STATE SPACE AND THE TIME-DOMAIN RESPONSE

Analytical and MATLAB® methods can be used to determine the time (forced) response of dynamic systems by means of state space models, which is the subject of the following sections.

8.3.1 Analytical approach: The State-Transition Matrix Method

As briefly mentioned at the beginning of this chapter, the time response of dynamic systems can be evaluated by an algorithm that consists of combined calculations in the Laplace and time domains. In essence, solving for the output $\{y\}$ based on a given input $\{u\}$ and a selected state vector $\{x\}$ can be performed as indicated in Figure 8.10. The state vector $\{x\}$ is first determined through integration from the state equation then substituted into the output equation, which yields the output vector $\{y\}$ as a function of the input vector $\{u\}$, the state vector $\{x\}$, and the state space model matrices $[A]$, $[B]$, $[C]$, and $[D]$. Two subcases are studied next: The first one analyzes the homogeneous state space model (with no forcing $\{u(t)\}$) and the other one focuses on the nonhomogeneous state space model where $\{u(t)\} \neq 0$. Both categories consider nonzero initial conditions. The *state-transition matrix method* is utilized, as explained next.

Homogeneous State Space Model

The homogeneous case corresponds to the absence of an input vector, whereby the system response is caused by initial conditions only. The state and output equations are given in Eqs. (8.2) and (8.4). The aim of this method is to express the state vector at any time instant in terms of the initial state vector as

$$\{x(t)\} = [\varphi(t)]\{x(0)\} \tag{8.104}$$

where $[\varphi(t)]$ is the *state transition matrix*. There are two methods of calculating the state transition matrix: The *Laplace transform method* is discussed next and the *matrix exponential method* is presented in the companion website Chapter 8.

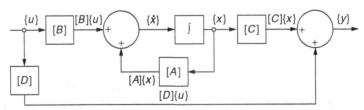

FIGURE 8.10

Block Diagram of Operations in a State Space Model.

By applying the Laplace transform to the state equation, Eq. (8.6), and considering a nonzero initial state vector $\{x(0)\}$, the following relationship is obtained:

$$\{X(s)\} = (s[I] - [A])^{-1}\{x(0)\} \tag{8.105}$$

whose inverse Laplace transform yields the time-domain state vector:

$$\{x(t)\} = \mathcal{L}^{-1}[(s[I] - [A])^{-1}]\{x(0)\} \tag{8.106}$$

By comparing Eqs. (8.104) and (8.106), it follows that the matrix connecting $\{x(t)\}$ to $\{x(0)\}$ is the state-transition matrix, which is therefore calculated as

$$[\varphi(t)] = \mathcal{L}^{-1}[(s[I] - [A])^{-1}] \tag{8.107}$$

The output equation, Eq. (8.6), becomes

$$\{y(t)\} = [C][\varphi(t)]\{x(0)\} = [C]\mathcal{L}^{-1}[(s[I] - [A])^{-1}]\{x(0)\} \tag{8.108}$$

Example 8.12

The homogeneous state space model of a dynamic system is defined by the matrices

$$[A] = \begin{bmatrix} 0 & 2 & 0 \\ -8 & -1 & 8 \\ 0 & 0 & 1 \end{bmatrix}; \ [C] = \{1 \ 0 \ 1\}$$

Determine the system's response $y(t)$ and plot it with respect to time if the following initial condition is used: $\{x(0)\} = \{0 \ 1 \ 0\}^t$.

Solution

The following MATLAB® commands

```
» a = [0,2,0;-8,-1,8;0,0,1];
» c = [1,0,1];
» syms s
» c*ilaplace(inv(s*eye(3)-a))*[0;1;0]
```

which are based on the analytical Eq. (8.108), yield the following approximate output:

$$y(t) = 0.5e^{-0.5t}\sin(4t) \tag{8.109}$$

The response curve is plotted in Figure 8.11.

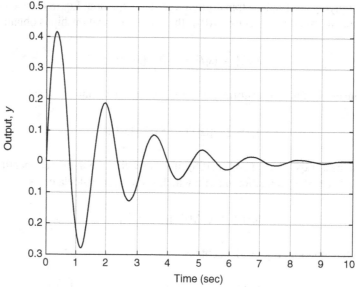

FIGURE 8.11

Time Response of a Homogeneous State Space Model.

Example 8.13

Use the state space approach to find the response of the hydraulic system shown in Figure 7.14 of the solved Example 7.7 by considering that there is no input flow rate. Instead, there is an initial value of the fluid head in the left tank, h_{10}, which acts as an initial condition. Plot the system's time response for $R_{l1} = 20$ s/m², $R_{l2} = 25$ s/m², $C_{l1} = 1$ m², $C_{l2} = 2$ m², and $h_{10} = 0.05$ m.

Solution

For the case of the free response, the differential equations are derived using the definitions of hydraulic capacitances, Eq. (5.27), and resistances, Eq. (5.39):

$$C_{l1}\frac{dh_1}{dt} = -q; \quad C_{l2}\frac{dh_2}{dt} = q - q_o; \quad R_{l1}q = h_1 - h_2; \quad R_{l2}q_o = h_2 \qquad (8.110)$$

After eliminating the intermediate and output flow rates q and q_o, the hydraulic system's mathematical model is

$$\begin{cases} C_{l1}\dfrac{dh_1}{dt} + \dfrac{1}{R_{l1}}h_1 - \dfrac{1}{R_{l1}}h_2 = 0 \\[3mm] C_{l2}\dfrac{dh_2}{dt} - \dfrac{1}{R_{l1}}h_1 + \left(\dfrac{1}{R_{l1}} + \dfrac{1}{R_{l2}}\right)h_2 = 0 \end{cases} \qquad (8.111)$$

Two state variables are needed for this two-DOF first-order system, and they can be selected as

$$x_1 = h_1; \; x_2 = h_2 \tag{8.112}$$

By using these state variables in conjunction with Eqs. (8.111), the following state equation is obtained:

$$\begin{Bmatrix} \dot{x}_1 \\ \dot{x}_2 \end{Bmatrix} = \begin{bmatrix} -\dfrac{1}{R_{11} C_{11}} & \dfrac{1}{R_{11} C_{11}} \\ \dfrac{1}{R_{11} C_{12}} & -\dfrac{1}{C_{12}}\left(\dfrac{1}{R_{11}} + \dfrac{1}{R_{12}}\right) \end{bmatrix} \begin{Bmatrix} x_1 \\ x_2 \end{Bmatrix} \tag{8.113}$$

Matrix [A] of the state space model is the one connecting the two vectors in Eq. (8.115). The output equation is obtained based on Eqs. (8.112):

$$\begin{Bmatrix} h_1 \\ h_2 \end{Bmatrix} = \begin{Bmatrix} y_1 \\ y_2 \end{Bmatrix} = \begin{bmatrix} 1 & 0 \\ 0 & 1 \end{bmatrix} \begin{Bmatrix} x_1 \\ x_2 \end{Bmatrix} \tag{8.114}$$

Matrix [C] of the state space model is the identity matrix [I]; therefore, {x(0)} = {y(0)}. Using the parameters of this example, Eq. (8.108) yields the following time-domain output:

$$\begin{cases} h_1 = 0.0035[14.177\cosh(0.035t) - \sinh(0.035t)]e^{-0.047t} \\ h_2 = 0.035\sinh(0.035t)e^{-0.047t} \end{cases} \tag{8.115}$$

Figure 8.12 contains the plots of the two response components. For $t \to \infty$, the heads become $h_1(\infty) = 0$ and $h_2(\infty) = 0$.

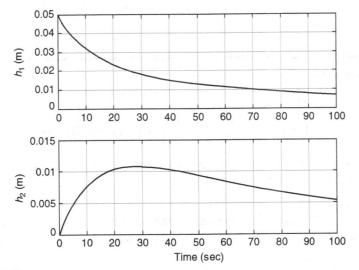

FIGURE 8.12

Time-Response Curves of Hydraulic Heads.

Nonhomogeneous State Space Model

The nonhomogeneous case implies intervention of the input (forcing) vector $\{u(t)\}$; therefore, the state and output equations are the ones originally expressed in Eqs. (8.2) and (8.4). Application of the Laplace transform to the state Eq. (8.4), considering there is a nonzero initial state vector $\{x(0)\}$, results in the following Laplace-domain equation:

$$\{X(s)\} = (s[I] - [A])^{-1}\{x(0)\} + (s[I] - [A])^{-1}[B]\{U(s)\} \tag{8.116}$$

The inverse Laplace transform is now applied to Eq. (8.116), which leads to

$$\{x(t)\} = [\varphi(t)]\{x(0)\} + \mathcal{L}^{-1}[(s[I] - [A])^{-1}[B]\{U(s)\}] \tag{8.117}$$

According to the convolution theorem, which is studied in Chapter 6, the second term on the right-hand side of Eq. (8.117) can be calculated as

$$\mathcal{L}^{-1}[(s[I] - [A])^{-1}[B]\{U(s)\}] = \int_0^t [\varphi(t - \tau)][B]\{u(\tau)\}\,d\tau \tag{8.118}$$

and, as a consequence, Eq. (8.117) becomes

$$\{x(t)\} = [\varphi(t)]\{x(0)\} + \int_0^t [\varphi(t - \tau)][B]\{u(\tau)\}d\tau \tag{8.119}$$

where $[\varphi(t)]$ is computed by means of Eq. (8.107). The output vector is found as

$$\{y(t)\} = [C]\left([\varphi(t)]\{x(0)\} + \int_0^t [\varphi(t - \tau)][B]\{u(\tau)\}d\tau\right) + [D]\{u(t)\} \tag{8.120}$$

Example 8.14

A dynamic system is described in state space form by the matrices

$$[A] = \begin{bmatrix} 0 & 1 \\ -60 & -5 \end{bmatrix}; [B] = \begin{Bmatrix} 0 \\ 1 \end{Bmatrix}; [C] = \{1 \ \ 0\}; [D] = 0$$

A step input $u = 3$ is applied to the system and the initial conditions are zero. Find the system response and plot it against time.

Solution

The matrices' dimensions show that the system is SISO and there are two state variables. For the particular case of this example, the output is expressed from Eq. (8.120) as

$$\{y(t)\} = [C]\int_0^t [\varphi(t - \tau)][B]u(\tau)d\tau \tag{8.121}$$

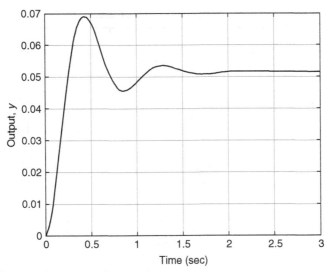

FIGURE 8.13

Time Response of a Nonhomogeneous State Space Model.

where $u(\tau) = 3$. The state transition matrix $[\varphi(t)]$ is first computed by using the definition Eq. (8.107) as

$$[\varphi(t)] = \begin{bmatrix} [\cos(7.3t) + 0.34\sin(7.3t)] & 0.14\sin(7.3t) \\ -8.18\sin(7.3t) & [\cos(7.3t) - 0.34\sin(7.3t)] \end{bmatrix} e^{-2.5t} \quad (8.122)$$

The variable $t - \tau$ is used instead of the variable t in Eq. (8.122), then the operations necessary are performed in Eq. (8.121), which yields the following time-domain response:

$$y(t) \simeq 0.05\{1 - [\cos(7.3t) + 0.35\sin(7.3t)]e^{-2.5t}\} \quad (8.123)$$

Figure 8.13 displays the output as a function of time.   ▬

8.3.2 MATLAB® Approach

This section presents the application of built-in MATLAB® functions that use state space modeling to determine the forced response (including nonzero initial conditions) of dynamic systems. The free response with nonzero initial conditions and the forced response are studied next.

Free Response with Nonzero Initial Conditions

With MATLAB®, it is possible to directly model the free response of a dynamic system when the initial conditions are different from zero. The basic command is

```
» initial (sys, x₀)
```

where x_0 is the initial state vector and sys is a previously formulated state space model.

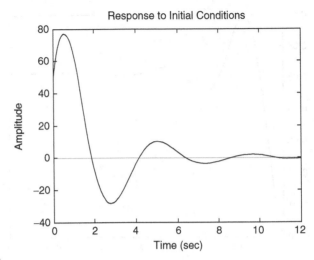

FIGURE 8.14

Example Plot for Free Response with Nonzero Initial Conditions.

Example 8.15

The free-response state space model of a dynamic system is represented by the matrices [A] and [C] and the initial state vector x_0, which are defined as

$$[A] = \begin{bmatrix} -1 & -1.5 \\ 1.5 & 0.1 \end{bmatrix}; \quad [C] = \{4 \quad 10\}; \quad \{x_0\} = \begin{Bmatrix} 10 \\ 1 \end{Bmatrix}$$

Use MATLAB® to determine the time response and plot it in terms of time.

Solution

The following MATLAB® command sequence

```
» a = [-1, -1.5; 1.5, 0.1]; b = [0; 0];
» c = [4, 10]; d = 0;
» x0 = [10; 1];
» sys = ss (a, b, c, d);
» initial (sys, x0)
```

is used to model the free response of this specific state space model and the result is plotted in Figure 8.14.

Several previously defined state space models can be plotted on the same graph and time interval and specified with the command

```
» initial (sys1, sys2,…,sysn, x0, t)
```

When using a command like

```
» [y,t,x] = initial (sys1, sys2,…,sysn, x₀, t)
```

no plot is returned but the output matrix and state matrix are formed, each having as many rows as time increments. The number of columns in y is equal to the number of inputs and the number of columns in x is equal to the number of state variables. Subsequent plotting is possible with a plot command. Instead of initial, initialplot can be used with the same specifications.

Forced Response

To find the forced time response, the state space models in MATLAB® employ the same predefined functions as the transfer function models, step, impulse, and lsim. For a state space model defined as

```
» sys = ss ([-1, -1; 1, 0], [1; 0], [1, 0; 0, 2], [0; 1]);
```

the command step(sys, 'k') (k indicates that the color of the plot should be black) produces the plots shown in Figure 8.15.

Because the matrices of this particular example define a one-input, two-output model, two plots result, each corresponding to a unit step input.

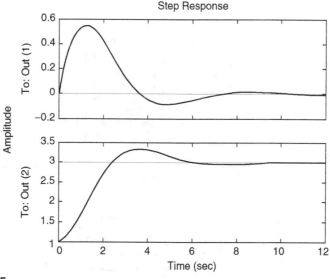

FIGURE 8.15

Response of a Two-Output State Space Model to Unit-Step Input.

Example 8.16

Consider the mechanical microsystem shown in Figure 8.16, which is formed of three identical serpentine springs, each having a spring constant of $k = 10$ N/m and two identical shuttle masses with $m = 30$ µg. A force $f = 5\sin(2t)$ µN acts on the body on the right, as shown in the figure. Also, an initial displacement $y_1(0) = -1.5 \times 10^{-11}$ m is applied to the body on the left of the figure. Consider that damping acts on both masses, with a damping coefficient $c = 0.1$ N-s/m. Find a state space model of the system and use MATLAB® to plot its time response. Use both the regular and the descriptor forms of the state equation.

Solution

To use MATLAB® for the time-domain solution, the problem is divided into two subproblems: it is considered first that the system is under the action of the force alone, then that only the initial conditions are applied. Since the system is linear, the two individual solutions are then added to obtain the total (actual) solution. The lumped-parameter model corresponding to the micromechanical system is shown in Figure 8.17, being formed of two masses m, three dampers c, and three springs k. The force f is also indicated in the figure.

The dynamic equations of this system are

$$\begin{cases} m\ddot{y}_1 = -c\dot{y}_1 - ky_1 - c(\dot{y}_1 - \dot{y}_2) - k(y_1 - y_2) \\ m\ddot{y}_2 = -c\dot{y}_2 - ky_2 - c(\dot{y}_2 - \dot{y}_1) - k(y_2 - y_1) + f \end{cases} \tag{8.124}$$

The following state variables are selected:

$$x_1 = y_1; \; x_2 = \dot{y}_1; \; x_3 = y_2; \; x_4 = \dot{y}_2 \tag{8.125}$$

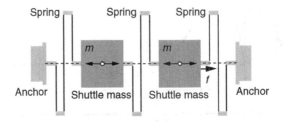

FIGURE 8.16

Two-Mass, Three-Spring Mechanical Microsystem.

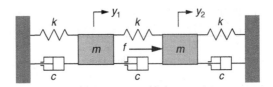

FIGURE 8.17

Lumped-Parameter Model of Two-DOF Translatory Mechanical System.

the input is the force, that is, $u = f$, whereas the output is formed of the two displacements y_1 and y_2. Combination of Eqs. (8.124) and (8.125) generates the standard-form state equation:

$$
\begin{Bmatrix} \dot{x}_1 \\ \dot{x}_2 \\ \dot{x}_3 \\ \dot{x}_4 \end{Bmatrix} = \begin{bmatrix} 0 & 1 & 0 & 0 \\ -2\dfrac{k}{m} & -2\dfrac{c}{m} & \dfrac{k}{m} & \dfrac{c}{m} \\ 0 & 0 & 0 & 1 \\ \dfrac{k}{m} & \dfrac{c}{m} & -2\dfrac{k}{m} & -2\dfrac{c}{m} \end{bmatrix} \begin{Bmatrix} x_1 \\ x_2 \\ x_3 \\ x_4 \end{Bmatrix} + \begin{Bmatrix} 0 \\ 0 \\ 0 \\ \dfrac{1}{m} \end{Bmatrix} u \qquad (8.126)
$$

which defines the [A] and [B] matrices. The descriptor-form state equation is also obtained from Eqs. (8.124) and (8.125):

$$
\begin{bmatrix} 1 & 0 & 0 & 0 \\ 0 & m & 0 & 0 \\ 0 & 0 & 1 & 0 \\ 0 & 0 & 0 & m \end{bmatrix} \begin{Bmatrix} \dot{x}_1 \\ \dot{x}_2 \\ \dot{x}_3 \\ \dot{x}_4 \end{Bmatrix} = \begin{bmatrix} 0 & 1 & 0 & 0 \\ -2k & -2c & k & c \\ 0 & 0 & 0 & 1 \\ k & c & -2k & -2c \end{bmatrix} \begin{Bmatrix} x_1 \\ x_2 \\ x_3 \\ x_4 \end{Bmatrix} + \begin{Bmatrix} 0 \\ 0 \\ 0 \\ 1 \end{Bmatrix} u \qquad (8.127)
$$

where the matrix $[A']$ of Eq. (8.5) is the one on the left-hand side of Eq. (8.127) and the matrix $[B']$ of Eq. (8.5) is the first one on the right-hand side of Eq. (8.127). The output equation is determined from the first and third of Eqs. (8.125):

$$
\begin{Bmatrix} y_1 \\ y_2 \end{Bmatrix} = \begin{bmatrix} 1 & 0 & 0 & 0 \\ 0 & 0 & 1 & 0 \end{bmatrix} \begin{Bmatrix} x_1 \\ x_2 \\ x_3 \\ x_4 \end{Bmatrix} + \begin{Bmatrix} 0 \\ 0 \end{Bmatrix} u \qquad (8.128)
$$

which defines the [C] and [D] matrices. The second sub-problem uses the initial-condition vector $\{x(0)\} = \{y_1(0) \ 0 \ 0 \ 0\}^t$ with the same state space model as of the first sub-problem (the one with forcing). With the numerical values of this problem, the MATLAB® code is given here:

```
» t = 0:0.0001:10;
» m = 30e-6;
» da = 0.1;
» k = 10;
» u = 5e-6*sin(2*t);
» a = [0,1,0,0;-2*k/m,-2*da/m,k/m,da/m;0,0,0,1;k/m,da/m,...
-2*k/m,-2*da/m];
» b = [0;0;0;1/m];
» c = [1,0,0,0;0,0,1,0];
» d = [0;0];
» sys = ss(a,b,c,d);
» % individual response 'yf' to forcing
» [yf,t,x] = lsim(sys,u,t);
» % individual response 'yic' to initial conditions
```

```
» y01 = -1.5e-11;
» x0 = [y01;0;0;0];
» [yic,t,x] = initial(sys,x0,t);
» % total response 'y' as superposition of individual…
responses 'yf' and 'yic'
» y = yf + yic;
» y1 = y;
» y1(:,2) = []; % deletes the second column and keeps…
the first column of 'y1 = y'
» y2 = y;
» y2(:,1) = []; % deletes the first column and keeps…
the second column of 'y2 = y'
» subplot(211);
» plot(t,y1)
» ylabel('y_1 (m)')
» grid on
» subplot(212);
» plot(t,y2)
» ylabel('y_2 (m)')
» xlabel('Time (sec)')
» grid on
```

Figure 8.18 shows the response curves of the mechanical microsystem of Figure 8.16.

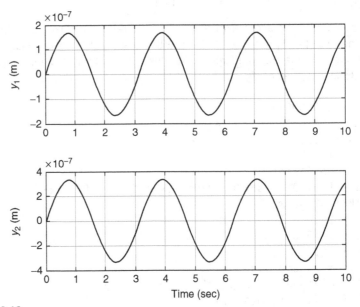

FIGURE 8.18

Plot of Displacement Outputs for the Two-DOF Mechanical System.

If the descriptor form of the state equation is used, then Eq. (8.127) is needed to define the [*E*] and [*A'*] matrices, and the following additional commands are inserted in the previous MATLAB® sequence:

```
» e = [1,0,0,0;0,m,0,0;0,0,1,0;0,0,0,m];
» ap = [0,1,0,0;-2*k,-2*da,k,da;0,0,0,1;k,da,-2*k,-2*da];
» bp = [0; 0; 0; 1];
```

and instead of the `sys = ss(a,b,c,d)` command, the command

```
» sys2 = dss(ap,bp,c,d,e)
```

is employed. The remaining part of the code is identical to the one employed in this example and the plots of Figure 8.18 are obtained as well.   ■

8.4 USING SIMULINK® FOR STATE SPACE MODELING

Simulink® offers an elegant environment for creating and simulating state space models. The minimum configuration of a Simulink® state space model is formed of an input block, the state space block, and an output (visualization) block. Let us analyze a few examples of formulating and solving state space problems by means of Simulink®. As a reminder, in order to open the Simulink® environment, just type `simulink` at the MATLAB® prompt, then click on `File`, `New`, and `Model` to open a new model window.

■

Example 8.17
A state space model is defined by the matrices

$$A = \begin{bmatrix} 0.001 & 80 \\ 1 & 0 \end{bmatrix}; \quad B = \begin{bmatrix} 1 & -1 \\ 0 & 0 \end{bmatrix}; \quad C = \{0\ 1\}; \quad D = \{0\ 0\}$$

The input to this system has two components: One is a pulse with an amplitude of 1, period of 5 s, and pulse width of 2.5 s; the second input is a random function with a mean value of 0 and a variance of 1. The initial conditions of the problem are $u(0) = \{1, 2\}^t$. Use Simulink® to plot the system response.

Solution
A new model window is opened and the following blocks are dragged to it from the `Library` window: a `Sine Wave` block from the `Sources` category, the `State Space` block from the `Continuous` category, and `Scope` from the `Sinks` category. The system is a MISO (multiple-input/single-output) system, as the dimensions of the matrices indicate. We therefore need to use two signals combined into a single input, since the Simulink® state space operator works as a single-input, single-output one. The Simulink® source library possesses a *concatenating operator* that accepts several vectors (or matrices) as the input, which it combines into a single vector (or matrix) at the output without modifying the original components. In Figure 8.19 the two input signals preserved their individuality, although

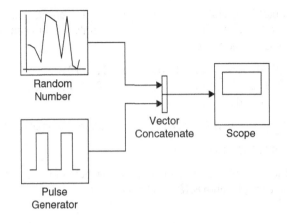

FIGURE 8.19

Simulink® Block Diagram of the Two Inputs.

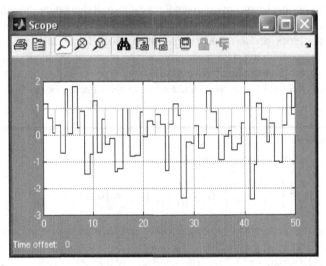

FIGURE 8.20

Simulink® Plot of the Pulse and Random Input Signals.

they are combined in one vector. Configuration of the state space block (which you have to double click) is done by filling in the following data:

```
Function Block Parameters: State-Space Window
   A: [0.001, 80; 1, 0]
   B: [1, -1; 0, 0]
   C: [0, 1]
   D: [0, 0]
   Initial conditions: [1; 2]
```

Configuring the other blocks is straightforward.

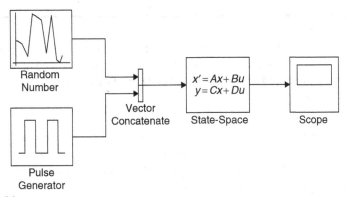

FIGURE 8.21

Simulink® Block Diagram of the Two-Input, One-Output State Space Model.

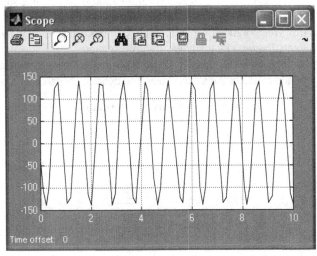

FIGURE 8.22

Simulink® Plot of the State Space Model Time-Domain Output.

The two inputs, the `Pulse Generator` and the `Random Number` shown in Figure 8.19, are selected from the `Source` block category, and the parameters of the example are entered after clicking the respective operators and opening of the corresponding windows. The `Vector Concatenate` box, which creates a contiguous output signal from several input vector-type signals, is dragged from the `Mathematical Operations` library group, whereas, as before, `Scope` is taken from the `Sink` library. Figure 8.20 is the plot resulting from the two input functions.

Figure 8.21 illustrates the block diagram of the simulation, which uses the `State-Space` block in addition to those already mentioned. The *A*, *B*, *C*, and *D* matrices elements are entered, and the initial condition vector is also inserted as `[1; 2]`. The result of the simulation is shown in Figure 8.22.

Example 8.18

A body of mass $m = 1$ kg slides on a horizontal surface with friction. The coefficient of static friction is $\mu_s = 0.3$ and the coefficient of kinematic friction is $\mu_k = 0.2$. A ramp force $f = 0.5t$ N acts on the body. Find a state space model of this mechanical system and determine its time response using Simulink®.

Solution

(a) The free-body diagram is sketched in Figure 8.23 where f is the active (ramp) force and f_f is the kinematic friction force. It is known that, when the ramp force is applied, the body will start moving only after the active force has reached the value of the opposing static friction force, which is

$$f_{f,s} = \mu_s mg \tag{8.129}$$

or $f_{f,s} = 0.3 \times 1$ kg $\times 9.8$ m/s² $= 2.94$ N and Figure 8.24(a) depicts this situation, where the motion starts after a delay time t_d. Another time-domain force definition is displayed in Figure 8.24(b), which indicates a sudden jump in the active force to the level of the static friction force, followed by the linear (ramp) variation.

Let us use the free-body diagram of Figure 8.23 to derive the mathematical model and then the state space model of the sliding body. The dynamic equation of motion is:

$$m\ddot{y} = f - f_f \quad \text{or} \quad m\ddot{y} = f - \mu_k mg \tag{8.130}$$

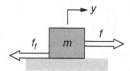

FIGURE 8.23

Free-Body Diagram for a Body under the Action of Active and Friction Forces.

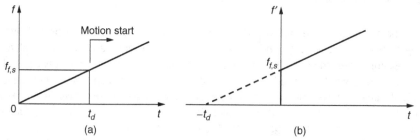

FIGURE 8.24

Force-Time Relationship Indicating (a) Linear Effects; (b) Nonlinear Effects Due to Coulomb Friction.

The state variables are

$$x_1 = y; x_2 = \dot{y} \tag{8.131}$$

The output is y and the input is formed by superimposing (adding) the active and friction forces as $u = u_1 + u_2$, where

$$u_1 = f, u_2 = -\mu_k mg \tag{8.132}$$

The state equation is therefore

$$\begin{Bmatrix} \dot{x}_1 \\ \dot{x}_2 \end{Bmatrix} = \begin{bmatrix} 0 & 1 \\ 0 & 0 \end{bmatrix} \begin{Bmatrix} x_1 \\ x_2 \end{Bmatrix} + \begin{Bmatrix} 0 \\ \frac{1}{m} \end{Bmatrix} u \tag{8.133}$$

The output equation is

$$y = \{1 \quad 0\} \begin{Bmatrix} x_1 \\ x_2 \end{Bmatrix} + 0 \times u \tag{8.134}$$

There are two different ways in Simulink® of modeling the time definition of Figure 8.24(b) and Figure 8.25 shows the block diagram of the two models described next. One model, which is displayed in the top area of Figure 8.25, combines the two inputs by means of an Add block, which operates similarly to a Sum block and is found in the Mathematical Operations of the Simulink library. The Ramp 1 input

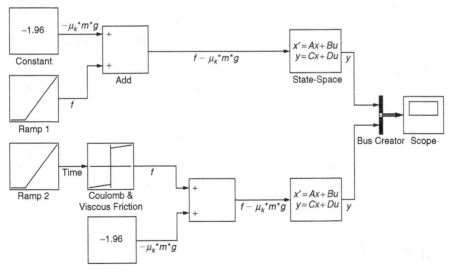

FIGURE 8.25

Simulink® Block Diagram of the State Space Model without and with the Nonlinear Effect of Coulomb Friction.

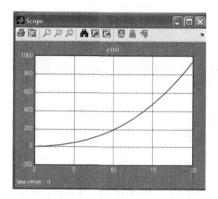

FIGURE 8.26

Simulink® Plot of State Space Model Time Response.

is defined with a Slope of 0.5 and a Start time of $t_d = f_{f,s} / K = 2.94$ N / 0.5 N/s = 5.88 s. This start time value, which is actually the delay time, has to be introduced with a negative sign, as shown in Figure 8.24(b). The Constant input is the kinematic friction force whose value is $f = \mu_k m g = 0.2 \times 1$ kg $\times$ 9.8 m/s² $= 1.96$ N—it has a minus sign to conform to Eq. (8.130). The addition result is fed to the State-Space block where A, B, C, and D are defined in Eqs. (8.133) and (8.134).

The other modeling procedure available in Simulink® is shown on the lower branch of Figure 8.25. The Ramp 2 block has a Slope of 1 and generates the time vector entering the Coulomb & Viscous Friction block. This block requires input of the Coulomb friction value (offset), which is equal to the static friction force $f_{f,s} = 2.94$ N, as well as of the Gain, which is the 0.5 slope. The output of this block is added to the output of the Constant block (the kinematic friction force), and the result is again fed to the State-Space model block. The outputs from the two state space blocks are mixed into the Bus Creator block, which functions similarly to a Mux block; the output of the block is a group (or bundle) formed of separate input signals, and this output is subsequently plotted by a Scope. Compared to the Vector Concatenate block, which has been utilized in the previous example and which accepts inputs of the same data type, such as vectors, the Bus Creator block accepts input signals of different data types. Because the two Simulink® models are identical, a single displacement plot is obtained, as pictured in Figure 8.26. ▮

SUMMARY

This chapter introduces the concept of state space as an approach that can be used to model and determine the response of dynamic systems in the time domain. Using a vector-matrix formulation, the state space procedure is an alternate method of characterizing mostly MIMO systems defined by a large number of coordinates (DOFs). Different state space algorithms are applied, depending on whether the input has time derivatives or no time derivatives. Methods of calculating the free response with nonzero initial conditions and the forced response using the state space approach also are presented. Methods of converting between state space and transfer function or zero-pole-gain models are presented, both analytically and by means of MATLAB® specialized commands. The chapter also studies the principles of linearizing nonlinear state space models. The material includes the application of specialized MATLAB® commands to solve state space formulated problems and examples of using Simulink® to model and solve system dynamics problems by the state space approach.

PROBLEMS

8.1 Find a state space model for the translatory mechanical system of Figure 8.27 considering that the input is the displacement u of the chain's free end and the output is the displacement z. Derive another state space model for an output formed of the displacement z and the velocity dz/dt.

8.2 Demonstrate that the voltage on the resistor, v_R, and the voltage on the capacitor, v_C, can be used as state variables to generate a valid state space model for the series electrical circuit of Figure 8.28.

8.3 **(a)** Find a state space representation for the electrical circuit shown in Figure 8.29 using only one state space variable.

  (b) If the resistor R_2 is substituted by an inductor L, formulate a state space model by using two state variables. Hint: Use the relationship between voltage v and magnetic flux Ψ: $v = d\Psi/dt$.

8.4 The pneumatic system of Figure 8.30 consists of a pneumatic resistance R_g and a gas container whose capacity is C_g. The input pressure is p_i and the output (container) pressure is p_o. Derive a state space model for this system.

8.5 Derive a state space model for the one-room thermal system of Figure 8.31. Consider the input is the outdoors temperature θ_2, whereas the output is the indoor temperature θ_1. The enclosed space has a thermal capacity C_{th} and the four identical walls each have a thermal resistance R_{th}.

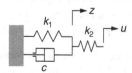

FIGURE 8.27

Translatory Mechanical System with Springs, Damper, and Displacement Input.

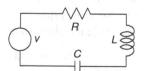

FIGURE 8.28

Series Electrical System with Resistor, Inductor, Capacitor, and Voltage Source.

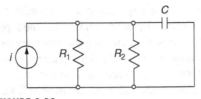

FIGURE 8.29

Electrical Circuit with Current Source.

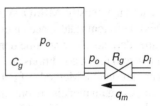

FIGURE 8.30

Pneumatic System with Resistance and Capacitance.

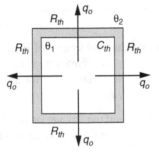

FIGURE 8.31

Four-Wall, One-Room Thermal System.

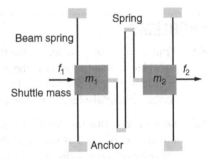

FIGURE 8.32

Linear-Motion MEMS Device.

8.6   The MEMS system of Figure 8.32 is formed of two shuttle masses m_1 and m_2 coupled by a serpentine spring and supported separately by two pairs of identical beam springs. The shuttle masses are subjected to viscous damping and acted upon by two electrostatic forces f_1 and f_2. Use a lumped-parameter model of this MEMS device and obtain a state space model for it by having an output vector formed of the two masses' displacements and velocities.

8.7   Find a state space model for the translatory mechanical system of Figure 8.33 considering that the input is the force f applied at the chain's free end and the output vector contains all relevant displacements.

8.8   Derive a state space model for the electrical system of Figure 8.34, where v_i is the applied (input) voltage. Consider the output is formed of the relevant currents in the circuit and the output voltage v_o. Known are R, L, and C.

8.9   The liquid system shown in Figure 8.35 is formed of two tanks of capacitances C_{l1} and C_{l2} and two valves of resistances R_{l1} and R_{l2}. Formulate a state space model for it by considering the input to the system is the volume flow rate q_i and the pressure p_3; the output consists of the pressures at the tanks' bottoms, p_1 and p_2.

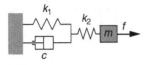

FIGURE 8.33

Translatory Mechanical System with
Springs, Damper, Mass, and Force Input.

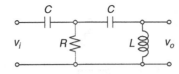

FIGURE 8.34

Two-Stage Electrical System.

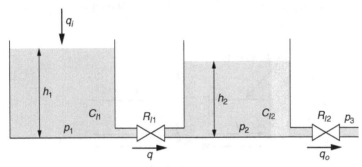

FIGURE 8.35

Two-Tank Liquid-Level System.

8.10 Find state space models for the SISO dynamic systems represented by the following differential equations, where y is the output and u is the input:

(a) $\dddot{y} + 6\ddot{y} + 2\dot{y} + 6y = 3\dot{u}$.

(b) $3\ddot{y} + \dot{y} + 5y = 3\ddot{u} + u$.

8.11 Derive a state space model for the rotary mechanical system of Figure 8.36, which is formed of a cylinder with a mass moment of inertia J, a spring of stiffness k, and two dampers defined by the damping coefficients c_1 and c_2. The input is the rotation angle θ_u and the output is the cylinder rotation angle.

8.12 Obtain a state space model for the electrical system of Figure 8.37, where the applied voltage v_i is the input and the voltage v_o is the output. Known are R, L, and C.

8.13 Use a lumped-parameter model for the single-DOF mechanical microacceler-ometer of Figure 8.38. Find a transfer function model of this system and con-vert it to a state space model. The five identical massless beams have a length $l = 100$ μm, their cross-section is square with a side of $a = 2$ μm, Young's modulus is $E = 150$ GPa, and the mass of the central plate is 200 μg. Use the state space model and MATLAB® to calculate and plot the mass displacement when a force $f = 10^{-8}\delta(t)$ N is applied to the plate. Ignore damping and inertia contributions from the beams.

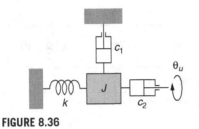

FIGURE 8.36

Rotary Mechanical System.

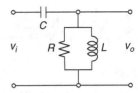

FIGURE 8.37

One-Stage Electrical System with Resistor, Inductor, and Capacitor.

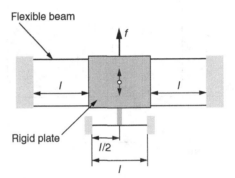

FIGURE 8.38

Microaccelerometer with Plate and Beam Springs.

8.14 Use complex impedances to determine the transfer function of the electrical system shown in Figure 8.39, and then convert the transfer function model into a SISO state space model using analytical derivation. Use MATLAB® to plot the output voltage as a function of time using the state space model. Known are $R = 450\ \Omega$, $L = 500$ mH, $C = 80\ \mu$F, and $v_i = 90$ V.

8.15 Transform the state space model of the electrical system of Problem 8.12 and pictured in Figure 8.37 into a transfer function for $R = 270\ \Omega$, $L = 6$ H, and $C = 300\ \mu$F. Use analytical and MATLAB® calculations.

8.16 Convert the state space model of the MEMS device of Problem 8.6 and shown in Figure 8.32 into the corresponding transfer function matrix model by means of analytical derivation. Use MATLAB® to confirm the analytical result. Known are $m_1 = 3 \times 10^{-11}$ kg, $m_2 = 4 \times 10^{-11}$ kg, damping coefficients $c_1 = c_2 = 0.1$ N-s/m, stiffness of beam springs $k_1 = 2$ N/m and stiffness of middle spring $k = 3$ N/m.

8.17 Derive a state-space model for the mass-spring lever system sketched in Figure 8.40, which vibrates in a vertical plane. Assume the elastic (spring) force is defined as $f_e = z + \frac{1}{2}z^2$ (where z represents the vertical displacement

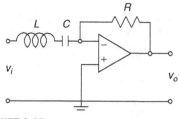

FIGURE 8.39

Single-Stage Operational-Amplifier
Electrical System.

FIGURE 8.40

Mass-Spring Lever System.

of the spring end attached to the lever) and the rod is massless. Linearize the
obtained model by considering small vibrations about the equilibrium position.
Hint: Formulate the dynamic equation of motion about the static equilibrium
position.

8.18 Consider an initial rotation angle of $5°$ is applied to the pivoting rod of the lever
system studied in Problem 8.17 and shown in Figure 8.40. Solve the linearized
state space model using the state transition matrix method; also use Simu-
link® to solve both the original nonlinear and the linearized state space models.
Known are $m = 0.6$ kg, $l = 0.4$ m, and $g = 9.8$ m/s^2.

8.19 An *RLC* series circuit with a voltage source v has a nonlinear resistor whose
voltage varies as $v_R = 300(i)^{3/2}$, where i is the current through the components.
Derive a nonlinear state space model for this system and obtain the linearized
state space model corresponding to an equilibrium point defined by $i_e \neq 0$ and
$(di/dt)_e = 0$.

8.20 For the electrical system of Problem 8.19 consider that the equilibrium point
corresponds to the initial moment where $i_e = 5$ mA. Known also are $L = 0.1$ H,
$C = 0.002$ F, and the source voltage, which is $v = te^{-3t}$ V. Solve the linearized
state space model using the state transition matrix; also use Simulink® to solve
both the original nonlinear and the linearized state space models.

8.21 The transfer function of a SISO dynamic system is $G(s) = (3s + 1)/(2s^2 +
s + 4)$. Use MATLAB® to determine the corresponding zero-pole-gain model,
then utilize the resulting zpk model to derive a state space model. Also obtain
a state space model directly from the transfer function model. If the two state
space models are different, explain the discrepancy.

8.22 Use the state space model derived for the mechanical system of Problem 8.7
and shown in Figure 8.33 together with the state transition matrix method to
solve for the mass displacement. Known are $m = 0.8$ kg, $c = 70$ N-s/m, $k_1 =
100$ N/m, $k_2 = 130$ N/m, and $f = 40\delta(t)$.

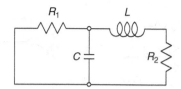

FIGURE 8.41

Two-Mesh Electrical System with Initial Charge on Capacitor.

8.23 Apply the state space approach and the state transition matrix to find the currents in the circuit of Figure 8.41 when an initial charge $q_0 = 0.4\ C$ is applied to the capacitor. Plot these currents in terms of time when known are $R_1 = 20\ \Omega$, $R_2 = 30\ \Omega$, $L = 0.6$ H, and $C = 0.01$ F.

8.24 The homogeneous state space model of a dynamic system is defined by the matrices

$$[A] = \begin{bmatrix} 0 & 1 \\ -120 & -0.1 \end{bmatrix}; \ [C] = \{1\ \ 0\}$$

Use MATLAB® to plot the system response $y(t)$ if the following initial condition is applied: $y(0) = 0.02$.

8.25 Consider the electrical system of Problem 8.3(a) and represented in Figure 8.29. Apply the state transition matrix method to the state space model derived for this system to calculate and plot the voltage that is relevant. Known are $R_1 = 100\ \Omega$, $R_2 = 80\ \Omega$, $C = 5$ mF, $i = 1$A and zero initial conditions.

8.26 Consider the two-room thermal system of Figure 8.42. Derive a state space model for the system and use MATLAB® to plot the two room temperatures θ_1 and θ_2 when the outside temperature is $\theta_o = 35°C$ and the initial room temperatures are $\theta_1(0) = 27°C$ and $\theta_2(0) = 16°C$. All walls are identical and of dimensions 8 m × 4 m × 0.2 m; the wall material thermal conductivity is $k = 0.05$ W/m-C. The rooms' thermal capacitance is $C_{th} = 220,000$ J/K.

8.27 A dynamic system is defined by the state space matrices

$$[A] = \begin{bmatrix} 1 & 0 \\ -3 & -80 \end{bmatrix}; [B] = \begin{Bmatrix} -1 \\ 0 \end{Bmatrix}; [C] = \{1\ \ 0\}; [D] = 2.$$

Find the system response and plot it against time when an input $u = 2\sin(20t)$ is applied to the system with zero initial conditions; use MATLAB® for that.

8.28 Find a state space model for the pneumatic system of Figure 8.43 knowing the input pressures $p_{i,1} = 30$ atm and $p_{i,2} = 20$ atm. Known also are the pneumatic resistances $R_{g1} = 2,000$ m^{-1}-s^{-1}, $R_{g2} = 3,000$ m^{-1}-s^{-1}, $R_{g3} = 1,000$ m^{-1}-s^{-1}, and tank capacitances $C_{g1} = 0.0005$ m-s^2, $C_{g2} = 0.0004$ m-s^2. Use the state transition

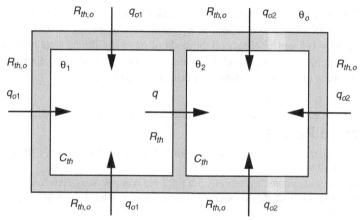

FIGURE 8.42

Two-Room Thermal System.

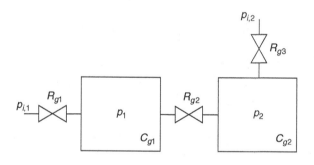

FIGURE 8.43

Pneumatic System with Two Containers and Ductwork.

matrix method , as well as Simulink®, to find and to plot the pressures in the two containers.

8.29 A parallel *RLC* circuit is connected to a voltage source that provides a sinusoidal voltage $v = 80 \sin(2t)$ V. It is also known that $L = 0.6$ H and $C = 380$ μF. Plot the currents in the three components by using Simulink®, when an initial charge $q_0 = 0.1$ C is applied to the capacitor and

(a) The resistor is linear with R $= 80$ Ω.

(b) The resistor defined at (a) has a saturation nonlinearity (discontinuity) defined by a limit voltage of 70 V.

8.30 Two bodies of masses $m_1 = 1.2$ kg and $m_2 = 1$ kg are connected by a dashpot (spring and damper) defined by $c = 90$ N-s/m and $k = 110$ N/m. The bodies slide on a horizontal surface with friction. The coefficient of static friction is $\mu_s = 0.35$ and the coefficient of kinematic friction is $\mu_k = 0.22$. A sinusoidal force $f = 0.5 \sin(5t)$ N acts on the body of mass m_2. Find a state space model of this mechanical system and determine its time response using Simulink® when nonlinear effects of Coulomb friction are considered.

Suggested Reading

H. Klee, *Simulation of Dynamic Systems with MATLAB® and Simulink®*, CRC Press, Boca Raton, FL, 2007.

N. S. Nise, *Control System Engineering*, 5th Ed. John Wiley & Sons, New York, 2008.

B. C. Kuo and F. Golnaraghi, *Automatic Control Systems*, 8th Ed. John Wiley & Sons, New York, 2003

P. Marchand and O. T. Holland, *Graphics and GUIs with MATLAB®*, 3rd Ed. Chapman & Hall/CRC, Boca Raton, FL, 2003.

Frequency-Domain Approach

OBJECTIVES

This chapter studies the frequency-domain behavior of dynamic systems by introducing and using the concept of complex transfer function, which is based on the transfer function of Chapter 7.
The following topics are analyzed:

- Calculation of natural frequencies for conservative dynamic systems.

- Evaluation of the amplitude and phase angle of the steady-state response of SISO and MIMO dynamic systems for harmonic input by using the complex transfer function.

- Harmonic vibration transmission, absorption, and isolation.

- Sensing of mechanical vibrations with harmonic input.

- Cascading unloading systems with sine input by means of the complex transfer function.

- Filtering applications by using electrical and mechanical systems.

- Utilization of specialized MATLAB® commands to model and solve frequency-domain problems.

INTRODUCTION

Two main themes are analyzed in this chapter, both based on a particular form of the transfer function, which is known as the *complex transfer function*. One topic utilizes the characteristic polynomial, which is the denominator of the transfer function for SISO systems and the denominator of the transfer function matrix for MIMO systems, to calculate the natural frequencies of conservative dynamic systems. The other important topic analyzes the steady-state response when the input (or forcing) to a dynamic system is harmonic (sinusoidal mainly). The output, in that particular case, can be determined from the input and the complex transfer function, which is

DOI: 10.1016/B978-0-240-81128-4.00009-X

obtained from the regular transfer function $G(s)$ for SISO systems (or transfer function matrix $[G(s)]$ for MIMO systems) using the substitution $s = j\omega$. The frequency-domain response consists of the output amplitude and phase angle defined in terms of frequency, which becomes the variable. The two resulting graphical representations are known as Bode diagrams, and MATLAB® enables plotting the magnitude (ratio of the output to the input amplitudes) and the phase angle between output and input by built-in functions. Vibration transmission, vibration reduction or isolation, as well as cascading nonloading dynamic systems and electrical/mechanical filters, are analyzed in the frequency domain. MATLAB® conversion between frequency-response data and zero-pole-gain, transfer-function, and state space models is also examined.

9.1 THE CONCEPT OF COMPLEX TRANSFER FUNCTION IN STEADY-STATE RESPONSE AND FREQUENCY-DOMAIN ANALYSIS

One section of this chapter is devoted to calculating the natural frequencies of conservative dynamic systems (of order two or higher) by means of a methodology relying on a particular expression of the transfer function that uses $j\omega$ instead of s as variable. Consider a single-DOF conservative translatory mechanical system formed of a mass m and a spring k. Its natural frequency, as seen in Chapter 2, is $\omega_n = \sqrt{k/m}$. Let us attempt to determine this natural frequency by utilizing the transfer function. When a "dummy" force $u(t)$ is considered to act on the mass aligned with the mass motion coordinate $y(t)$, the following transfer function results, as discussed in Chapter 7:

$$G(s) = \frac{Y(s)}{U(s)} = \frac{1}{ms^2 + k} \tag{9.1}$$

The variable s is complex, as seen in Chapter 6, and for the particular value $s = j\omega$ (where $j = \sqrt{-1}$), the function of Eq. (9.1) becomes

$$G(j\omega) = \frac{1}{-m\omega^2 + k} \tag{9.2}$$

which is known as the *complex transfer function*. Its denominator is the *characteristic polynomial*; its roots are found by solving the *characteristic equation*, which, for this example, is

$$-m\omega^2 + k = 0 \tag{9.3}$$

The root of Eq. (9.3) is the natural frequency of this single-DOF conservative mechanical system.

Another section of this chapter uses the same transfer function expression with $j\omega$ instead of s to evaluate the steady-state response (when time $\to \infty$) of dynamic systems under sinusoidal input. The *total solution* (or *time response*) $y(t)$ of a dynamic system to an input $u(t)$ is the sum of the *complementary solution* (the solution of the homogeneous equation) $y_c(t)$ and a *particular solution* $y_p(t)$; that is, $y(t) = y_c(t) +$ $y_p(t)$. The complementary solution is related to the inherent properties of a system, whereas the particular (or steady-state) solution is connected to the specific input. Our focus in this chapter falls on the *steady-state solution*, which is $y(\infty) = \lim_{t \to \infty} y(t)$, under the assumption that the *input* is *harmonic* (sinusoidal, in particular). We demonstrate that the complementary solution is zero when time $\to \infty$; therefore, the total solution consists of only the particular solution, the latter one having a harmonic form. Before generalizing, let us solve the example of a first-order system.

Example 9.1

Determine the steady-state response of a first-order system for a sinusoidal input $u(t) = U\sin(\omega t)$.

Solution

A first-order system, as introduced in Chapter 1, is defined by the differential equation

$$\tau \dot{y}(t) + y(t) = Ku(t) \tag{9.4}$$

with τ being the time constant and K the static sensitivity. As shown in Appendix A, the complementary solution to the homogeneous equation attached to Eq. (9.4) has the form

$$y_c(t) = c_1 e^{\lambda t} \tag{9.5}$$

where λ is the root of the *characteristic equation*:

$$\tau \lambda + 1 = 0 \tag{9.6}$$

which gives $\lambda = -1/\tau$; therefore, the particular solution of Eq. (9.2) becomes

$$y_c(t) = c_1 e^{-t/\tau} \tag{9.7}$$

As the time constant is a positive quantity and c_1 is a constant, it can be seen that

$$y_c(\infty) = \lim_{t \to \infty} y_c(t) = 0 \tag{9.8}$$

therefore, the total solution is identical to the particular solution when time goes to infinity:

$$y(\infty) = \lim_{t \to \infty} y(t) = y_p(\infty) = \lim_{t \to \infty} y_p(t) \tag{9.9}$$

There are several methods of determining the particular solution, such as the *method of undetermined coefficients*, which is used here. The particular solution is a linear combination of the input and its time derivative:

$$y_p(t) = Y_1 \sin(\omega t) + Y_2 \cos(\omega t) \tag{9.10}$$

Equation (9.10) is substituted into the differential Eq. (9.4) where $u = U\sin(\omega t)$. After identifying the sine and cosine factors in the resulting equation, the unknown coefficients Y_1 and Y_2 are found as

$$Y_1 = \frac{KU}{1 + \tau^2\omega^2}; \; Y_2 = -\frac{KU\tau\omega}{1 + \tau^2\omega^2} \qquad (9.11)$$

Using basic trigonometry, Eq. (9.10) can be reformulated by means of Eqs. (9.11) as

$$y_p(t) = \frac{KU}{\sqrt{1 + \tau^2\omega^2}}\sin[\omega t + \tan^{-1}(-\tau\omega)] \qquad (9.12)$$

which indicates that the particular solution is a sinusoidal function that has the same frequency as the input function, a phase angle with respect to the input, and an amplitude that is a multiplier of the input amplitude.

Let us check that the response amplitude and phase angle can be obtained from a specific form of the system's transfer function. The transfer function corresponding to Eq. (9.4) is

$$G(s) = \frac{Y(s)}{U(s)} = \frac{K}{\tau s + 1} \qquad (9.13)$$

When the substitution $s = j\omega$ is used, Eq. (9.13) changes to

$$G(j\omega) = \frac{Y(j\omega)}{U(j\omega)} = \frac{K}{1 + \tau\omega j} = \frac{K}{1 + \tau^2\omega^2} - \frac{K\tau\omega}{1 + \tau^2\omega^2}j \qquad (9.14)$$

$G(j\omega)$ of Eq. (9.14) is a complex number whose modulus $|G(j\omega)|$ and phase angle $\angle G(j\omega)$ are

$$|G(j\omega)| = \frac{K}{\sqrt{1 + \tau^2\omega^2}}; \; \angle G(j\omega) = \varphi = \tan^{-1}(-\tau\omega) \qquad (9.15)$$

If we now compare Eqs. (9.12) and (9.15), it follows that

$$y_p(t) = |G(j\omega)| \, U\sin[\omega t + \angle G(j\omega)] \qquad (9.16)$$

In other words, the steady-state output characteristics can be obtained from the amplitude and phase angle of $G(j\omega)$, which is known as the *complex* (or *sinusoidal*) *transfer function.* ■

Figure 9.1 illustrates the main calculation steps involved in frequency-domain modeling and that enable finding the natural frequencies of conservative systems and the steady-state response of dynamic systems under sinusoidal input. As

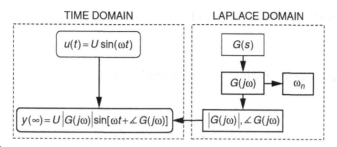

FIGURE 9.1

Calculation of the Natural Frequencies and Steady-State Response Using the Complex Transfer Function Model and Frequency-Domain Analysis.

Figure 9.1 indicates, the fundamental to this approach is the transfer function $G(s)$ for SISO systems (or the transfer function matrix $[G(s)]$ for MIMO systems), which was the subject of Chapter 7. It is noteworthy that evaluating the behavior at infinity of a time-domain response to harmonic input shifts the analysis in the frequency domain because the amplitude and the phase angle of the sinusoidal steady-state response depend on the input frequency solely.

9.2 CALCULATION OF NATURAL FREQUENCIES FOR CONSERVATIVE DYNAMIC SYSTEMS

Using the characteristic polynomial of the complex transfer function (for SISO or single-DOF systems) and the characteristic polynomial of the complex transfer function matrix (for MIMO or multiple-DOF systems), the natural frequencies of conservative dynamic systems are calculated here analytically and by means of MATLAB®.

9.2.1 Analytical Approach

In the previous section, the use of the complex transfer function to calculate the natural frequency of a conservative mechanical single-DOF system is examined. That particular approach can be generalized as follows.

The free undamped behavior of second-order single-DOF systems is described by the differential equation

$$a_2 \ddot{y}(t) + a_0 y(t) = 0 \tag{9.17}$$

Chapter 2 shows that the natural frequency resulting from Eq. (9.17) is

$$\omega_n = \sqrt{\frac{a_0}{a_2}} \tag{9.18}$$

Let us consider the forced undamped second-order system

$$a_2\ddot{y}(t) + a_0 y(t) = u(t) \tag{9.19}$$

The Laplace transform of Eq. (9.19) with zero initial conditions results in the transfer function

$$G(s) = \frac{Y(s)}{U(s)} = \frac{1}{a_2 s^2 + a_0} \tag{9.20}$$

During the modal motion, a single-DOF dynamic system undergoes harmonic (sinusoidal or cosinusoidal) vibrations at the natural frequency ω_n. It is mentioned in Chapter 7 that, even though it relates the input to the output, the transfer function is solely descriptive of a system's properties. The following *characteristic equation* results from the *characteristic polynomial* of Eq. (9.20), which is the denominator of $G(s)$:

$$a_2 s^2 + a_0 = 0 \tag{9.21}$$

Using the substitution $s = j\omega$ in Eq. (9.21) results in

$$-\omega^2 a_2 + a_0 = 0 \tag{9.22}$$

and this is actually the equation of the natural frequency as in Eq. (9.18). As a rule of thumb, we conclude that, for a free, undamped, single-DOF system the natural frequency can be determined by solving the *characteristic equation* (which corresponds to the characteristic polynomial, the denominator of the transfer function) and using the substitution $s = j\omega$.

The equation of the free response of multiple-DOF, second-order undamped systems is of the form

$$[a_2]\{\ddot{y}(t)\} + [a_0]\{y(t)\} = \{0\} \tag{9.23}$$

where $\{y(t)\}$ is the output vector. The harmonic free response is of the form

$$\{y(t)\} = \{Y\}\sin(\omega t) \tag{9.24}$$

which substituted in Eq. (9.23) leads to

$$(-\omega^2[a_2] + [a_0])\{Y\} = \{0\} \tag{9.25}$$

Regarded as a system of algebraic equations with the unknown being the amplitude vector $\{Y\}$, nonzero (nontrivial) solutions are possible only when the system's determinant is zero:

$$\det(-\omega^2[a_2] + [a_0]) = 0 \tag{9.26}$$

Equation (9.26), the *characteristic equation*, needs to be solved to find the natural frequencies of the free undamped response.

Let us turn our attention to modeling the forced response of undamped systems using the complex transfer function matrix approach, which means starting from the equation

$$[a_2]\{\ddot{y}(t)\} + [a_0]\{y(t)\} = \{u(t)\} \tag{9.27}$$

where $\{u(t)\}$ is the input or forcing vector. The Laplace transform with zero initial conditions is applied to Eq. (9.27), taking into account that the matrices $[a_2]$ and $[a_0]$ have constant components, which results in

$$([a_2]s^2 + [a_0])\{Y(s)\} = \{U(s)\} \tag{9.28}$$

The s-domain output vector $\{Y(s)\}$ is found by left-multiplying Eq. (9.28) by the inverse of the $[a_2]s^2 + [a_0]$ matrix:

$$\{Y(s)\} = ([a_2]s^2 + [a_0])^{-1}\{U(s)\} = [G(s)]\{U(s)\} \tag{9.29}$$

where

$$[G(s)] = ([a_2]s^2 + [a_0])^{-1} \tag{9.30}$$

is the *transfer function matrix*. It is known that the denominator of an inverse matrix is the determinant of the original matrix, but $[G(s)]$ is the transfer function matrix; therefore, its denominator is the characteristic polynomial. As a consequence, the characteristic equation corresponding to the transfer function matrix is

$$\det([a_2]s^2 + [a_0])^{-1} = \det([a_2]s^2 + [a_0]) = 0 \tag{9.31}$$

Equation (9.31) employs the property that a nonsingular matrix (one whose determinant is nonzero) and its inverse have the same determinant. As is the case with single-DOF systems, let us use $s = j\omega$ in Eq. (9.31), which produces Eq. (9.26), the known form of the characteristic equation for multiple-DOF systems.

Example 9.2

For the mechanical system of Figure 9.2(a) known are $m_1 = 1\,\text{kg}$, $m_2 = 1.8\,\text{kg}$, $k_1 = 200\,\text{N/m}$, $k_2 = 250\,\text{N/m}$. Assuming the two levers are massless and the motions are small, determine the natural frequencies of this system using the complex transfer function matrix and its characteristic equation.

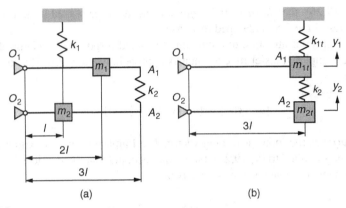

FIGURE 9.2

Lever Mechanical System: (a) Original System; (b) System with Relocated Components.

Solution

As discussed in Chapter 3, the masses and springs can be relocated conveniently on a lever, for instance, by having them all placed at the ends of the levers, at points A_1 and A_2, as shown in Figure 9.2(b). In doing so, the originally rotational system is transformed into a translational one, since the displacements are assumed small and all mechanical components are aligned to a vertical axis. In this situation, the coordinates y_1 and y_2 are the parameters defining the state of the system, which is therefore a two-DOF system. The transformed mass and spring parameters are

$$m_{1t} = \left(\frac{2l}{3l}\right)^2 m_1 = \frac{4}{9}m_1; \; m_{2t} = \left(\frac{l}{3l}\right)^2 m_2 = \frac{1}{9}m_2; \; k_{1t} = \left(\frac{l}{3l}\right)^2 k_1 = \frac{1}{9}k_1 \qquad (9.32)$$

The dynamic equations of motion for the two bodies are

$$\begin{cases} m_{1t}\ddot{y}_1 = -k_{1t}y_1 - k_2(y_1 - y_2) \\ m_{2t}\ddot{y}_2 = -k_2(y_2 - y_1) \end{cases} \qquad (9.33)$$

Laplace transforming Eqs. (9.33) results in the following vector-matrix equation:

$$\begin{bmatrix} m_{1t}s^2 + k_{1t} + k_2 & -k_2 \\ -k_2 & m_{2t}s^2 + k_2 \end{bmatrix} \begin{bmatrix} Y_1(s) \\ Y_2(s) \end{bmatrix} = \begin{bmatrix} 0 \\ 0 \end{bmatrix} \qquad (9.34)$$

Using $s = j\omega$, Eq. (9.34) becomes

$$\begin{bmatrix} -m_{1t}\omega^2 + k_{1t} + k_2 & -k_2 \\ -k_2 & -m_{2t}\omega^2 + k_2 \end{bmatrix} \begin{bmatrix} Y_1(j\omega) \\ Y_2(j\omega) \end{bmatrix} = \begin{bmatrix} 0 \\ 0 \end{bmatrix} \qquad (9.35)$$

Compared to the known format of a transfer function matrix, Eq. (9.35) indicates that the complex transfer function matrix is

$$[G(j\omega)] = \begin{bmatrix} -m_{1t}\omega^2 + k_{1t} + k_2 & -k_2 \\ -k_2 & -m_{2t}\omega^2 + k_2 \end{bmatrix}^{-1}$$

$$= \begin{bmatrix} \dfrac{-m_{2t}\omega^2 + k_2}{d} & \dfrac{k_2}{d} \\ \dfrac{k_2}{d} & \dfrac{-m_{1t}\omega^2 + k_{1t} + k_2}{d} \end{bmatrix} \tag{9.36}$$

with

$$d = m_{1t}m_{2t}\omega^4 - (m_{1t}k_2 + m_{2t}k_{1t} + m_{2t}k_2)\omega^2 + k_{1t}k_2 \tag{9.37}$$

being the characteristic polynomial. Using the numerical values of this example, the natural frequencies are $\omega_{n1} = 5.85$ rad/s and $\omega_{n2} = 42.8$ rad/s as obtained by solving the equation $d = 0$. ■

9.2.2 MATLAB® Approach

As shown in previous chapters, the MATLAB® command eig yields the eigenvalues and eigenvectors of a square matrix, either in numeric or symbolic format. However, another command furnishes the natural frequencies associated with a characteristic polynomial, in addition to providing the damping ratios. Let us analyze the following example, which introduces the damp(sys) command (in the Control System Toolbox™), where sys is an LTI object (zero-pole-gain, transfer function, or state space) previously defined.

Example 9.3
Use MATLAB® and the damp command to calculate the natural frequencies of the mechanical lever system analyzed in Example 9.2 and sketched in Figure. 9.2.

Solution
After introducing the numerical values of the parameters, the following commands can be used to calculate the natural frequencies of the two-DOF mechanical system of Example 9.2 and shown in Fig. 9.2:

```
s=tf('s');
g11=m1t*s^2+k1t+k2;
g12=-k2;
g22=m2t*s^2+k2;
ginv=[g11,g12;g12,g22];
g=inv(ginv);
damp(g)
```

and the return is

```
Eigenvalue                        Damping          Freq. (rad/s)
-2.22e-016 + 5.85e+000i           3.80e-017        5.85e+000
-2.22e-016 - 5.85e+000i           3.80e-017        5.85e+000
-2.22e-016 + 5.85e+000i           3.80e-017        5.85e+000
-2.22e-016 - 5.85e+000i           3.80e-017        5.85e+000
 5.33e-015 + 4.28e+001i          -1.25e-016        4.28e+001
 5.33e-015 - 4.28e+001i          -1.25e-016        4.28e+001
 5.33e-015 + 4.28e+001i          -1.25e-016        4.28e+001
 5.33e-015 - 4.28e+001i          -1.25e-016        4.28e+001
```

It can be seen that the natural frequencies (the last column of the table) are indeed the ones obtained in Example 9.2, as expected. ◼

9.3 STEADY-STATE RESPONSE OF DYNAMIC SYSTEMS TO HARMONIC INPUT

As presented in this section, analytical methods and MATLAB® can be used to evaluate the steady-state response of dynamic systems being subjected to sinusoidal input (forcing) as shown in the following.

9.3.1 Analytical Approach

The steady-state response of SISO and MIMO dynamic systems are formulated and studied analytically.

SISO Systems

This section studies SISO systems in relation to the problem of harmonic excitation and the corresponding steady-state solution. It also introduces the main frequency response parameters for the second-order systems.

Steady-State Solution Under Harmonic (Sinusoidal) Input

The results obtained for the particular case of a first-order system (see Example 9.1) can be extended to systems of higher order, as shown in the following. When the input is of harmonic form, $u = U \sin(\omega t)$, its relationship to the output and the transfer function into the s-domain is

$$Y(s) = G(s)\frac{U\omega}{s^2 + \omega^2} \tag{9.38}$$

Equation (9.38) indicates that the total response contains contributions from the system itself, represented by the transfer function $G(s)$, and from the forcing (input), identified by the Laplace transform of $U \sin(\omega t)$. The partial fraction expansion on the right-hand side of Eq. (9.38) leads to

$$Y(s) = \sum_i G_i(s) + \frac{c_1}{s - j\omega} + \frac{c_2}{s + j\omega} = Y_c(s) + Y_p(s) \tag{9.39}$$

where the sum is denoted by $Y_c(s)$, the Laplace transform of the complementary solution, and contains partial fractions resulting from $G(s)$. The two-fraction sum denoted by $Y_p(s)$, the Laplace transform of the particular solution, is the expansion of the fraction on the right-hand side of Eq. (9.38). For a *stable system* (more on this topic in the website Chapter 11), the inverse Laplace transform of $Y_c(s)$ is zero when time goes to infinity for a system with losses; therefore, the steady-state response consists of only the particular solution, as demonstrated for first-order systems. As a consequence, we are interested in working with only

$$G(s)\frac{U\omega}{s^2 + \omega^2} = \frac{c_1}{s - j\omega} + \frac{c_2}{s + j\omega} \tag{9.40}$$

The constants c_1 and c_2 are determined using the cover-up method (see Chapter 6) as

$$\begin{cases} c_1 = \left[G(s)\frac{U\omega}{s + j\omega} \right]_{s = j\omega} = \frac{U}{2j}G(j\omega) \\ c_2 = \left[G(s)\frac{U\omega}{s - j\omega} \right]_{s = -j\omega} = -\frac{U}{2j}G(-j\omega) \end{cases} \tag{9.41}$$

A complex number can be written in *phasor form*, which is obtained using the trigonometric definition of the complex number and Euler's equations:

$$G(j\omega) = |G(j\omega)| \left[\cos(\angle G(j\omega)) + j\sin(\angle G(j\omega)) \right]$$

$$= |G(j\omega)| \left[\frac{e^{j\angle G(j\omega)} + e^{-j\angle G(j\omega)}}{2} + j\frac{e^{j\angle G(j\omega)} - e^{-j\angle G(j\omega)}}{2j} \right]$$

$$= |G(j\omega)| \, e^{j\angle G(j\omega)} \tag{9.42}$$

At the same time, the numbers $G(j\omega)$ and $G(-j\omega)$ in Eqs. (9.41) are complex conjugates and therefore have the same modulus; as a result, Eqs. (9.38), (9.39), and (9.40) result in

$$Y_p(s) = U\,|G(j\omega)| \left(\frac{1}{2j} \times \frac{e^{j\angle G(j\omega)}}{s - j\omega} - \frac{1}{2j} \times \frac{e^{-j\angle G(j\omega)}}{s + j\omega} \right) \tag{9.43}$$

Application of the inverse Laplace transform to Eq. (9.43) results in

$$y_p(t) = U\,|G(j\omega)| \times \frac{e^{j[\omega t + \angle G(j\omega)]} - e^{-j[\omega t + \angle G(j\omega)]}}{2j}$$

$$= U\,|G(j\omega)| \sin\left[\omega t + \angle G(j\omega) \right] \tag{9.44}$$

Equation (9.44) shows that the steady-state response can be expressed as a sinusoidal function of time:

$$y_p(t) = Y_p \sin\left[\omega t + \varphi \right] \tag{9.45}$$

where, by comparison to Eq. (9.44),

$$Y_p = U \,|\, G(j\omega) \,|; \quad \varphi = \angle G(j\omega) \tag{9.46}$$

In other words,

- The steady-state amplitude is obtained from the input amplitude U multiplied by the modulus of the complex transfer function $|G(j\omega)|$ (when $s = j$ in $G(s)$).
- The phase angle between the output and the input is the angle of the complex transfer function, $\angle G(j\omega)$.

The *frequency response* therefore consists of the two functions, the *modulus of the complex transfer function* and the *phase* of the same function expressed in *terms of the frequency* ω. Alternatively, these amounts are called the *frequency response magnitude*, denoted by $M(\omega)$ and the *frequency response phase*, denoted by $\varphi(\omega)$. Performing a *frequency-domain analysis* means studying the two functions $M(\omega)$ and $\varphi(\omega)$. As it results from the transfer function $G(s)$, the complex transfer function $G(j\omega)$ is the ratio of two complex-valued polynomials. Because a complex number can be formulated in phasor form, as in Eq. (9.42), the complex transfer function modulus is the ratio of the numerator to the denominator modulii, whereas the complex transfer function phase angle is the difference between the numerator and denominator phase angles.

Example 9.4

Determine the frequency response of the operational amplifier electrical system of Figure 9.3 for sinusoidal input voltage and $R = 220\ \Omega$, $L = 0.8$ H, and $C = 360\ \mu$F.

Solution

As seen in Chapter 7, the transfer function of this system is calculated based on imped-ances as

$$G(s) = \frac{V_o(s)}{V_i(s)} = -\frac{Z_2}{Z_1} = -\frac{\dfrac{1}{Cs}}{R + Ls} = -\frac{1}{LCs^2 + RCs} \tag{9.47}$$

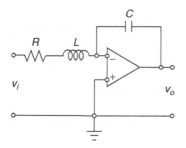

FIGURE 9.3

Operational Amplifier System with Input Resistor-Inductor and Feedback Capacitor.

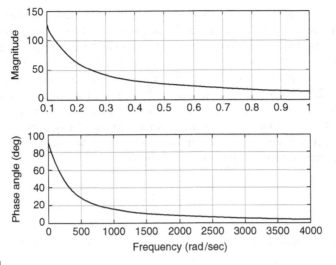

FIGURE 9.4

Magnitude and Phase Angle Plots for the Electrical System of Figure 9.3.

and the corresponding complex transfer function is

$$G(j\omega) = \frac{1}{LC\omega^2 - RC\omega j} \tag{9.48}$$

which is a complex number in standard algebraic form, whose modulus is

$$M(\omega) = |G(j\omega)| = \frac{1}{|LC\omega^2 - RC\omega j|} = \frac{1}{\omega C\sqrt{L^2\omega^2 + R^2}} \tag{9.49}$$

The phase of the complex transfer function of Eq. (9.48) is

$$\varphi = \angle G(j\omega) = \tan^{-1}(0) - \tan^{-1}\left(-\frac{R}{L\omega}\right) = \tan^{-1}\left(\frac{R}{L\omega}\right) \tag{9.50}$$

For the numerical values of the electrical components, the magnitude and phase angle are plotted in Figure 9.4. Equations (9.49) and (9.50) indicate that both the magnitude and the phase angle become zero when $\omega \to \infty$. As shown later in this chapter, the magnitude behavior displayed in the frequency response of Figure 9.4 is that of a *low-pass filter*, which magnifies the input signal for a narrow low-frequency range and blocks it (or diminishes it) for higher frequencies. ■

Frequency Response Parameters of Second-Order Systems
The typical frequency response for second- or higher-order dynamic systems displays a maximum in magnitude, which is known as *resonant peak*, denoted by

M_r, and occurs at the *resonant frequency* ω_r. The magnitude generally decreases for frequencies larger than a threshold frequency ω_c (the *cutoff frequency*, also named *bandwidth*), where the output power amplitude becomes half the input power amplitude. Since the power amplitude of a sinusoidal signal is proportional to the square of the amplitude of that signal, the cutoff frequency corresponds to a value of $M(\omega) = \sqrt{1/2}$ or 0.707 of the magnitude. The slope of the magnitude curve around the cutoff frequency is the *cutoff rate*. Figure 9.5 plots the typical magnitude and phase angle curves in terms of frequency for a second-order system.

For a second-order system with unitary gain ($K = 1$), whose mathematical model is introduced in Chapter 1, the transfer function is

$$G(s) = \frac{\omega_n^2}{s^2 + 2\xi\omega_n s + \omega_n^2} \tag{9.51}$$

The magnitude and phase angle are obtained from Eq. (9.51) by taking $s = j\omega$:

$$|G(j\omega)| = M(\beta) = \frac{1}{\sqrt{(1 - \beta^2)^2 + (2\xi\beta)^2}}; \ \varphi(\beta) = -\tan^{-1}\frac{2\xi\beta}{1 - \beta^2} \tag{9.52}$$

where $\beta = \omega/\omega_n$. The resonant frequency is found by equating the derivative of $M(\beta)$ to zero:

$$-\frac{4\beta(\beta^2 - 1 + 2\xi^2)}{2\left[\sqrt{(1 - \beta^2)^2 + (2\xi\beta)^2}\right]^3} = 0 \tag{9.53}$$

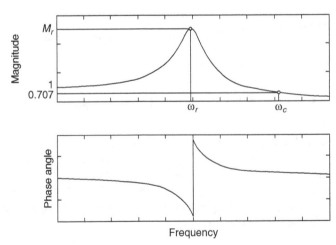

FIGURE 9.5

Magnitude and Phase Angle Plots Illustrating the Frequency Response of a Typical Second-Order System.

The valid solution of Eq. (9.53) is

$$\beta_r = \sqrt{1 - 2\xi^2} \tag{9.54}$$

therefore, the *resonant frequency* is

$$\omega_r = \omega_n \sqrt{1 - 2\xi^2} \tag{9.55}$$

Real values of the resonant frequency are possible only for $1 - 2\xi^2 \geq 0$, which means that $\xi \leq 1/\sqrt{2} = 0.707$. For damping ratios that are larger than 0.707, there is no resonance at all. For all $\xi \leq 0.707$, the resonant peak is found by combining Eqs. (9.52) and (9.54):

$$M_r = \frac{1}{2\xi\sqrt{1 - \xi^2}} \tag{9.56}$$

It can be seen that the resonant frequency depends on both the natural frequency and damping ratio, while the resonant peak depends only on the damping ratio. Figure 9.6 plots $M(\beta)$ for three values of the damping coefficient: $\xi_1 = 0.1$, $\xi_2 = 0.4$, and $\xi_3 = 0.707$.

The cutoff frequency is obtained by taking $M = 1/\sqrt{2}$ in Eq. (9.52), which results in

$$\omega_c = \omega_n \sqrt{1 - 2\xi^2 + \sqrt{4\xi^4 - 4\xi^2 + 2}} \tag{9.57}$$

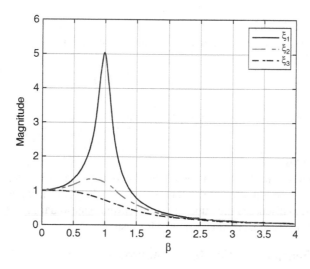

FIGURE 9.6

Magnitude as a Function of the Frequency Ratio.

Note: The natural frequency and damping ratio are properties intrinsic to second-order systems, which are defined through the characteristic polynomial of Eq. (9.51). As a consequence, ω_n and ξ can be determined from transfer functions with numerators that are different from the one of Eq. (9.51).

Example 9.5

A microcantilever of constant rectangular cross-section has a thickness $h = 200$ nm, a width $w = 5$ μm, and a length $l = 50$ μm. The microcantilever undergoes harmonic out-of-plane bending vibrations. Use lumped-parameter modeling to determine the cantilever mass density ρ as well as the damping coefficient c of the gaseous environment. Consider that known are the following parameters: $\omega_n = 5.65 \times 10^5$ rad/s, $\omega_r = 5.1 \times 10^5$ rad/s, and $E = 160$ GPa.

Solution

The damping ratio is determined from Eq. (9.55) as

$$\xi = \sqrt{\frac{1}{2}\left(1 - \frac{\omega_r^2}{\omega_n^2}\right)} \tag{9.58}$$

which yields $\xi = 0.3043$. As shown in Chapter 3, the lumped parameter model of a cantilever consists of a spring of equivalent stiffness k_e, given in Table 3.2, and a pointlike equivalent mass m_e, provided in Table 3.1, with the natural frequency $\omega_n = \sqrt{k_e/m_e}$. This latter equation yields

$$\rho = \frac{35Eh^2}{33\omega_n^2 l^4} \tag{9.59}$$

thus, $\rho = 3402.2$ kg/m³. The damping coefficient is calculated as

$$c = 2\xi\omega_n m_e = 2\xi \times \frac{33}{140}\rho l w h \omega_n \tag{9.60}$$

whose numerical value is $c = 1.38 \times 10^{-8}$ N-s/m.

MIMO Systems

MIMO systems can also be rendered into frequency-domain models using either the transfer function matrix or the state space modeling approach, as shown in the following. Also discussed are superimposing several sinusoidal inputs and calculating the total steady-state response of MIMO systems.

Complex Transfer Function Matrix Approach

For MIMO systems under the action of harmonic input, it is necessary to firstly determine the transfer function matrix $|G(s)|$, which creates a relationship between the input $\{U(s)\}$ and output $\{Y(s)\}$ Laplace-transformed vectors, such as the ones studied in Chapter 7. The next step is to use the substitution $s = j\omega$ to obtain the complex transfer function matrix $|G(j\omega)|$. Let us analyze an example then generalize the frequency response of MIMO systems.

Example 9.6

Determine the frequency response of the two-vessel pneumatic system shown in Figure 5.19 of Example 5.12, evaluating the pressures in the two vessels when the input pressure p_i is sinusoidal. Known are the following pneumatic system properties: $R_{g1} = 500$ m^{-1}s^{-1}, $R_{g2} = 600$ m^{-1}s^{-1}, $C_{g1} = 0.004$ m-s^2, $C_{g2} = 0.005$ m-s^2.

Solution

By combining the second and fourth of Eqs. (5.119), the following equation is produced:

$$R_{g2}C_{g2}\frac{dp_o(t)}{dt} + p_o(t) - p(t) = 0 \tag{9.61}$$

A similar combination of the first, third, and fourth of Eqs. (5.119) results in

$$R_{g1}C_{g1}\frac{dp(t)}{dt} + \left(1 + \frac{R_{g1}}{R_{g2}}\right)p(t) - \frac{R_{g1}}{R_{g2}}p_o(t) = p_i(t) \tag{9.62}$$

Equations (9.61) and (9.62) can be assembled in vector-matrix form:

$$\begin{bmatrix} R_{g1}C_{g1} & 0 \\ 0 & R_{g2}C_{g2} \end{bmatrix}\begin{Bmatrix} \dfrac{dp(t)}{dt} \\ \dfrac{dp_o(t)}{dt} \end{Bmatrix} + \begin{bmatrix} 1+\dfrac{R_{g1}}{R_{g2}} & -\dfrac{R_{g1}}{R_{g2}} \\ -1 & 1 \end{bmatrix}\begin{Bmatrix} p(t) \\ p_o(t) \end{Bmatrix} = \begin{Bmatrix} p_i(t) \\ 0 \end{Bmatrix} \tag{9.63}$$

which indicates the pneumatic system is a single-input, two-output system, so a MIMO system. The input is p_i and the output components are the two pressures in the vessels, p and p_o. The form of this system is

$$[a_1]\{\dot{y}(t)\} + [a_0]\{y(t)\} = \{u(t)\} \tag{9.64}$$

By applying the Laplace transform with zero initial conditions to Eq. (9.64), a relationship $\{Y(s)\} = [G(s)]\{U(s)\}$ is obtained; the transfer function matrix $[G(s)]$ is

$$[G(s)] = (s\,[a_1] + [a_0])^{-1} \tag{9.65}$$

By using the numerical values of this example and MATLAB®, the following transfer matrix is obtained:

$$[G(s)] = \begin{bmatrix} \dfrac{6s + 2}{12s^2 + 15s + 2} & \dfrac{1.67}{12s^2 + 15s + 2} \\ \dfrac{2}{12s^2 + 15s + 2} & \dfrac{4s + 3.67}{12s^2 + 15s + 2} \end{bmatrix} \tag{9.66}$$

If $G_{11}(s)$, $G_{12}(s)$ are the elements of the transfer function matrix on the first row, and $G_{21}(s)$, $G_{22}(s)$ are the elements on the second row, the following relationship can be written in the Laplace domain:

$$\begin{Bmatrix} P(s) \\ P_o(s) \end{Bmatrix} = \begin{bmatrix} G_{11}(s) & G_{12}(s) \\ G_{21}(s) & G_{22}(s) \end{bmatrix} \begin{Bmatrix} P_i(s) \\ 0 \end{Bmatrix} \tag{9.67}$$

which can be reduced to the following scalar equations:

$$\begin{cases} P(s) = G_{11}(s) P_i(s) \\ P_o(s) = G_{21}(s) P_i(s) \end{cases} \tag{9.68}$$

The transfer functions of Eqs. (9.68) are defined in Eq. (9.66). Each equation in Eq. (9.68) represents a SISO system, which can be treated separately. The complex transfer functions corresponding to $G_{11}(s)$ and $G_{21}(s)$ are needed first. They are

$$G_{11}(j\omega) = \frac{2 + 6\omega j}{2 - 12\omega^2 + 15\omega j}; \quad G_{21}(j\omega) = \frac{2}{2 - 12\omega^2 + 15\omega j} \tag{9.69}$$

The following modulii and phase angles are obtained from Eqs. (9.69):

$$\begin{cases} |G_{11}(j\omega)| = \sqrt{\dfrac{4 + 36\omega^2}{144\omega^4 + 177\omega^2 + 4}}; \angle G_{11}(j\omega) = \varphi_{11} \\ \quad = \tan^{-1}(3\omega) - \tan^{-1}\dfrac{15\omega}{2 - 12\omega^2} \\ |G_{21}(j\omega)| = \dfrac{2}{\sqrt{144\omega^4 + 177\omega^2 + 4}}; \angle G_{21}(j\omega) = \varphi_{21} \\ \quad = -\tan^{-1}\dfrac{15\omega}{2 - 12\omega^2} \end{cases} \tag{9.70}$$

and these amounts are plotted in Figure 9.7.

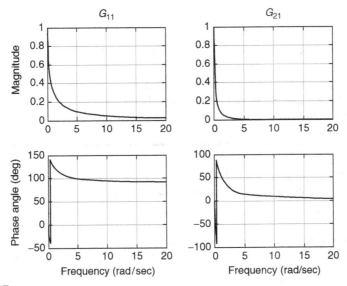

Magnitude and Phase Angle Plots for a Pneumatic System with One Input and Two Outputs.

The previous example with one input and two outputs needs only two transfer functions of the 2×2 transfer function matrix to obtain the two outputs. For the general MIMO case, where there might be m inputs and p outputs, the general relationship between the input vector and output vector through the transfer function matrix is written directly with $s = j\omega$ as

$$\{Y(j\omega)\} = [G(j\omega)]\{U(j\omega)\} \tag{9.71}$$

The complex transfer matrix $[G(j\omega)]$ contains $p \times m$ transfer functions connecting every component of the input vector to every component of the output vector. To graphically qualify the frequency response of such a system, $p \times m$ magnitude plots and an equal number of phase angle plots are needed.

Linear Superposition for Time Response

The *time response of a MIMO system* under *multiple sinusoidal input* can be found through *the linear superposition of individual responses*, since any component of the output vector results as the sum of all input components multiplied by their corresponding transfer functions. The output component i, for instance, is computed as

$$y_i(t) = \sum_{k=1}^{m} |G_{ik}(j\omega_k)| \, U_k \sin(\omega_k t + \varphi_{ik}), \text{ for } i = 1, 2, \ldots, p \tag{9.72}$$

with

$$|G_{ik}(j\omega_k)| = \sqrt{[\mathrm{Re}\,G_{ik}(j\omega_k)]^2 + [\mathrm{Im}\,G_{ik}(j\omega_k)]^2}; \quad \varphi_{ik} = \tan^{-1}\frac{\mathrm{Im}\,G_{ik}(j\omega_k)}{\mathrm{Re}\,G_{ik}(j\omega_k)} \qquad (9.73)$$

where Re and Im are the real and imaginary parts of the complex transfer function G_{ik}, which connects the output i to the input k. In Eq. (9.72), U_k is the amplitude of the sinusoidal input k and φ_{ik} is the phase angle between the output i and the input k.

Example 9.7

Consider the mechanical system of Figure 9.8 with $m_1 = m_2 = m_3 = 1$ kg, $c_1 = c_2 = 2$ N-s/m, $k_1 = k_2 = k_3 = 1$ N/m is acted upon by the forces $u_1 = 5 \sin t$; $u_2 = 2 \sin(3t)$. Both forces are expressed in Newtons.

a. Determine and plot the magnitude and phase angle that define the frequency response connecting the output y_1 and the input u_1.

b. Express the steady-state outputs (time responses) of this system under the given force inputs.

Solution

a. The mechanical system is a two-input, three-output one and its equations of motion are

$$\begin{cases} m_1\ddot{y}_1 = u_1 - c_1(\dot{y}_1 - \dot{y}_2) - k_1(y_1 - y_2) \\ m_2\ddot{y}_2 = u_2 - c_1(\dot{y}_2 - \dot{y}_1) - k_1(y_2 - y_1) - c_2(\dot{y}_2 - \dot{y}_3) - k_2(y_2 - y_3) \\ m_3\ddot{y}_3 = -c_2(\dot{y}_3 - \dot{y}_2) - k_2(y_3 - y_2) - k_3 y_3 \end{cases} \qquad (9.74)$$

Application of the Laplace transform to Eqs. (9.74) with zero initial conditions generates the equation

$$\begin{bmatrix} m_1 s^2 + c_1 s + k_1 & -(c_1 s + k_1) & 0 \\ -(c_1 s + k_1) & m_2 s^2 + (c_1 + c_2)s + k_1 + k_2 & -(c_2 s + k_2) \\ 0 & -(c_2 s + k_2) & m_3 s^2 + c_2 s + k_2 + k_3 \end{bmatrix} \begin{bmatrix} Y_1(s) \\ Y_2(s) \\ Y_3(s) \end{bmatrix}$$

$$= \begin{bmatrix} U_1(s) \\ U_2(s) \\ 0 \end{bmatrix} \qquad (9.75)$$

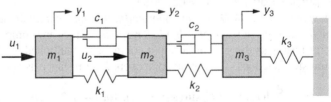

FIGURE 9.8

Lumped-Parameter Mechanical System with Harmonic Input Forces.

Equation (9.75) enables solving for $Y_1(s)$, $Y_2(s)$ and $Y_3(s)$ in terms of $U_1(s)$ and $U_2(s)$ as

$$\begin{bmatrix} Y_1(s) \\ Y_2(s) \\ Y_3(s) \end{bmatrix} = \begin{bmatrix} G_{11}(s) & G_{12}(s) \\ G_{21}(s) & G_{22}(s) \\ G_{31}(s) & G_{32}(s) \end{bmatrix} \begin{bmatrix} U_1(s) \\ U_2(s) \end{bmatrix} \tag{9.76}$$

The transfer functions of the matrix in Eq. (9.76) are, for the numerical values of this example,

$$\begin{cases} G_{11}(s) = \dfrac{s^4 + 6s^3 + 8s^2 + 8s + 3}{s^6 + 8s^5 + 17s^4 + 18s^3 + 10s^2 + 4s + 1} \\[3mm] G_{12}(s) = G_{21}(s) = \dfrac{2s^3 + 5s^2 + 6s + 2}{s^6 + 8s^5 + 17s^4 + 18s^3 + 10s^2 + 4s + 1} \\[3mm] G_{22}(s) = \dfrac{s^4 + 4s^3 + 7s^2 + 6s + 2}{s^6 + 8s^5 + 17s^4 + 18s^3 + 10s^2 + 4s + 1} \\[3mm] G_{31}(s) = \dfrac{4s^2 + 4s + 1}{s^6 + 8s^5 + 17s^4 + 18s^3 + 10s^2 + 4s + 1} \\[3mm] G_{32}(s) = \dfrac{2s^3 + 5s^2 + 4s + 1}{s^6 + 8s^5 + 17s^4 + 18s^3 + 10s^2 + 4s + 1} \end{cases} \tag{9.77}$$

The magnitudes and phase angles of the transfer functions in Eq. (9.76) are determined in the regular manner using $s = j\omega$. The magnitude of each is then the ratio of the numerator modulus to the denominator modulus and the phase angle is the difference between the numerator's phase angle and the denominator's phase angle. Figure 9.9 shows the frequency domain plots corresponding to $G_{11}(s)$.

b. By following the generic Eq. (9.72), the three outputs are expressed as

$$\begin{cases} y_1 = U_1|G_{11}(j)|\sin(t + \varphi_{11}) + U_2|G_{12}(3j)|\sin(3t + \varphi_{12}) \\ y_2 = U_1|G_{21}(j)|\sin(t + \varphi_{21}) + U_2|G_{22}(3j)|\sin(3t + \varphi_{22}) \\ y_3 = U_1|G_{31}(j)|\sin(t + \varphi_{31}) + U_3|G_{32}(3j)|\sin(3t + \varphi_{32}) \end{cases} \tag{9.78}$$

with $U_1 = 5$ and $U_2 = 2$ being the amplitudes of the inputs, and the six complex transfer functions $G_{11}(j\omega)$ through $G_{32}(j\omega)$ can be obtained from the corresponding ones listed in the transfer function matrix of this example for G_{11}, G_{21}, and G_{31} when $\omega = 1$ rad/s, as well as for G_{12}, G_{22}, and G_{32} when $\omega = 3$ rad/s. After determining the modulii and phase angles of all the complex transfer functions, they are substituted correspondingly into Eqs. (9.78). The input forces are plotted in Figure 9.10 and the steady-state time responses are plotted in Figure 9.11.

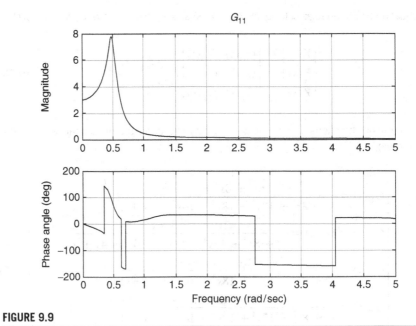

FIGURE 9.9

Magnitude and Phase Angle Plots for the $G_{11}(j\omega)$ Transfer Function.

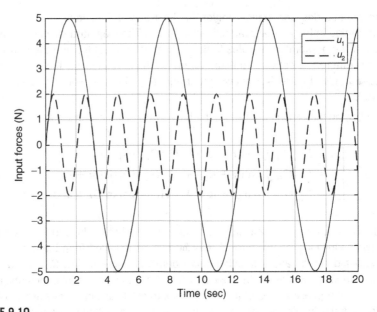

FIGURE 9.10

Sinusoidal Input Forces.

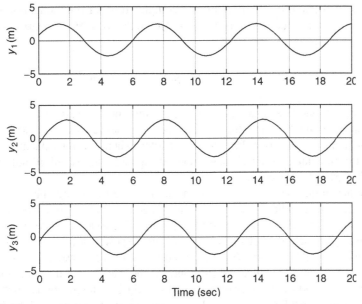

FIGURE 9.11

Steady-State Response of Two-Input, Three-Output Dynamic System.

9.3.2 **Using MATLAB® for Frequency Response Analysis**

MATLAB® can use built-in functions either in the core program or in the Control System Toolbox™ that generate frequency-domain calculations and plots quite simply. Specifically, it is possible to determine the frequency response of dynamic systems through *Bode plots* using either the transfer function or state space models. Discrete frequency data can also be used to obtain frequency plots. Conversions from zero-pole-gain, transfer function, or state space models to a frequency-response data model can be performed using MATLAB®.

The Logarithmic Scale and Bode Plots

Frequency plots similar to the ones shown thus far in this chapter but that utilize logarithmic scales for magnitude and frequency are known as *Bode plots*. The Control System Toolbox™ of MATLAB® has the bode(sys) command to obtain the two plots (magnitude and phase versus frequency) once a linear time-invariant model is available, as defined by sys. Traditionally, Bode plots utilize a logarithmic scale for the frequency, because using $\log_{10}\omega$ instead of simply ω provides a wider dynamic range (for instance, when $\omega = 1000$ rad/s, the logarithmic counterpart is only $\log_{10}\omega = 3$). Another reason for using a logarithmic scale for magnitudes is that the output, input, and complex transfer function amplitudes are connected in a product form as $Y(j\omega) = G(j\omega)U(j\omega)$, which, logarithmic scale, becomes

$$\log_{10}|Y(j\omega)| = \log_{10}|G(j\omega)| + \log_{10}|U(j\omega)| \qquad (9.79)$$

Therefore, an additive relationship between amplitudes (magnitudes) is obtained using the log scale. For cascading systems (serially connected systems without interstage loading), as is shown later in this chapter, where the total magnitude is the product of individual systems' magnitudes, using addition instead of multiplication is particularly advantageous.

The actual unit being used for the magnitude $M(\omega)$ in a Bode plot is known as *decibel*, and the number of decibels is calculated as $20 \log_{10}M(\omega)$. The magnitude is the ratio of two signal amplitudes, and power is proportional to the square of the amplitude (such as in electrical systems where power = resistance $\times$ current^2); in other words, $|G(j\omega)|^2 = M(\omega)^2$ is proportional to the power ratio. A factor of 10 in a power ratio is named *bel* (in honor of the telephone's inventor, Alexander Graham Bell), and the number of bels corresponding to the magnitude is # bells = $\log_{10}|G(j\omega)|^2$. Considering that 1 bel has 10 decibels, the number of decibels is # decibels = $10 \log_{10}|G(j\omega)|^2 = 20 \log_{10}|G(j\omega)|$.

In MATLAB®, MIMO systems can also be handled by means of the transfer function matrix, and the frequency response can readily be determined as illustrated in the following example.

Example 9.8

Use MATLAB® to plot the Bode diagrams corresponding to the transfer function matrix

$$[G(s)] = \begin{bmatrix} \dfrac{1}{s^2 + 2s + 10} & \dfrac{s + 2}{s^2 + 2s + 10} \\ -\dfrac{2}{s^2 + 2s + 10} & \dfrac{2s}{s^2 + 2s + 10} \end{bmatrix}$$

Solution

The following MATLAB® command sequence can be used

```
>> g11 = tf(1,[1,2,10]);
>> g12 = tf([1,2],[1,2,10]);
>> g21 = tf(-2,[1,2,10]);
>> g22 = tf([2,0],[1,2,10]);
>> m = [g11,g12;g21,g22];
>> bode(m)
```

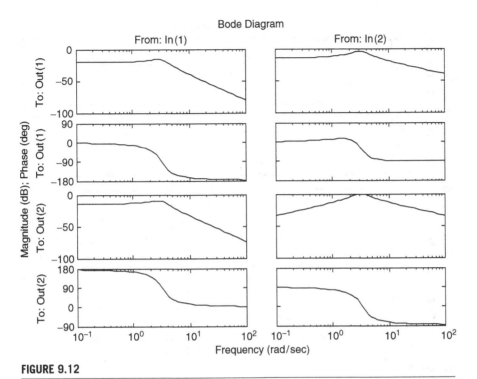

FIGURE 9.12

Bode Plots for a Two-Input, Two-Output Dynamic System.

to generate the Bode diagrams of Figure 9.12. The 2×2 transfer function matrix indicates that the system is a two-input, two-output one; therefore, four plot pairs resulted. ■

Plotting the Frequency Response from a State Space Model

The MATLAB® bode command can be used in conjunction with a state space model directly, unlike any previous conversion to transfer function matrix model. Let us study the following example.

Example 9.9

A state space model is defined by the following matrices:

$$[A] = \begin{bmatrix} 1 & 0 \\ -3 & 2 \end{bmatrix}; [B] = \begin{Bmatrix} 1 \\ -1 \end{Bmatrix}; [C] = \begin{bmatrix} 0 & 1 \\ -1 & 4 \end{bmatrix}; [D] = \begin{Bmatrix} -2 \\ 0.5 \end{Bmatrix}$$

Plot the frequency response corresponding to this system model.

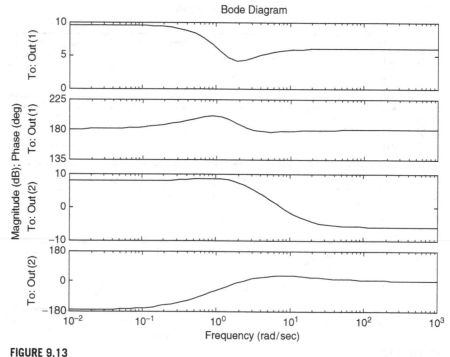

FIGURE 9.13

Bode Plots for a Three-Input, Two-Output Dynamic System.

Solution

According to the matrices' dimensions, the system has two outputs (the number of rows of [C]) and one input (number of columns of [B] and [D]). We can use the following MATLAB® command sequence

```
>> a = [1,0;-3,2];
>> b = [1;-1];
>> c = [0,1;-1,4];
>> d = [-2;0.5];
>> sys = ss(a,b,c,d);
>> bode(sys)
```

to obtain the plots of Figure 9.13. Two pairs of plots, as expected, are produced by relating the single input to the two outputs. ■

Frequency Response Data Handling

MATLAB® possesses the capability of handling frequency response data by creating a linear time invariant (LTI) model. The command frd (r,f)—which stands for frequency response data, with *r* being a vector that contains the frequency response values (the values of $G(j\omega)$) and f being the vector comprising the corresponding frequency values—can be used in situations when discrete experimental or numerically generated data available in the frequency domain need to be further analyzed. For instance, Bode plots can be drawn for such data, as shown in the following example.

Example 9.10

Frequency response data are furnished by a frequency analyzer corresponding to the frequency range from 0 to 100 rad/s with an increment of 10 rad/s as follows:

```
0.1000, -0.0246-0.0031i, -0.0052-0.0003i, -0.0023-0.0001i,
-0.0013, -8.0021e-004-1.6123e-005i, -5.5038e-004-9.3211e-
006i, -4.0675e-004-5.8120e-006i, -3.1004e-004-3.9342e-006i,
-2.5089e-004-2.7127e-006i, -2.0038e-004-2.0078e-006i
```

where $i = j = \sqrt{-1}$. Plot the magnitude of these data in terms of frequency and determine $G(j\omega)$ for $\omega = 88$ rad/s.

Solution

We need to define the frequency response vector r as well as the frequency vector f first to be able to call the frd function. Once the frd LTI model has been created, the command bodemag can be used to plot just the magnitude of the complex transfer function. Retrieving data from the frequency response, other than the discrete values that have been provided, can be done by means of the freqresp command. The following MATLAB® code can be used to solve this problem:

```
>> f = 0:10:100;
>> r = []; % input-given values of G(jω) - not included here
>> g = frd(r, f);
>> bodemag (g)
>> freqresp (g,88)
```

The last command returns the value of -2.6068e-004-2.9914e-006i, whereas the bodemag command gives the plot of Figure 9.14.

Frequency Response Model Conversion

MATLAB® allows conversion to an frd model from any of the other three (previously defined) linear time invariant (LTI) models: zpk (zero-pole-gain), tf (transfer function), and ss (state space) models. Conversion to a frequency response model is possible by means of the command frd (sys, f), where sys is an LTI object (zpk, tf, or ss) and f is a frequency vector. The frd vector returns frequency response data for the specified frequencies. It should be noted that conversion from an existing frd model into any of the zpk, tf, or ss models is not possible in MATLAB®.

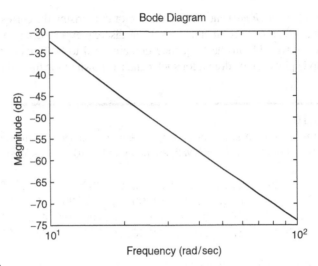

FIGURE 9.14

Bode Plot of Magnitude from Frequency Response Data.

Example 9.11

Obtain the frequency response data model when known are

a. A zero-pole-gain model for which: the zeroes are 1 and 2; the poles are 0, 3 (of the second order of multiplicity), 5; the gain is 1.5.

b. A transfer function model defined by

$$G(s) = \frac{1.5s^2 - 4.5s + 3}{s(s^3 - 11s^2 + 39s - 45)}$$

Plot the magnitude and phase angle corresponding to the [0; 1000] rad/s frequency range for each model.

Solution

a. The following MATLAB® program results in the Bode plots of Figure 9.15:

```
>> z = [1,2];
>> p = [0,3,3,5];
>> f = zpk(z,p,1.5);
>> om = 0:1:1000;
>> g = frd(f,om);
>> bode(g)
```

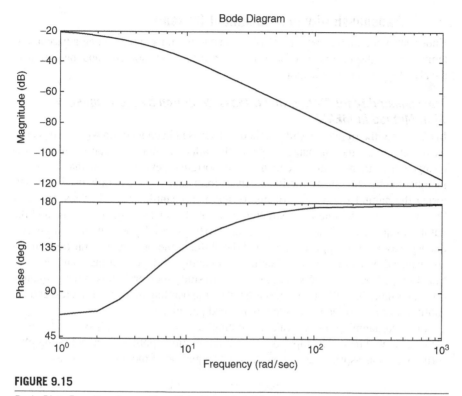

FIGURE 9.15

Bode Plots Resulting from a Zero-Pole-Gain Model Data.

b. The following MATLAB® commands

```
>> f = tf([1.5,-4.5,3],[1,-11,39,-45,0]);
>> om = 0:1:1000;
>> g = frd(f,om);
>> bode(g)
```

generate the same plot of Figure 9.15; the reader is encouraged to check the reason for that. ■

9.4 **FREQUENCY-DOMAIN APPLICATIONS**

This section studies several applications where frequency-domain analysis is utilized: vibration transmission in mechanical systems, serially connected systems, and electrical/mechanical filters.

9.4.1 Transmissibility in Mechanical Systems

Under harmonic input, mechanical vibration can be transmitted among adjacent systems by either displacement or force. Vibration absorption, isolation, and measurement are also discussed in this section.

Transmissibility for Motion Input, Mass Detection by the Frequency Shift Method in MEMS

In Chapter 3, the frequency shift method is discussed in connection with the possibility of detecting small amounts of matter that attach to microresonators through the change (shift) in the natural frequency of the original device. We saw that variations in either the bending or the torsional natural frequencies are generally used by monitoring the natural frequencies of micro- and nano-cantilevers or bridges. In either of the cases, the flexible microstructure is excited over its frequency range and the natural frequencies of interest are measured. After mass deposition takes place, frequency excitation is applied again, and the shifts in the original natural frequencies are detected, which enables evaluating the quantity of deposited mass and its position. Lumped-parameter models can be used to study such problems in the frequency domain. Figure 9.16 illustrates the case where the bending mode is of interest and the cantilever or the bridge is modeled by lumped parameters.

Lumped damping (or equivalent damping) as accounting for all system losses is also considered here. The input is a sinusoidal motion applied by the vibrating platform to the mass-spring-damper system. The equation of motion of the mass is

$$m\ddot{y} = -c(\dot{y} - \dot{u}) - k(y - u) \tag{9.80}$$

which can be rewritten as

$$m\ddot{y} + c\dot{y} + ky = c\dot{u} + ku \tag{9.81}$$

The following transfer function results after applying the Laplace transform to Eq. (9.81):

$$G(s) = \frac{Y(s)}{U(s)} = \frac{cs + k}{ms^2 + cs + k} \tag{9.82}$$

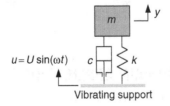

FIGURE 9.16

Lumped-Parameter Mechanical System with Input Harmonic Motion as Physical Model of a Microcantilever or Bridge Undergoing Out-of-Plane Bending Vibrations.

Using $s = j\omega$, as well as the equations for natural frequency ω_n, damping ratio ξ, and frequency ratio β, which have previously been defined as

$$\omega_n = \sqrt{\frac{k}{m}}; \ 2\xi\omega_n = \frac{c}{m}; \ \beta = \frac{\omega}{\omega_n}$$

changes Eq. (9.82) to

$$G(j\omega) = \frac{Y(j\omega)}{U(j\omega)} = \frac{1 + 2\xi\beta j}{1 - \beta^2 + 2\xi\beta j} \tag{9.83}$$

The modulus of the ratio of two complex numbers is equal to the ratio of the modulii of the same numbers. At the same time, $Y(j\omega)$ and $U(j\omega)$ are complex numbers; therefore, their modulii are also their amplitudes. It follows from Eq. (9.83) that

$$TR = |G(j\omega)| = \frac{|Y(j\omega)|}{|U(j\omega)|} = \frac{Y}{U} = \frac{\sqrt{1 + 4\xi^2\beta^2}}{\sqrt{(1 - \beta^2)^2 + 4\xi^2\beta^2}} \tag{9.84}$$

The modulus of the transfer function, which was shown to be equal to the ratio of the output amplitude Y to the input amplitude, is known as *transmissibility* and denoted by *TR*. Figure 9.17 plots the transmissibility corresponding to Eq. (9.84) in terms of the frequency ratio for several values of the damping ratio, such that the underdamped ($\xi < 1$), overdamped ($\xi > 1$), and critically damped ($\xi = 1$) cases are illustrated.

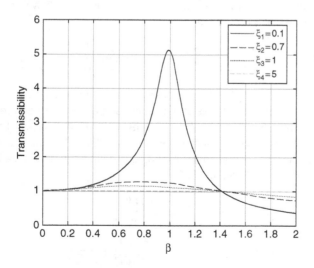

FIGURE 9.17

Transmissibility as a Function of the Frequency Ratio for Various Damping Ratios.

As Figure 9.17 shows, all curves pass through the points $\beta = 0$, $TR = 1$ and $\beta = 2^{1/2}$, $TR = 1$ (these values of β are found by solving $TR = 1$). For these two frequency ratios, the amplitude transmitted to the body of mass m is equal to the input amplitude. The spike in transmissibility is visible for the small (underdamped) damping ratio of $\xi = 0.1$. It can be checked that when $\xi \to 0$, the transmissibility $\to \infty$. For larger values of the damping ratio, including the critically damped and overdamped cases, the transmissibility curve flattens almost to superposition over $TR = 1$ for overdamping. It can also be seen in Figure 9.17 that for $\beta > 2^{1/2}$, $TR < 1$ irrespective of the damping ratio, which indicates that the output amplitude is less than the input amplitude.

Example 9.12

An external mass equal to $\Delta m = 5 \times 10^{-16}$ kg deposits at the free end of a rectangular cross-section microcantilever. Analyze the frequency spectrum before and after mass deposition by plotting the two transmissibility curves when the cantilever is supported on a vibratory table, as shown in Figure 9.16, and determine the value of the frequency shift. The cantilever is 90 μm long, 10 μm wide, and 120 nm thick. It is also known that $E = 160$ GPa, $\rho = 2500$ kg/m³. The damping coefficient is $c = 1.55 \times 10^{-9}$ N-s/m.

Solution

Similar to the generic derivation of the transfer function of Eq. (9.82), when the total mass that vibrates is m (the effective mass of the cantilever) plus Δm, the new transfer function is

$$G'(s) = \frac{Y'(s)}{U(s)} = \frac{cs + k}{(m + \Delta m)s^2 + cs + k} \tag{9.85}$$

Using the substitutions utilized in the generic transmissibility derivation, the following modified transmissibility is obtained:

$$|G'(j\omega)| = \frac{|Y'(j\omega)|}{|U(j\omega)|} = \frac{Y'}{U} = \frac{\sqrt{1 + 4\xi^2\beta^2}}{\sqrt{[1 - (1 + r_m)\beta^2]^2 + 4\xi^2\beta^2}} \tag{9.86}$$

where $r_m = \Delta m/m$ is the ratio of the deposited mass to the cantilever's effective mass. The effective mass m and stiffness of the cantilever are calculated according to Tables 3.2 and 3.1 as $m = 6.364 \times 10^{-14}$ kg, $k = 9.48 \times 10^{-4}$ N/m. It follows that the natural bending-related frequency of the cantilever is $\omega_n = 1.22 \times 10^5$ rad/s. The damping ratio is $\xi = 0.1$ and the mass ratio becomes $r_m = 0.0079$. With these values, the two transmissibility functions of Eqs. (9.84) and (9.86) are

$$\begin{cases} |G(j\omega)| = \dfrac{\sqrt{1 + 0.04\beta^2}}{\sqrt{(1 - \beta^2)^2 + 0.04\beta^2}} \\[4mm] |G'(j\omega)| = \dfrac{\sqrt{1 + 0.04\beta^2}}{\sqrt{[1 - 1.0079\beta^2]^2 + 0.04\beta^2}} \end{cases} \tag{9.87}$$

and Figure 9.18 shows the plot of the transmissibilities in terms of the frequency ratio.

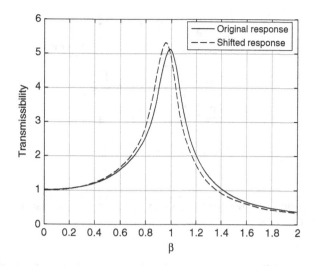

FIGURE 9.18

Transmissibility and Frequency Ratio Change as a Result of Mass Addition on a Vibrating Nanocantilever.

The precise frequency shift is computed by evaluating the difference between the original and altered natural frequency:

$$\Delta\omega_n = \omega_n - \omega'_n = \sqrt{\frac{k}{m}} - \sqrt{\frac{k}{m + \Delta m}} \tag{9.88}$$

For the numerical values of this problem, the difference of Eq. (9.88) is $\Delta\omega_n = 476.65$ rad/s. Approximate frequency values can also be obtained by right clicking the peaks of the two plots of Figure 9.18. ∎

Transmissibility for Force Input

In some situations, the input to a mechanical system is a harmonic force or torque instead of displacement, and the interest lies on how that force (or moment) is transmitted to other parts of the mechanical system as either force (moment) or displacement; this connection is known as *transmissibility for force input*. Consider the mechanical system of Figure 9.19(a), where a sinusoidal force acts on a body of mass m connected to a fixed base through a damper spring with damping coefficient c and stiffness k. We want to compare the amplitude of the input force F to the amplitude F_b, which is transmitted to the base.

Based on the free-body diagram of the mass and base, Figure 9.19(b), the equation of motion of the mass is

$$m\ddot{y} = f - f_d - f_e \tag{9.89}$$

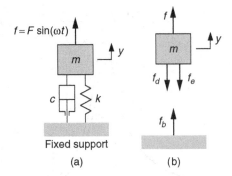

FIGURE 9.19

Mechanical System with Input Harmonic Force: (a) Lumped-Parameter Model; (b) Free-Body Diagrams.

where f_d is the damping force and f_e is the elastic force. Using known relationships, Eq. (9.89) becomes

$$m\ddot{y} + c\dot{y} + ky = f \tag{9.90}$$

The force transmitted to the base f_b is equal to the sum of the damping and elastic forces as the damper and the spring connect the moving body to the fixed base; therefore,

$$f_b = c\dot{y} + ky \tag{9.91}$$

Laplace-transforming Eqs. (9.90) and (9.91) with zero initial conditions results in the following transfer function after substituting $Y(s)$ from one equation into the other:

$$G(s) = \frac{F_b(s)}{F(s)} = \frac{cs + k}{ms^2 + cs + k} \tag{9.92}$$

which is identical to Eq. (9.82), which expressed the transfer function between the input motion of the base $U(s)$ and the output motion of the body $Y(s)$. Since the input to this system is sinusoidal, the output is sinusoidal as well; therefore, the transmissibility can be expressed as

$$TR = |G(j\omega)| = \frac{|F_b(j\omega)|}{|F(j\omega)|} = \frac{F_b}{F} = \frac{\sqrt{1 + 4\xi^2\beta^2}}{\sqrt{(1 - \beta^2)^2 + 4\xi^2\beta^2}} \tag{9.93}$$

where F_b is the amplitude of the force transmitted to the base, this force being expressed as

$$f_b = F_b \sin(\omega t + \varphi) \tag{9.94}$$

Example 9.13

Two identical electric motors that are synchronized and rotate in opposite directions are placed symmetrically on a rigid platform connected to the ground by means of two identical dashpot supports, as shown in Figure 9.20(a). The mass center of each rotor has an eccentricity *e* with respect to its rotation axis, as sketched in Figure 9.20(b). Determine the amplitude of the force transmitted to the ground when the rotation frequency is one half the system's natural frequency. Known are the mass of the platform and the motor stators, which is $m_p = 80$ kg, the mass of one rotor $m_r = 0.5$ kg, the damping coefficient $c = 80$ N-s/m, the spring stiffness $k = 200$ N/m, and the eccentricity $e = 4$ mm.

Solution

Figure 9.20(b) is an enlarged schematic view of the rotor's center of mass; it rotates eccentrically about the bearings axis, which coincides with the geometric center of the motor placed on the right of Figure 9.20(a). The eccentricity produces a centrifugal force, whose vertical projection is

$$f_y = f_c \sin(\omega t + \varphi) = m_r \omega^2 e \sin(\omega t + \varphi) \tag{9.95}$$

Due to synchronization between the two motors, a force identical to the one of Eq. (9.95) acts on the rotor on the left in Figure 9.20(a). The horizontal projections of the centrifugal forces on the two rotors cancel out because of opposite rotation directions, whereas the vertical projections are identical. As a consequence, Figure 9.21 shows the simplified, lumped-parameter physical model of the mechanical system.

The amplitude of the force acting on the mass *m* (which is $m = m_p + 2m_r$) is

$$F = 2m_r \omega^2 e \tag{9.96}$$

The following equations can be written:

$$\begin{cases} m\ddot{y} + 2c\dot{y} + 2ky = f \\ f_b = 2c\dot{y} + 2ky \end{cases} \tag{9.97}$$

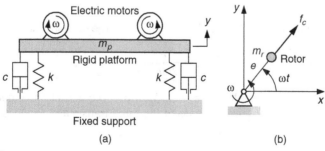

FIGURE 9.20

Rigid Platform with Eccentric Motors and Damper-Spring Supports: (a) Schematic of the System; (b) Enlarged Sketch of the Right Motor Showing Eccentricity and Centrifugal Force.

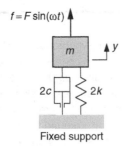

FIGURE 9.21

Lumped-Parameter Model of the Platform-Motors Mechanical System.

Application of Laplace transform with zero initial conditions to Eqs. (9.97) produces the following transfer function:

$$G(s) = \frac{F_b(s)}{F(s)} = \frac{2(cs + k)}{ms^2 + 2cs + 2k} \tag{9.98}$$

which results in a transmissibility:

$$TR = |G(j\omega)| = \frac{|F_b(j\omega)|}{|F(j\omega)|} = \frac{F_b}{F} = \frac{2\sqrt{1 + 4\xi^2\beta^2}}{\sqrt{(2 - \beta^2)^2 + 16\xi^2\beta^2}} \tag{9.99}$$

Combining Eqs. (9.96) and (9.99) yields

$$F_b = \frac{4m_r\beta^2\omega_n^2 e\sqrt{1 + 4\xi^2\beta^2}}{\sqrt{(2 - \beta^2)^2 + 16\xi^2\beta^2}} \tag{9.100}$$

The natural frequency of the mechanical system is

$$\omega_n = \sqrt{\frac{2k}{m}} \tag{9.101}$$

which results in a numerical value of $\omega_n = 2.22$ rad/s. The damping ratio is calculated as

$$\xi = \frac{c}{2m\omega_n} \tag{9.102}$$

Numerically, it is obtained that $\xi = 0.22$. With $\beta = 0.5$ and all the other numerical values of this example, the amplitude of the force transmitted to the base is $F_b = 0.0056$ N.

Vibration Absorption and Vibration Isolation

When a harmonic force acts on a body that is placed on a fixed base through a damper spring, it is possible to reduce the amplitude of the force transmitted to the support, but it is impossible to completely eliminate it, as can be seen by reformulating Eq. (9.99):

$$F_b = \frac{2F\sqrt{1 + 4\xi^2\beta^2}}{\sqrt{(1 - \beta^2)^2 + 16\xi^2\beta^2}} \qquad (9.103)$$

because $1 + 4\xi^2\beta^2 \neq 0$. One way of isolating (eliminating) the base force is adding a mass-spring system to the existing system, as shown in the next example.

Example 9.14
A sinusoidal force acts on the mass-damper-spring mechanical system of Figure 9.22(a). Study the possibility of reducing to zero the force transmitted to the fixed base by adding a mass-spring subsystem to the existing system, as suggested in Figure 9.22(b).

Solution
The equations of motion for the two-DOF mechanical system of Figure 9.22(b) are

$$\begin{cases} m_a\ddot{y}_2 = -k_a(y_2 - y_1) \\ m\ddot{y}_1 = f - k_a(y_1 - y_2) - c\dot{y}_1 - ky_1 \end{cases} \qquad (9.104)$$

The force transmitted to the base results from the damper and spring contact and is expressed as

$$f_b = c\dot{y}_1 + ky_1 \qquad (9.105)$$

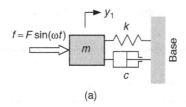

(a)

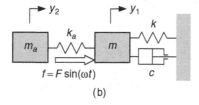

(b)

FIGURE 9.22

Mechanical System with Sinusoidal Input: (a) Original System; (b) System with Added Mass and Spring.

Laplace transforming Eqs. (9.104) and (9.105) results in

$$\begin{cases} (m_a s^2 + k_a) Y_2(s) = k_a Y_1(s) \\ (ms^2 + cs + k_a + k) Y_1(s) = F(s) + k_a Y_2(s) \\ F_b(s) = (cs + k) Y_1(s) \end{cases} \quad (9.106)$$

Successive substitution of $Y_2(s)$ and $Y_1(s)$ from the first two Eqs. (9.106) into the third of Eqs. (9.106) results in

$$\frac{F_b(s)}{F(s)} = \frac{(cs + k)(m_a s^2 + k_a)}{(m_a s^2 + k_a)(ms^2 + cs + k_a + k) - k_a^2} \quad (9.107)$$

The modulus of the base force can be now determined by means of the transmissibility as

$$F_b = |F_b(j\omega)| = TR\,|F(s)| = \frac{|k + c\omega j|\,(k_a - m_a \omega^2)}{|(m_a s^2 + k_a)(ms^2 + cs + k_a + k) - k_a^2|} F \quad (9.108)$$

Equation (9.108) indicates that when

$$\frac{k_a}{m_a} = \omega^2 \quad (9.109)$$

the amplitude of the force transmitted to the base is zero. In other words, when the natural frequency of the added mass-spring subsystem becomes equal to the frequency of the sinusoidal force, the transmitted vibration is annihilated. This additional mass-spring system behaves as a *vibration isolator*. ■

Measuring Vibration Displacement and Acceleration Amplitudes

The transmissibility principle can be used to measure the amplitude of either the displacement or the acceleration of a vibratory input, as studied in the following example.

Example 9.15
A massless platform undergoes a sinusoidal motion, as shown in Figure 9.23. Using a mass-damper-spring system as a sensor, analyze the situations when displacement or acceleration amplitudes can be detected using the relative motion of the body with respect to the platform.

Solution
The equation of motion for the mass of Fig. 9.23 is:

$$m\ddot{x} = -c(\dot{x} - \dot{u}) - k(x - u) \quad (9.110)$$

which is rewritten as

$$m(\ddot{x} - \ddot{u}) = -m\ddot{u} - c(\dot{x} - \dot{u}) - k(x - u) \quad (9.111)$$

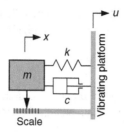

FIGURE 9.23

Mechanical Instrument to Measure Vibration.

The motion of interest is the one of the mass relative to the vibrating platform; therefore, the following relative motion coordinate is introduced:

$$y = x - u \tag{9.112}$$

Equation (9.111) becomes

$$m\ddot{y} + c\dot{y} + ky = -m\ddot{u} \tag{9.113}$$

The Laplace transform is applied to Eq. (9.113), which results in the transfer function

$$G(s) = \frac{Y(s)}{U(s)} = \frac{-ms^2}{ms^2 + cs + k} \tag{9.114}$$

The magnitude corresponding to $G(j\omega)$, which derives from $G(s)$, can be expressed as

$$|G(j\omega)| = \frac{|Y(j\omega)|}{|U(j\omega)|} = \frac{Y}{U} = \frac{m\omega^2}{\sqrt{(k - m\omega^2)^2 + (c\omega)^2}}$$

$$= \frac{\omega^2}{\sqrt{(\omega_n^2 - \omega^2)^2 + (2\xi\omega_n\omega)^2}} \tag{9.115}$$

where the definitions of the natural frequency ω_n and damping ratio ξ have been used.

Depending on the relationship between ω and ω_n, two cases are possible, as discussed next.

System with Small Natural Frequency and High Excitation Frequency: $\omega \gg \omega_n$
Equation (9.115) can be changed to

$$Y = \frac{1}{\sqrt{\left(\frac{\omega_n^2}{\omega^2} - 1\right)^2 + \left(2\xi\frac{\omega_n}{\omega}\right)^2}} U \tag{9.116}$$

The square of the natural-to-excitation frequency ratio is a very small positive quantity, which can be neglected. As a consequence, Eq. (9.116) simplifies to

$$Y \approx U \tag{9.117}$$

Equation (9.95) indicates that the instrument measurement, y, is approximately equal to the amplitude of the input motion. As a result, the system can be used as a *displacement sensor*.

System with Large Natural Frequency and Small Excitation Frequency: $\omega \ll \omega_n$
In this situation, Eq. (9.115) can be written as

$$\frac{\omega_n^2}{\omega^2} Y = \frac{1}{\sqrt{\left(1 - \frac{\omega^2}{\omega_n^2}\right)^2 + \left(2\xi \frac{\omega}{\omega_n}\right)^2}} U \tag{9.118}$$

The square of the excitation-to-natural frequency ratio is a very small positive quantity this time, and as a consequence, the denominator of the fraction on the right-hand side of Eq. (9.118) is equal to 1 approximately. Therefore, Eq. (9.118) simplifies to

$$Y \approx \frac{1}{\omega_n^2} (\omega^2 U) \tag{9.119}$$

Because the input motion is sinusoidal, the acceleration also is sinusoidal:

$$a = -\omega^2 U \sin(\omega t) = -A \sin(\omega t) \tag{9.120}$$

where $A = \omega^2 U$ is the acceleration amplitude. As a consequence, and as per Eq. (9.119), the amplitude of the relative motion of the mass (which is the measured parameter) is proportional to the amplitude of the input acceleration; therefore, the instrument can be used as an *acceleration sensor*.

9.4.2 Cascading Nonloading Systems

In many instances dynamic systems interact in a cascading (series) manner, such that the output from one system becomes the input to the next system, as sketched in Figure 9.24. If there is no interstage loading (meaning that the output of one stage converts fully into the input of the subsequent stage) and the input to the first system is sinusoidal with an amplitude U and a frequency ω, the output from the second system is

$$y(t) = U |G(j\omega)| \sin(\omega t + \varphi) \tag{9.121}$$

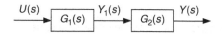

FIGURE 9.24

Two-Stage Dynamic System with Elements Connected in Series and without Interstage Loading.

where

$$\begin{cases} |G(j\omega)| = |G_1(j\omega)||G_2(j\omega)|; \\ \varphi = \varphi_1 + \varphi_2 \end{cases} \tag{9.122}$$

Based on Figure 9.24, the Laplace transform of the first-stage output is

$$Y_1(s) = G_1(s)U(s) \tag{9.123}$$

Because there is no interstage loading, the output from the second stage system is calculated similarly:

$$Y(s) = G_2(s)Y_1(s) = G_1(s)G_2(s)U(s) \tag{9.124}$$

Let us use $s = j\omega$ in Eq. (9.124), which leads to

$$Y(j\omega) = G_1(j\omega)G_2(j\omega)U(j\omega) \tag{9.125}$$

The right-hand side of Eq. (9.125) can be written in phasor form:

$$\begin{aligned} y(t) = Y(j\omega) &= |G_1(j\omega)|e^{j\varphi_1}|G_2(j\omega)|e^{j\varphi_2}U(j\omega) \\ &= |G_1(j\omega)||G_2(j\omega)|e^{j(\varphi_1 + \varphi_2)}U(j\omega) \end{aligned} \tag{9.126}$$

where $U(j\omega) = U\sin(\omega t)$ is the input to the system. Equations (9.121) and (9.122) have thus been demonstrated. This result can be generalized to a system made up of n subsystems cascading so that the output from one subsystem is input to the next one: *The output from a serially connected system under sinusoidal input is of sinusoidal form; its magnitude is the product of individual magnitudes and its phase angle is the sum of all individual phase angles.*

Example 9.16

A cascading dynamic system is formed of two subsystems; one is first order and the next one is second order, such that the output from the first-order system is the output to the second-order system. The two systems are defined by the following differential equations: $3\dot{y}_1 + y_1 = 20\sin(3t)$; $\ddot{y} + 2\dot{y} + 0.01y = 5y_1$ (see Figure 9.24). Determine the steady-state response of the system and plot the corresponding Bode plots.

Solution

The transfer functions corresponding to the differential equations of the two coupled systems are found by applying Laplace transforms to the differential equations:

$$G_1(s) = \frac{1}{3s + 1}; \; G_2(s) = \frac{5}{s^2 + 2s + 0.01} \tag{9.127}$$

The magnitudes and phase angles are found from the $G_1(j\omega)$ and $G_2(j\omega)$ functions as

$$\begin{cases} M_1(\omega) = |G_1(j\omega)| = \dfrac{1}{\sqrt{1 + (3\omega)^2}}; \; \varphi_1 = -\tan^{-1}(3\omega) \\[4mm] M_2(\omega) = |G_2(j\omega)| = \dfrac{5}{\sqrt{(0.01 - \omega^2)^2 + (2\omega)^2}}; \; \varphi_2 = -\tan\dfrac{2\omega}{0.01 - \omega^2} \end{cases} \tag{9.128}$$

Numerically, the magnitudes and phase angles of Eq. (9.128) are $M_1(\omega) = 0.11$, $M_2(\omega) = 0.46$, $\varphi_1 = -83.66°$, $\varphi_2 = 33.72°$. The steady-state response is therefore

$$y(t) = |G_1(j\omega)||G_2(j\omega)| \, A \sin(\omega t + \varphi_1 + \varphi_2) \tag{9.129}$$

and, with the numerical values of this example, the output is

$$y(t) = \sin(\omega t - 49.94°) \tag{9.130}$$

The combined transfer function, which is necessary for the Bode plots, is obtained by multiplying the transfer functions of Eqs. (9.127):

$$G(s) = G_1(s)G_2(s) = \frac{5}{3s^3 + 7s^2 + 2.03s + 0.01} \tag{9.131}$$

and the Bode diagrams are shown in Figure 9.25. ■

9.4.3 Filters

Electrical or mechanical filter systems are utilized to remove the input signal over certain frequency ranges and transmit it in its original magnitude over the other intervals. The most utilized filter types are sketched in Figure 9.26: Filters can be *low pass*, *high pass*, *band pass* or *notch*. The best way to analyze filter behavior, as shown in Figure 9.26, is to have their magnitude plotted against the frequency. Ideally, the magnitude is equal to 1 over the *passband* and equal to 0 over the *stopband*. The frequency that separates the passband and the stopband is the *cutoff frequency* (or *corner frequency*), denoted by ω_c.

Actual filter circuits display a behavior closer to the schematic representation of Figure 9.26(e), which shows a real low-pass filter with a *transition band* between the

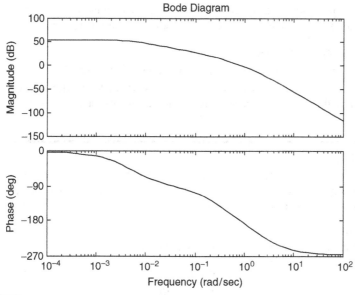

FIGURE 9.25

Magnitude and Phase Angle Plots of a Two-Stage Cascading System under Sinusoidal Input without Interstage Loading.

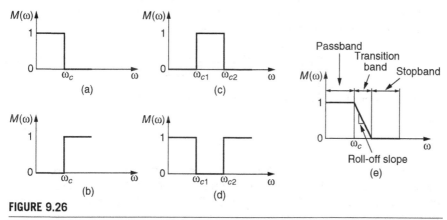

FIGURE 9.26

Filter Magnitude Ratio versus Frequency: (a) Ideal Low Pass; (b) Ideal High Pass; (c) Ideal Band Pass; (d) Ideal Notch; (e) Real Low Pass.

passband and the stopband. The cutoff frequency is defined as the frequency where the input power amplitude is reduced by half, and since power is proportional to the square of amplitude, the cutoff frequency is given by the equation

$$M(\omega_c) = \sqrt{\frac{1}{2}} = 0.707 \qquad (9.132)$$

A couple of examples of electrical and mechanical filters are presented next, but more filter examples are contained in the complementary Chapter 9.

Electrical Filter Systems

Electrical filters can be *passive* when they contain only passive elements such as resistors, capacitors, or inductors, or *active*, when they incorporate elements that need external energy sources, such as actuators, sensors, or operational amplifiers.

Example 9.17

Analyze the frequency response of the low-pass (*Butterworth*) filter of Figure 9.27 using the following sets of values: $R_1 = 100\ \Omega$, $C_1 = 0.001$ F, and $R_2 = 1000\ \Omega$, $C_2 = 0.01$ F. Determine the cutoff frequency for each case.

Solution

The transfer function of the electrical circuit of Figure 9.27 is

$$G(s) = \frac{V_o(s)}{V_i(s)} = \frac{1}{RCs + 1} \tag{9.133}$$

The two plots of Figure 9.28 have been obtained for the combinations shown in the legend using the `tf` and `bodemag` MATLAB® commands. The conclusion is that the transition band (and the roll-off slope) cannot be changed by changing the values of either R or C. However, a shift to the left is achieved by increasing the values of either R or C.

The magnitude is derived from Eq. (9.133) as

$$M(\omega) = |G(j\omega)| = \frac{1}{\sqrt{1 + (RC\omega)^2}} \tag{9.134}$$

therefore, the cutoff frequency is obtained based on Eq. (9.132) as

$$\omega_c = \frac{1}{RC} \tag{9.135}$$

Numerically, the cutoff frequency values are $\omega_{c1} = 10$ rad/s and $\omega_{c2} = 0.1$ rad/s for the two cases under analysis. The companion website Chapter 9 shows that, by adding two or more stages identical to the one of Figure 9.27, it is possible to reduce the transition band.

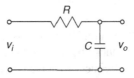

FIGURE 9.27

Electrical Low-Pass RC (Butterworth) Filter.

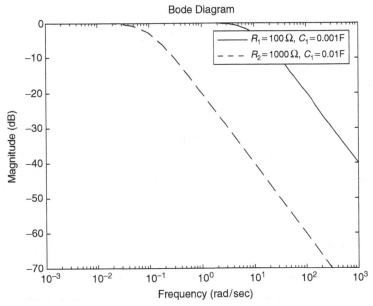

Bode Diagram

$R_1 = 100\,\Omega,\ C_1 = 0.001\text{F}$
$R_2 = 1000\,\Omega,\ C_1 = 0.01\text{F}$

FIGURE 9.28

Magnitude Ratio versus Frequency for an RC Low-Pass (Butterworth) Filter.

Mechanical Filters

Mechanical systems, too, can function as filters with respect to their input and output signals, as illustrated in the following example.

Example 9.18

Verify whether the mechanical system of Figure 9.29 produces any filtering effects when the input is the displacement u and the output is the displacement y. Consider the following numerical values: $m = 1$ kg, $c = 2$ N-s/m, and $k = 200$ N/m.

Solution

The dynamic equation of motion for the body of mass m is

$$m\ddot{y} = -c(\dot{y} - \dot{u}) - ky \qquad (9.136)$$

which results in the transfer function

$$G(s) = \frac{Y(s)}{U(s)} = \frac{cs}{ms^2 + cs + k} \qquad (9.137)$$

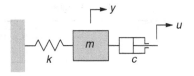

FIGURE 9.29

Translatory Mechanical System as a Filter.

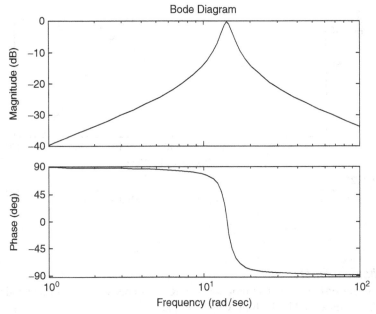

FIGURE 9.30

Magnitude and Phase Plots of a Mechanical Filter.

The magnitude and phase angle corresponding to $G(j\omega)$ are

$$
\begin{cases}
M(\omega) = |G(j\omega)| = \dfrac{Y(j\omega)}{U(j\omega)} = \dfrac{c\omega}{\sqrt{(k - m\omega^2)^2 + c^2\omega^2}} \\[2ex]
\varphi(\omega) = \dfrac{\pi}{2} - \tan^{-1}\dfrac{c\omega}{k - m\omega^2}
\end{cases}
\tag{9.138}
$$

and Figure 9.30 shows the Bode plots associated with Eqs. (9.138). The mechanical system works as a band-pass filter, as can be seen in Figure 9.26(c). The cutoff frequency is obtained by solving the equation $M(\omega_c) = 0.707$, which actually has two valid roots: $\omega_{c1} = 13.18$ rad/s and $\omega_{c2} = 15.18$ rad/s. This is illustrated in the magnitude plot of Figure 9.30, where a line parallel to the frequency axis at -3dB (which is equal to $1/\sqrt{2}$) intersects the magnitude curve at two points, whose abscissas are the cutoff frequencies calculated previously. ■

SUMMARY

The concepts of complex transfer function and complex transfer function matrix are introduced and applied in this chapter as modeling tools of the natural response and the steady-state response of SISO and MIMO dynamic systems under sinusoidal input. The frequency response of dynamic systems under sinusoidal excitation consists of an output amplitude and a phase angle that depend on the input frequency. The topics of transmission, absorption, and elimination of mechanical vibrations generated by sinusoidal input are also studied as well as the principles of mechanical vibration sensing, the cascading unloading systems, and electrical/mechanical filter systems. Specialized MATLAB® commands are utilized to model and solve several applications of these topics.

PROBLEMS

9.1 Use the transfer function approach and a lumped-parameter model to determine a relationship between the geometrical and material parameters of the nanobridge shown in Figure 9.31 so that the natural frequency corresponding to the y translation is twice the one corresponding to the θ_x rotation. The rigid plate has a mass m and a moment of inertia J about the longitudinal axis. The identical side rods are flexible and massless; they have circular cross-sections and their length is l; the material has a Poisson's ratio μ. Hint: Apply separate adequate input for each of the two motions (a force for the y translation and a moment for the θ_x rotation).

9.2 Determine the natural frequencies of the electrical system of Figure 9.32 by formulating the complex transfer function matrix between a (dummy) input voltage and the two independent output currents. Consider that $L_1 = 12$ mH, $L_2 = 16$ mH, $C_1 = 24$ μF, and $C_2 = 30$ μF. Use MATLAB® to verify the analytical results.

9.3 Calculate the resonant frequency and resonant peak of a dynamic system whose transfer function is $G(s) = (3s + 2)/(s^2 + s + 10)$. Determine and plot the magnitude and the phase angle corresponding to a sinusoidal input being

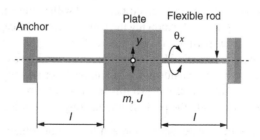

FIGURE 9.31

Top View of Paddle Nanobridge.

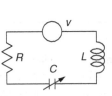

FIGURE 9.32

Two-Stage Electrical System.

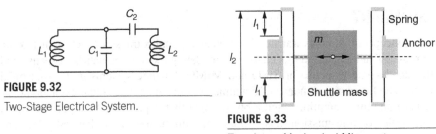

FIGURE 9.33

Translatory Mechanical Microsystem.

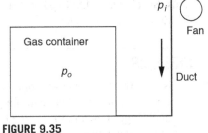

FIGURE 9.34

Electrical Microcircuit with Resistor, Inductor, Voltage Source, and Variable Capacitor.

FIGURE 9.35

Pneumatic System with Container, Duct, and Supply Fan.

applied to this system by means of the analytical approach and also using MATLAB®.

9.4 Use a lumped-parameter model and the analytical transfer function approach to calculate the frequency-domain response of the mechanical microsystem sketched in Figure 9.33 when a sinusoidal input force is applied to the shuttle mass in the motion direction. Use MATLAB® to verify the analytical results. Each of the two identical end springs has three flexible beams of circular cross-section with $d = 1$ μm, $l_1 = 10$ μm, $l_2 = 30$ μm, and $E = 1.5 \times 10^{11}$ N/m². The mass of the shuttle mass is $m = 200$ μg and the coefficient of viscous damping between the mass and the surrounding gas is $c = 0.004$ N-s/m.

9.5 A fixed-gap, longitudinal motion, variable-capacitor actuator (as the one of Figure 4.11) is connected in an electrical circuit with a sinusoidal voltage source, as sketched in Figure 9.34. Knowing the gap $g = 2$ μm, the capacitor plate width $w = 30$ μm, the air permittivity $\varepsilon = 8.8 \times 10^{-12}$ F/m, the resistance $R = 0.2$ Ω, and the inductance $L = 10$ μH, plot the frequency response using MATLAB® and applying the complex transfer function approach for two superposition lengths of the capacitor armatures: $x_1 = 10$ μm and $x_2 = 15$ μm. Calculate analytically and evaluate graphically the peak magnitude response for the two cases.

9.6 Apply the complex impedance approach to the fan-container-duct pneumatic system of Figure 9.35 and plot the frequency response both analytically and with MATLAB® considering the Hagen-Poiseuille pneumatic losses in the duct.

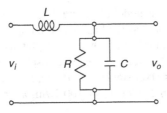

FIGURE 9.36

Electrical System with Inductor, Resistor, and Capacitor.

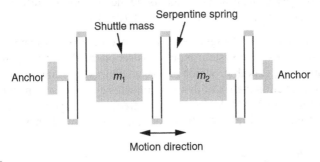

FIGURE 9.37

Mechanical Microsystem with Shuttle Masses and Serpentine Springs.

Known are the following parameters: diameter and length of duct are $d_i = 0.01$ m and $l = 12$ m, container diameter $d = 0.5$ m, gas density $\rho = 1.2$ kg/m³, and gas dynamic viscosity $\mu = 1.9 \times 10^{-5}$ N-s/m².

9.7 A constant rectangular cross-section nanocantilever vibrates in out-of-plane bending, driven by a sinusoidal input displacement applied to the anchor point. Find the equivalent viscous damping ratio of the environment when the cantilever has a length of 60 µm, a width of 5 µm, and a thickness of 120 nm. Young's modulus is 150 GPa and the mass density is 6000 kg/m³. The experimentally determined bandwidth is 30,000 Hz.

9.8 Use the complex transfer function approach to determine the values of R and C in the electrical system of Figure 9.36 knowing the resonant frequency is 21 kHz, the resonant peak is 1.3, and $L = 0.4$ H.

9.9 Use a lumped-parameter model for the mechanical microsystem of Figure 9.37, where the serpentine springs are identical; they are illustrated in Figure 3.6 with $l = 20$ µm and a square cross-section of 300 nm per side. The material Young's modulus is 170 GPa. The masses of the shuttles are $m_1 = 200$ µg and $m_2 = 280$ µg. Determine all the magnitudes and phase angles corresponding to a sinusoidal force applied on m_1 in the motion direction. Consider an overall damping coefficient $c = 0.05$ applying to both shuttle masses' motions.

9.10 The lumped-parameter model of Figure 9.38, is defined by the chassis mass m and moment of inertia J as well as by two damper-spring suspensions. Determine the system's number of DOFs and plot its frequency response. Also calculate and plot the steady-state time response of the system when $u_1 = 0.05 \sin(2t)$ m, $u_2 = 0.05 \sin(2t - \pi/10)$ m, $m = 1$ kg, $J = 0.01$ kg-m², $c_1 = 0.4$ N-s/m, $c_2 = 0.6$ N-s/m, $k_1 = 100$ N/m, $k_2 = 160$ N/m, $l_1 = 1$ m, and $l_2 = 2$ m.

9.11 Use the transfer function modeling approach to obtain the Bode plots for the electrical system represented in Figure 9.39. Consider that the inputs are the sine voltages v_1 and v_2 and the outputs are the meaningful currents. The electrical components are $L = 10$ H, $R = 20$ Ω, $C = 1$ F.

9.12 The liquid system of Figure 9.40 is formed of two tanks and two identical pipe segments. The system input consists of the flow rate q_i and the external pressure p_o, both of which are assumed to vary sinusoidally. Derive the transfer function matrix of this system, whose outputs are the tank pressures p_1 and p_2. Use MATLAB® to plot the frequency response when known are $d_1 = 3$ m, $d_2 = 3$ m (tank diameters), $d_i = 0.03$ m (diameter of the pipe),

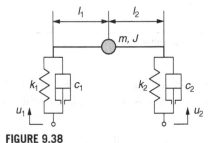

FIGURE 9.38

Mechanical System with Two Sinusoidal Inputs.

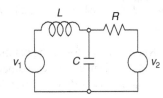

FIGURE 9.39

Electrical System with Voltage Input.

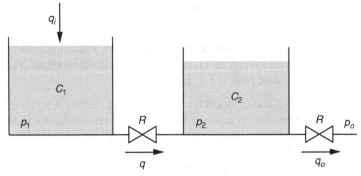

FIGURE 9.40

Tank-Pipe Liquid System under Sinusoidal Input.

$l = 40$ m (length of pipe between tanks and also length of pipe at the second tank outflow), $\mu = 0.001$ N-s/m^2, and $\rho = 1000$ kg/m^3.

9.13 Derive a state space model and plot the Bode diagrams for the mechanical system of Figure 9.41, where u_1, u_2 are the input sinusoidal force components and y_1, y_2 are the output displacements. Known are $m_1 = 2.5$ kg, $m_2 = 3$ kg, $c = 20$ N-s/m, $k_1 = 200$ N/m, $k_2 = 150$ N/m, $k_3 = 400$ N/m. Also plot the steady-state time-domain solution of this system for $u_1 = 90 \sin(2t)$ and $u_2 = 100 \sin(5t)$.

9.14 Consider the inputs are the voltages v_1, v_2 and the output is the voltage v_o for the operational amplifier electrical system of Figure 9.42. Use the complex transfer function approach to calculate and plot the steady-state time-domain response of this system for $R_1 = 200$ Ω, $R_2 = 180$ Ω, $L = 0.22$ H, $C = 3$ mF, $v_1 = 60 \sin(8t)$ V, and $v_2 = 80 \sin(2t)$ V.

9.15 A frequency analyzer provided the following experimental frequency response data:

0.2, $-0.2 + 0.01i$, $-0.3 + 0.1i$, $-0.02 + 0.03i$, $0.1 + 0.04i$, $0.1i$, $-0.015i$, $0.002 - 0.6i$

for the 10–60 rad/s frequency range. Plot the magnitude and phase angle corresponding to this data set and calculate $G(j\omega)$ for $\omega = 25$ rad/s and $\omega = 43$ rad/s.

9.16 A zero-pole-gain (zpk) model has the zeroes 1, 1, 3; the poles 2, 5, 6, 6; and a gain of 120. Plot the Bode diagrams corresponding to this system for the

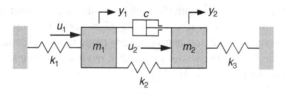

FIGURE 9.41

Translatory Mechanical System.

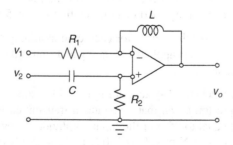

FIGURE 9.42

Operational Amplifier System with Two Input Voltages and One Output Voltage.

0–8000 rad/s frequency range using the conversion from the (zpk) model to the frequency response data (frd) model.

9.17 A dynamic system is defined by the differential equation: $3\dddot{y} + 2\ddot{y} + \dot{y} + 5y = 100\sin(\omega t)$. Use the conversion from a transfer function model to the frequency response data model to plot the Bode diagrams corresponding to this system for the 0–3000 rad/s frequency range.

9.18 A dynamic system is described by the following state space matrices:

$$[A] = \begin{bmatrix} 1 & 5 \\ 2 & 1 \end{bmatrix}; [B] = \begin{bmatrix} 1 \\ 2 \end{bmatrix}; [C] = \{-3 \ 0\}; D = 4.$$

Use MATLAB® to convert the model to a frequency response model and find its natural frequencies. Considering that a sine input is applied to the system, plot the Bode diagrams of the system response for the 0−1500 rad/s frequency range.

9.19 What is the minimum quantity of mass that can be detected by deposition at the free end of a microcantilever if the detection equipment has a resolution of 50 Hz? For that minimum mass, plot the original and shifted transmissibilities in terms of the frequency ratio when a sinusoidal displacement is applied to the cantilever base. Known are the dimensions of the cantilever, $l = 50$ μm, $w = 30$ μm, $h = 2$ μm, also the material properties, $E = 160$ GPa, $\rho = 6200$ kg/m³, and the viscous damping coefficient is $c = 5 \times 10^{-10}$ N-s/m.

9.20 A nanobridge is formed of a rigid platform having the dimensions of 200 μm × 200 μm × 2 μm and two identical nanowires of diameter 120 nm and length 30 μm made of a material with $E = 140$ GPa, $G = 100$ GPa, and $\rho = 5400$ kg/m³ (assume both the plate and nanowires are made of the same material). Decide which of the following two resonance shift methods is more sensitive in detecting a mass quantity of 2×10^{-14} kg:

(a) Bending mode with mass deposited at the center of the plate; the damping coefficient in translation is $c_t = 4 \times 10^{-9}$ N-s/m;

(b) Torsional mode with mass deposited off the plate center at a distance of 100 μm; the damping coefficient in rotation is $c_r = 5 \times 10^{-6}$ N-m-s.

Draw the transmissibility curve for each case assuming sinusoidal input displacement of the base is applied in bending and sinusoidal input rotation of the base is applied in torsion.

9.21 For the rotary mechanical system of Figure 9.43, evaluate the frequency of the sinusoidal input torque m_i that generates a transmitted torque amplitude on the right base equal to 85% of the input amplitude. Known are $J_1 = 2$ kg-m², $J_2 = 3$ kg-m², $c = 45$ N-m-s, $k_1 = 320$ N-m, $k_2 = 280$ N-m.

9.22 A sinusoidal force acts on the mechanical system illustrated in Figure 9.44. Considering that the mass is $m = 0.5$ kg and the stiffness is $k = 200$ N/m,

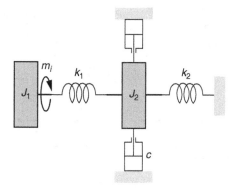

FIGURE 9.43

Rotary Mechanical System with Torque Transmission.

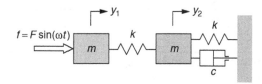

FIGURE 9.44

Mechanical System with Sinusoidal Input.

analyze the transmissibility of the input force to the force on the wall for $\xi = 0.3$ and $\xi = 0.7$.

9.23 A constant rectangular cross-section microcantilever needs to be designed to sense an input base acceleration of $100g$ (g is the gravitational acceleration constant). Known are the thickness $h = 5$ μm, the mass density $\rho = 5600$ kg/m³, and Young's modulus $E = 160$ GPa. Calculate the microcantilever length if the tip deflection cannot exceed 1 μm.

9.24 The microbridge of Figure 9.45 operates as a sensor for the sinusoidal input rotary displacement of its base. To be effective for this purpose, the natural frequency needs to be least 100 times smaller than the input frequency. It is also expected that the actual input frequency be larger than 2000 rad/s. Find the flexible rods' diameter d for $J = 3 \times 10^{-19}$ kg-m², $l = 500$ μm, and $G = 100$ GPa. Calculate the damping coefficient c corresponding to the viscous interaction between the microbridge plate and its environment knowing that the maximum value of this system's magnitude is 1.0001.

9.25 Two dynamic systems form a nonloading cascade defined by the following differential equations: $\ddot{y}_1 + 3\dot{y}_1 + 6y_1 = 5\sin(8t)$; $3\ddot{y} + 2\dot{y} = 4\dot{y}_1 + y_1$. Evaluate the steady-state response of this system and plot the corresponding Bode diagrams.

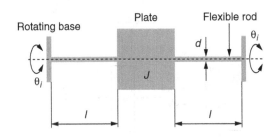

FIGURE 9.45

Microbridge with Base Rotary Input.

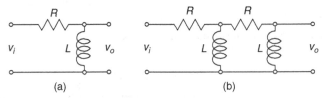

FIGURE 9.46

Electrical Filters: (a) Single-Stage; (b) Two-Stage.

9.26 A sinusoidal input $u = 10 \sin(100t)$ is provided to an unknown first-order system, which serially connects to a first-order transducer system defined by a sensitivity of 1.2 and a time constant of 0.01 s. The output from the transducer is further fed into a filter-amplifier device, which acts as a second-order system with a sensitivity of 3, a natural frequency of 300 rad/s, and a damping ratio of 0.3. Knowing that the signal output from the filter-amplifier system has an amplitude of 200 and a phase angle of $-70°$ with respect to the sinusoidal input, evaluate the unknown first order system parameters.

9.27 Study the frequency response of the electrical systems of Figure 9.46(a) and 9.46(b), for which $R = 1200 \, \Omega$ and $L = 0.8$ H. Demonstrate that both systems behave as high-pass filters.

9.28 Determine whether the electrical circuit of Figure 9.47 can function as a filter and establish its type in case it is a filter. The electrical components are $R_1 = 4 \, \Omega, R_2 = 40 \, \Omega, L = 1$ H, $C = 0.125$ mF. Calculate the corresponding cutoff frequency.

9.29 Check whether the mechanical system of Figure 9.48 operates as a filter when connecting the input displacement u to the output displacement y. Draw the Bode plots for the following pairs of parameter values: $k_1 = 10$ N/m, $c_1 = 1$ N-s/m and $k_2 = 15$ N/m, $c_2 = 2$ N-s/m.

9.30 Use the complex transfer function that connects the input displacement u to the output displacement y to predict the filter behavior of the translatory mechanical system of Figure 9.49 when the following amounts are known: $m = 1$ kg, $c_1 = 100$ N-s/m, $c_2 = 85$ N-s/m, $k_1 = 140$ N/m, $k_2 = 200$ N/m.

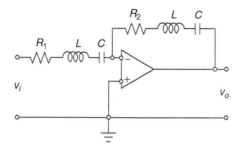

FIGURE 9.47

Operational-Amplifier Filter Circuit with Resistors, Inductors, and Capacitors.

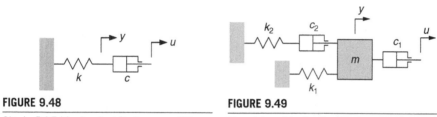

FIGURE 9.48

Single-DOF Mechanical System as a Filter.

FIGURE 9.49

Translatory Mechanical System as a Filter.

Suggested Reading

J. F. Lindsay and S. Katz, *Dynamics of Physical Circuits and Systems*, Matrix Publishers, Champaign, IL, 1978.

A. V. Oppenheim and A. S. Willsky, *Signal and Circuits*, 2nd Ed. Prentice Hall, Upper Saddle River, NJ, 1997.

N. S. Nise, *Control Systems Engineering*, 5th Ed. Wiley, New York, 2008.

K. Ogata, *System Dynamics*, 4th Ed. Pearson Prentice Hall, Upper Saddle River, NJ, 2004.

B. J. Kuo and F. Golnaraghi, *Automatic Control Systems*, 8th Ed. Wiley, New York, 2003.

N. Lobontiu, *Dynamics of Microelectromechanical Systems*, Springer, New York, 2007.

R. S. Figliola and D. E. Beasley, *Theory and Design for Mechanical Measurements*, 4th Ed. Wiley, New York, 2006.

Coupled-Field Systems

10

OBJECTIVES

This chapter introduces the notion of dynamic system coupling, which studies the interaction between systems belonging to different domains (fields) and generates mathematical models reflecting that interaction. Specifically, you learn about

- System analogies based on differential equations and transfer functions.
- Actuation and sensing by means of coupled-field systems.
- Electromechanical and electromagnetomechanical coupled-field examples.
- Principles of piezoelectricity and applications of piezoelectric coupling.
- Thermomechanical coupling by means of bimetallic strips.
- Nonlinear electrothermomechanical coupling.
- Utilization of MATLAB® and Simulink® to model and solve the mathematical models of coupled-field systems.

INTRODUCTION

Dynamic systems seldom operate in isolation but commonly interact with other systems. Interaction between dynamic systems of different physical nature is known as domain or field coupling, and some examples of the many coupled-field possibilities are presented in this chapter, such as electromechanical (including strain-gauge macro-scale and MEMS applications), electromagnetomechanical (with or without optical detection), piezoelectric, and thermomechanical or electrothermomechanical; more coupled-field dynamic system examples are studied in the companion website Chapter 10. Analogies among systems are also analyzed in this chapter using differential equations or transfer functions. Various algebraic and numeric examples are used to illustrate linear and nonlinear system coupling that apply MATLAB® and Simulink® to determine the solutions.

DOI:10.1016/B978-0-240-81128-4.00010-6

10.1 CONCEPT OF SYSTEM COUPLING

We dealt with *system coupling* implicitly in almost all the previous chapters, but the coupling between coordinates was restricted to a system of a single physical type (mechanical, electrical, fluid, or thermal). Consider, for instance, the two separate mass-spring mechanical systems of Figure 10.1(a); they are physically isolated and therefore *uncoupled*. Once a spring of stiffness k is added between the masses m_1 and m_2, as in Figure 10.1(b), the two systems become *coupled* and the resulting mathematical model is represented by the equations

$$\begin{cases} m_1\ddot{x}_1 + (k_1 + k)x_1 - \boxed{kx_2} = 0 \\ m_2\ddot{x}_2 + (k_2 + k)x_2 - \boxed{kx_1} = 0 \end{cases} \qquad (10.1)$$

The presence of the boxed term with x_2 in the first Eq. (10.1) indicates that this is a coupled system. This equation describes the motion of m_1, which is defined by coordinate x_1, and conversely, the fact that the boxed x_1 term is included in the second Eq. (10.1), the equation formulating the x_2 motion. Because the elements of the system sketched in Figure 10.1(b) are from one field, this system is a single-field coupled system.

Single-field coupled systems, such as the one just discussed, are not within the scope of this chapter but the companion website Chapters 3 and 4 study mechanical and electrical single-field coupled examples. In basic terms, *a multiple-field coupled system* (or, simply, a *coupled-field system*) is defined by at least one (differential) equation with variables (coordinates or DOFs) from at least two domains of different physical nature. The notion of coupled-field systems is introduced in Chapter 1 by means of an electromechanical dc motor example, which is discussed qualitatively. Another basic couple-field example is the one of a bar of original length l subjected to a temperature increase $\Delta\theta$ and that l deforms axially by the quantity:

$$\Delta l = \alpha l \Delta\theta \qquad (10.2)$$

Evaluated as an input-output relationship, Eq. (10.2) shows that the input (the temperature increase $\Delta\theta$) is from the thermal domain whereas the output (the bar elongation Δl) is from the mechanical domain. The coupling between the two system

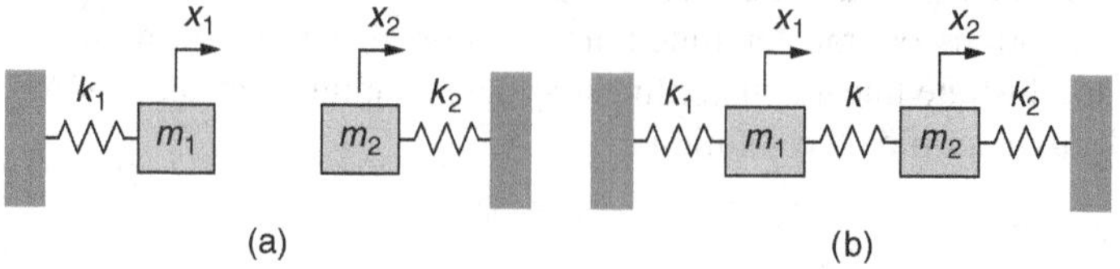

(a)   (b)

FIGURE 10.1

(a) Two Uncoupled Mechanical Systems; (b) Coupled Mechanical System.

variables is realized by the coefficient of thermal expansion α, which is measured in units of length per units of temperature (m/deg). This example indicates that two variables from systems of different physical types can be connected only by means of parameters that are defined over both fields.

Equation (10.2) described a zero-order coupled system, but coupled higher-order systems (with time derivatives of the output variable) also exist, as we see in this chapter. One convenient way of representing the mathematical model of a coupled MIMO system is by using the Laplace domain and the transfer function matrix. Consider two systems, denoted by 1 and 2, each being formulated in a different field, see Figure 10.2, where $\{U(s)\} = \{U_1(s)\ U_2(s)\}'$ is the input vector and $\{Y(s)\} = \{Y_1(s)\ Y_2(s)\}'$ is the output vector. When at least one of the output components is a function of both inputs, the systems are *coupled*.

Coupled systems, as shown in Figure 10.3, have nonzero terms on the secondary diagonal of the transfer function matrix. When these terms are zero, the two systems are *uncoupled* (as illustrated in Figure 10.3) because the two outputs, $Y_1(s)$ and $Y_2(s)$, depend on the input of only their own systems.

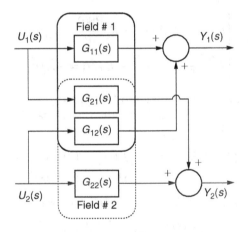

FIGURE 10.2

MIMO Coupled-Field System Studied in the Laplace Domain.

Uncoupled systems

$$\begin{Bmatrix} Y_1(s) \\ Y_2(s) \end{Bmatrix} = \begin{bmatrix} G_{11}(s) & 0 \\ 0 & G_{22}(s) \end{bmatrix} \begin{Bmatrix} U_1(s) \\ U_2(s) \end{Bmatrix}$$

Coupled systems

$$\begin{Bmatrix} Y_1(s) \\ Y_2(s) \end{Bmatrix} = \begin{bmatrix} G_{11}(s) & G_{12}(s) \\ G_{21}(s) & G_{22}(s) \end{bmatrix} \begin{Bmatrix} U_1(s) \\ U_2(s) \end{Bmatrix}$$

FIGURE 10.3

MIMO Uncoupled and Coupled-Field Systems.

Example 10.1

Consider the electromechanical system represented by the dc motor of Figure 10.4, studied in Chapter 1. Assume that, in addition to the given parameters, a load torque m_l acts on the mechanical cylinder.

a. Discuss the coupling of this system based on the transfer function matrix that connects the input vector $\{v_a, m_l\}$ to the output vector $\{i_a, \omega\}$, where v_a, i_a, and ω are the armature voltage and current and angular velocity of the mechanical system.

b. Use MATLAB® to plot the system response for the following numerical parameters: $J_l = 0.02$ N-m-s², $c = 0.1$ N-m-s, $m_l = 0.02$ N-m, $R_a = 6\,\Omega$, $L_a = 2$ H, $v_a = 80$ V, $K_e = 0.015$ N-m/A (1 N-m/A = 1 V-s/rad), and $K_t = 0.2$ N-m/A.

Solution

a. Equation (1.19) is the only one that changes due to the presence of the load torque:

$$J_l \frac{d^2\theta(t)}{dt^2} = m_a(t) - m_l(t) - c\frac{d\theta(t)}{dt} \tag{10.3}$$

Combining Eqs. (1.18), (10.3), and (1.20) results in

$$\begin{cases} L_a \dfrac{di_a(t)}{dt} + R_a i_a(t) + K_e \dfrac{d\theta(t)}{dt} = v_a(t) \\ J_l \dfrac{d^2\theta(t)}{dt^2} + c\dfrac{d\theta(t)}{dt} - K_t i_a(t) = -m_l(t) \end{cases} \tag{10.4}$$

This system is field coupled as the two Eqs. (10.4) comprise variables from both the electrical and the mechanical fields. The Laplace transformation of Eqs. (10.4) results in

$$\begin{bmatrix} L_a s + R_a & K_e \\ -K_t & J_l s + c \end{bmatrix}\begin{Bmatrix} I_a(s) \\ \Omega(s) \end{Bmatrix} = \begin{Bmatrix} V_a(s) \\ -M_l(s) \end{Bmatrix} \tag{10.5}$$

where $\Omega(s)$ is the Laplace transform of $\omega(t) = d\theta(t)/dt$, the angular velocity. The relationship

$$\begin{Bmatrix} I_a(s) \\ \Omega(s) \end{Bmatrix} = [G(s)]\begin{Bmatrix} V_a(s) \\ -M_l(s) \end{Bmatrix} \tag{10.6}$$

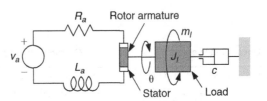

FIGURE 10.4

Schematic of a dc Motor as an Electromechanical System.

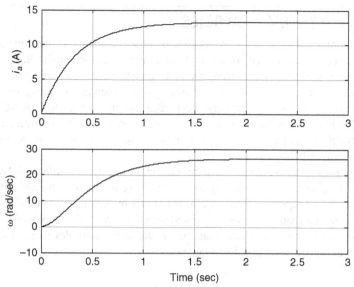

FIGURE 10.5

Time-Domain Plots for a Two-Input, Two-Output Coupled-Field Electromechanical dc-Motor System.

indicates, by comparison to Eq. (10.5), that the transfer function matrix [G(s)] is

$$[G(s)] = \begin{bmatrix} L_a s + R_a & K_e \\ -K_t & J_l s + c \end{bmatrix}^{-1} = \begin{bmatrix} \dfrac{J_l s + c}{D(s)} & -\dfrac{K_e}{D(s)} \\ \dfrac{K_t}{D(s)} & \dfrac{L_a s + R_a}{D(s)} \end{bmatrix} \qquad (10.7)$$

with the denominator $D(s)$ being

$$D(s) = J_l L_a s^2 + (J_l R_a + c L_a)s + K_e K_t + c R_a \qquad (10.8)$$

Field coupling is also indicated by the nondiagonal transfer function matrix.

b. Once the transfer function matrix of Eq. (10.7) is obtained, the Laplace-transformed output vector of Eq. (10.6) can be determined, and then the time-domain counterpart vector (which is not given here) is obtained symbolically using MATLAB®'s ilaplace command. The armature current and shaft angular velocity are plotted in Figure 10.5 for the numerical values of this example. Both output amounts become constant after approximately 2 s. ∎

10.2 SYSTEM ANALOGIES

Two or more dynamic systems are *analogous* if they have similar mathematical models expressed by differential equations, transfer functions (or transfer function matrices), or state space models. System analogies enable testing a system that is simpler to design and monitor (usually an electrical system), then extrapolating the obtained results to the analogous system (say, mechanical, for instance). The expectation for a coupled-field system therefore is to enable conversion of parameters from one field to their counterparts in the analogous system; this subject is discussed in the companion website Chapter 10.

Analogous systems are studied in this chapter based on their *differential equations* and *transfer functions*. Two analogous systems have the same order and structure of their differential equations; they are also defined by similar transfer functions (or transfer function matrices) as is illustrated next for first- and second-order systems. An example of zero-order system analogy is the one relating levers (as mechanical systems) to transformers (as electrical systems).

10.2.1 First-Order Systems

The following example studies analogous examples of first-order systems from several fields.

Example 10.2

Demonstrate that the mechanical, electrical, pneumatic, and operational amplifier systems of Figure 10.6 and the liquid-level system of Figure 5.15 are analogous.

Solution

For the mechanical system of Figure 10.6(a), consisting of the damper of coefficient c and the spring of stiffness k, the equation of motion is formulated for coordinate y as

$$0 = -c\dot{y} - k(y - u) \tag{10.9}$$

which can be rewritten as

$$\frac{c}{k}\dot{y} + y = u \tag{10.10}$$

The transfer function of the electrical circuit sketched in Figure 10.6(b) is determined by means of the complex impedance approach as

$$G(s) = \frac{V_o(s)}{V_i(s)} = \frac{Z_2}{Z_1 + Z_2} = \frac{1}{RCs + 1} \tag{10.11}$$

whose inverse Laplace transform yields

$$RC\dot{v}_o + v_o = v_i \tag{10.12}$$

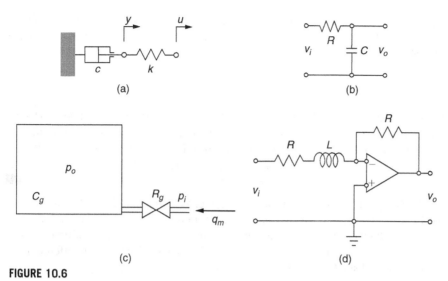

FIGURE 10.6

Analogous First-Order Systems: (a) Mechanical; (b) Electrical; (c) Pneumatic;
(d) Operational Amplifier.

with R and C being the electrical resistance and capacitance and v_i and v_o being the input and output voltages. The companion website Chapter 5 demonstrates that the equation governing the dynamic behavior of the pneumatic system of Figure 10.6(c) is

$$R_g C_g \dot{p}_o + p_o = p_i \tag{10.13}$$

where R_g and C_g are the pneumatic (gas) resistance and capacitance, p_i and p_o are the input and output pressures. Similarly, the differential equation governing the dynamic behavior of the liquid system of Figure 5.15 is given in Eq. (5.91) and rewritten here:

$$R_{l,p} C_{l,p} \dot{q}_o + q_o = q_i \tag{10.14}$$

with $R_{l,p}$ and $C_{l,p}$ being the pressure-defined liquid resistance and capacitance, and q_i, q_o being the input and output volume flow rates.

By using the complex impedance approach, the transfer function of the electrical system of Figure 10.6(d) is

$$G(s) = \frac{V_o(s)}{V_i(s)} = -\frac{Z_2}{Z_1} = -\frac{R}{R + Ls} \tag{10.15}$$

Equation (10.15) can be reformulated as

$$\frac{V_o(s)}{V_i(s)} = -\frac{1}{\dfrac{L}{R}s + 1} \tag{10.16}$$

and, if we ignore the minus sign of Eq. (10.16) or if we add another op amp basic inverting stage with two identical components, the inverse Laplace transform with zero initial
conditions yields

$$\frac{L}{R} \times \dot{v}_o + v_o = v_i \tag{10.17}$$

Equations (10.10), (10.12), (10.13), (10.14), and (10.17) are similar first-order differential equations. As a consequence, the four systems of Figure 10.6 together with the one
of Figure 5.15 are analogous. The quantities c/k in Eq. (10.10), RC in Eq. (10.12), $R_g C_g$
in Eq. (10.13), $R_{l,p} C_{l,p}$ in Eq. (10.14), and L/R in Eq. (10.17) are all *time constants* (they
are measured in seconds), as they relate to *first-order systems*.

By comparing Eqs. (10.11) and (10.16), it can be seen that the analogous systems
of Figures 10.6(b) and 10.6(d) have similar transfer functions, which proves that
system analogy can also be pursued using transfer functions instead of time-domain
equations; more examples on system analogy by transfer functions are offered in the
companion website Chapter 10.

10.2.2 Second-Order Systems

Force-voltage or *force-current analogies* can be established between second-order
mechanical and electrical systems as discussed next.

For the mechanical system of Figure 10.7(a), the equation of motion is

$$m\ddot{y} + c\dot{y} + ky = f \tag{10.18}$$

By using known voltage-current relationships, the dynamic equation of the electrical
system shown in Figure 10.7(b) is

$$L\ddot{q} + R\dot{q} + \frac{1}{C}q = v \tag{10.19}$$

The analogy between the mechanical system of Figure 10.7(a) and the electrical system
of Figure 10.7(b) is illustrated by Eqs. (10.18) and (10.19). Table 10.1 shows the pairing
of physical quantities according to this particular analogy. It should be noted that rotary
mechanical systems are also analogous to the electrical system of Figure 10.7(b), where

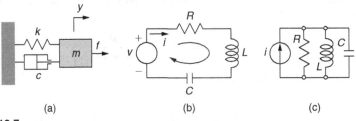

(a)   (b)   (c)

FIGURE 10.7

Analogous Second-Order Systems: (a) Mechanical; (b) Series Electrical; (c) Parallel Electrical.

Table 10.1 Force-Voltage Analogy Pairs

Mechanical	f	y	m	c	k
Electrical	v	q	L	R	$1/C$

Table 10.2 Force-Current Analogy Pairs

Mechanical	f	y	m	c	k
Electrical	i	Ψ	C	$1/R$	$1/L$

angular displacement, velocity, and acceleration replace the translational counterparts and moments (torques) are needed instead of forces (so the *moment-voltage analogy* is the correct denomination in this case).

For the electrical system of Figure 10.7(c), the source current is equal to the sum of currents through the resistor, inductor, and capacitor, which means that

$$C\frac{dv(t)}{dt} + \frac{1}{R}v(t) + \frac{1}{L}\int v(t)\,dt = i(t) \tag{10.20}$$

where v is the voltage across each of the three electrical components. When using the relationship between voltage and electrical flux, $v = d\Psi(t)/dt$, Eq. (10.20) changes to

$$C\ddot{\Psi} + \frac{1}{R}\dot{\Psi} + \frac{1}{L}\Psi = i \tag{10.21}$$

which is similar to Eq. (10.18) and therefore indicates the analogy between the mechanical system of Figure 10.7(a) and the electrical one of Figure 10.7(c). This *force-current analogy* can be summarized by the pairs illustrated in Table 10.2. Again, rotary mechanical systems can be used instead of translatory ones with the corresponding changes in displacement, velocity, and accelerations (angular instead of linear) and forcing (moment instead of force), which results in a *moment-current analogy*.

Example 10.3
For the electrical system of Figure 10.8, determine an analogous rotary mechanical system using the differential equations of the two systems. Also demonstrate that the transfer function matrices of the two systems are similar.

Solution
By using Kirchhoff's second law, the following equations are obtained that represent the mathematical model of the two-mesh electrical system of Figure 10.8:

$$\begin{cases} R_1 i_1(t) + \dfrac{1}{C}\int [i_1(t) - i_2(t)]\,dt = v_1(t) \\ L\dfrac{di_2(t)}{dt} + R_2 i_2(t) - \dfrac{1}{C}\int [i_1(t) - i_2(t)]\,dt = v_2(t) \end{cases} \tag{10.22}$$

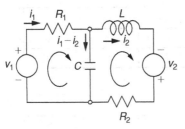

FIGURE 10.8

Two-Mesh Electrical Network.

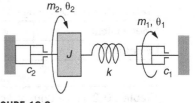

FIGURE 10.9

Two-DOF Rotary Mechanical System as an Analogous to the Electrical System of Figure 10.8.

Equations (10.22) can be written by using charges instead of currents as:

$$
\begin{cases}
R_1 \dot{q}_1 + \dfrac{1}{C}(q_1 - q_2) = v_1 \\[2mm]
L \ddot{q}_2 + R_2 \dot{q}_2 - \dfrac{1}{C}(q_1 - q_2) = v_2
\end{cases}
\tag{10.23}
$$

Application of the force-voltage analogy of Table 10.1 (with rotary-motion parameters instead of translatory-motion ones) results in the following equations for the analogous mechanical system:

$$
\begin{cases}
c_1 \dot{\theta}_1 + k(\theta_1 - \theta_2) = m_1 \\[2mm]
J \ddot{\theta}_2 + c_2 \dot{\theta}_2 - k(\theta_1 - \theta_2) = m_2
\end{cases}
\tag{10.24}
$$

A candidate for a rotational mechanical system described by Eqs. (10.24) is the one of Figure 10.9; it can easily be checked that Eqs. (10.24) are indeed the ones describing the dynamic response of this mechanical system, which confirms this is the mechanical analogous of the electrical system of Figure 10.8. It should be noted that the number of DOFs of two analogous systems must be the same.

Application of the Laplace transform with zero initial conditions to Eqs. (10.23) leads to a relationship of the type $\{Q(s)\} = [G_e(s)]\{V(s)\}$, where the transfer function matrix of the electrical system is

$$
[G_e(s)] =
\begin{bmatrix}
R_1 s + \dfrac{1}{C} & -\dfrac{1}{C} \\[3mm]
-\dfrac{1}{C} & L s^2 + R_2 s + \dfrac{1}{C}
\end{bmatrix}^{-1}
\tag{10.25}
$$

and $\{Q(s)\} = \{Q_1(s), Q_2(s)\}^t$, $\{V(s)\} = \{V_1(s), V_2(s)\}^t$ contain the Laplace transforms of q_1, q_2, v_1, and v_2, respectively. Similarly, by applying the Laplace transform with zero initial conditions to Eqs. (10.24), the following vector-matrix is obtained: $\{\Theta(s)\} = [G_m(s)]\{M(s)\}$, with

$$
[G_m(s)] =
\begin{bmatrix}
c_1 s + k & -k \\[2mm]
-k & J s^2 + c_2 s + k
\end{bmatrix}^{-1}
\tag{10.26}
$$

being the transfer function matrix of the mechanical system. $\{\Theta(s)\} = \{\Theta_1(s),\ \Theta_2(s)\}^t$, $\{M(s)\} = \{M_1(s),\ M_2(s)\}^t$ are formed of the Laplace transforms of θ_1, θ_2, m_1, and m_2, respectively. The similarity between the source transfer matrices of Eqs. (10.25) and (10.26) is noticed, as well as compliance with the pairs of Table 10.1.   ■

10.3 ELECTROMECHANICAL COUPLING

Systems defined by at least one equation comprising electrical and mechanical variables are known as *electromechanical systems*. Usually, differential equations can be formulated for the mechanical and electrical parts separately, complemented by one or several coupling equations connecting amounts from the two fields, as shown in the dc motor example. Several examples of electromechanically coupled systems are studied next, including electromagnetomechanical and piezoelectric ones.

10.3.1 Mechanical Strain, Electrical Voltage Coupling

Mechanical strain can be related to electrical voltage by means of sensors, for instance. Consider the block of Figure 10.10, which has a rectangular cross-section (width is w and thickness is h) and a length l being acted upon by two axial forces f. The block has an electric resistance R that can be defined along its length. It is demonstrated in the companion website Chapter 10 that the following relationship exists between the electrical resistance variation and the mechanical strain:

$$\frac{dR}{R} = \frac{\Delta R}{R} = (1 + 2\nu)\varepsilon_a = K\varepsilon_a \qquad (10.27)$$

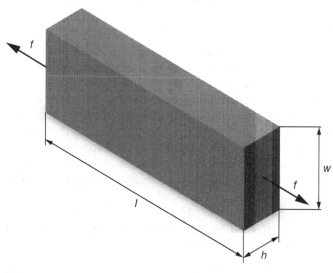

FIGURE 10.10

Mechanical Block under Axial Load.

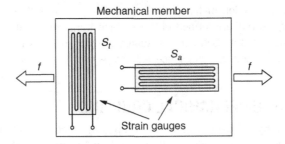

FIGURE 10.11

Wire Strain-Gauge Sensing for Mechanical Deformations of Member under Load.

where the symbol Δ indicates small but finite resistance variation whereas d points at infinitesimal variations; ν is *Poisson's ratio*, a material constant (e.g., 0.3 for steel) and K is the *material sensitivity*. Equation (10.27) is the base for a lot of sensors that measure mechanical deformation (strain) through electrical resistance variation. As known from *mechanics of materials* (see also Appendix D), the ratio of a length variation to the initial length is the *mechanical strain*:

$$\begin{cases} \varepsilon_a = \dfrac{\Delta l}{l} \\ \varepsilon_t = \dfrac{\Delta w}{w} = \dfrac{\Delta h}{h} \end{cases} \tag{10.28}$$

where ε_a is the *axial strain* (the strain aligned with the external force f), and ε_t is the *transverse strain* (the strain perpendicular to the force direction).

It is also known that the relationship between the two strains is

$$\varepsilon_t = -\nu\varepsilon_a \tag{10.29}$$

Equation (10.29) indicates that, when the axial strain is positive (the length l increases), the transverse strain is negative (the dimensions w and h decrease under the action of f).

The *strain gauge* is a practical way of implementing the use of resistance pickup for mechanical strain, and an example is discussed in the companion website Chapter 4. Figure 10.11 is the top view of a mechanical member under the action of two axial loads f. Two strain gauges, S_a and S_t, are affixed (glued) to the member. The strain gauge that has its long wire segments parallel to the forces f picks up the axial strain (this is the reason of being denoted by S_a), whereas the strain gauge identified as S_t, which has its longitudinal dimension perpendicular to the force line of action, senses transverse deformations.

For strain gauges, an equation similar to Eq. (10.27) connects the resistance variation to the mechanical strain, so that the two strain gauges of Figure 10.11 sense the following resistance-strain relationships:

$$\begin{cases} \left(\dfrac{\Delta R}{R}\right)_a = K_g \varepsilon_a \\ \left(\dfrac{\Delta R}{R}\right)_t = K_g \varepsilon_t = -K_g \nu \varepsilon_a \end{cases} \tag{10.30}$$

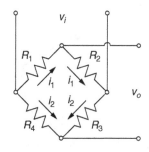

FIGURE 10.12

Wheatstone Full Bridge with Four Resistors.

where K_g is the *strain gauge sensitivity*. Strain gauges are usually connected in a *Wheatstone-bridge* electrical system, such as the one shown in Figure 10.12.

If v_i is the input voltage to the bridge, then assuming all four strain gauges are identical, the following relationship between the input voltage v_i and the variation of the output voltage Δv_o is demonstrated to be valid in the companion website Chapter 10:

$$\Delta v_o = \frac{1}{4}\left(\frac{\Delta R_1}{R_1} - \frac{\Delta R_2}{R_2} + \frac{\Delta R_3}{R_3} - \frac{\Delta R_4}{R_4}\right)v_i \tag{10.31}$$

where the subscripts 1, 2, 3, and 4 have been kept only to indicate the positions of the four strain gauges in the Wheatstone bridge. For a strain-gauge Wheatstone bridge, which is used to monitor the variation of a mechanical amount, say y, the *sensor sensitivity* K relates to the bridge output voltage as

$$\Delta v_o = Ky \tag{10.32}$$

Example 10.4

Design a displacement sensor to detect a sinusoidal motion using a microcantilever beam, such as the one shown in Figure 10.13, and four identical strain gauges connected in a Wheatstone bridge. Indicate the location and position of the strain gauges on the cantilever, as well as their connection in the Wheatstone bridge. Express the output voltage variation as a function of the mechanical strain.

Solution

The sinusoidal motion of the support generates out-of-plane oscillatory bending of the microcantilever. It is known from *Mechanics of Materials* that the strains are maximum on the two faces (face 1 is shown in Figure 10.13 and face 2 is the opposite face), one is positive and the other one is negative, and at the microcantilever's root. They are calculated by means of *Hooke's law* (stress = Young's modulus × strain or $\sigma = E\varepsilon$, see Appendix D) as

$$\varepsilon = \pm\frac{\sigma}{E} = \pm\frac{\dfrac{m_b h}{2I}}{E} = \pm\frac{\dfrac{flh}{2}\dfrac{wh^3}{12}}{E} = \pm\frac{6fl}{Ewh^2} \tag{10.33}$$

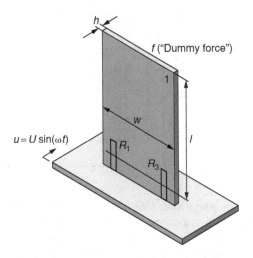

FIGURE 10.13

Cantilever Beam under Sinusoidal Motion.

In Eq. (10.33), m_b is the bending moment at a length l from the free end, $m_b = fl$ (f being a "dummy" force applied at the free end of the beam that would bend the microcantilever identically to the actual support motion), and I is the moment of inertia of the cross-section. Two strain gauges need to be placed on one face (such as on face 1 in Figure 10.13) and the other two on the opposite face, as close as possible to the root (fixed end). Let us assume that face 1 is the one that extends under bending and the opposite face 2 compresses. The strain gauges denoted by R_1 and R_3 are connected in the place of resistances R_1 and R_3 in the Wheatstone bridge of Figure 10.12. The other two strain gauges, which are not shown in Figure 10.13 but are placed on face 2 of the cantilever, are connected as R_2 and R_4 in the Wheatstone bridge. The relative resistance variations are

$$\frac{\Delta R_1}{R_1} = \frac{\Delta R_3}{R_3} = -\frac{\Delta R_2}{R_2} = -\frac{\Delta R_4}{R_4} = K_g \varepsilon \tag{10.34}$$

As a consequence, Eq. (10.31) simplifies to

$$\Delta v_o = K_g v_i \varepsilon \tag{10.35}$$

Example 10.5

For the microsystem of Example 10.4 and sketched in Figure 10.13, analyze the sensor sensitivity K (connecting the output voltage amplitude to the input motion amplitude) as a function of the subresonant excitation frequency. The output voltage amplitude variation is the sensitivity times the cantilever's tip deflection amplitude, as defined in Eq. (10.32).

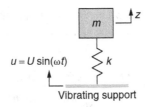

FIGURE 10.14

Lumped-Parameter Model of a Microcantilever with Harmonic Input Motion.

Known are $l = 200\ \mu m$, $h = 100\ nm$, $w = 40\ \mu m$, $\rho = 5200\ kg/m^3$, $E = 1.6 \times 10^{11}\ N/m^2$, $K_g = 2.2$, and $v_i = 20\ V$.

Solution

The lumped-parameter model of the microcantilever, where m and k are effective mass and stiffness (the subscript e has been dropped) connected to the free end of the cantilever, is shown in Figure 10.14. A similar application is studied in Chapter 9 when discussing transmissibility.

The dynamic equation of motion of the mass is

$$m\ddot{z} = -k(z - u) \tag{10.36}$$

This equation can also be written as

$$m\ddot{y} + ky = -m\ddot{u} \tag{10.37}$$

where

$$y = z - u \tag{10.38}$$

is the relative displacement of the microcantilever free end with respect to the moving support. The transfer function corresponding to Eq. (10.37) is

$$G(s) = \frac{Y(s)}{U(s)} = -\frac{ms^2}{ms^2 + k} \tag{10.39}$$

and the corresponding $G(j\omega)$ function is

$$G(j\omega) = \frac{m\omega^2}{k - m\omega^2} \tag{10.40}$$

which is a real number, so its modulus is equal to the function itself. The amplitude of the relative motion, Y, can therefore be expressed in terms of the input amplitude U:

$$Y = G(j\omega)U = \frac{m\omega^2}{k - m\omega^2}U \tag{10.41}$$

Working with amplitudes, it is known that a force F applied at the end of the microcantilever produces a deflection Y according to the *mechanics of materials* equation (see also Appendix D):

$$Y = \frac{Fl^3}{3EI} = \frac{Fl^3}{3E\frac{wh^3}{12}} = \frac{4Fl^3}{Ewh^3} \qquad (10.42)$$

By using Eq. (10.41), Eq. (10.42) can be written as

$$F = \frac{Ewh^3}{4l^3}Y = \frac{Ewh^3 m\omega^2}{4l^3(k - m\omega^2)}U \qquad (10.43)$$

At the same time, the maximum strain (amplitude), E, which is recorded at the length l from the free end, can be related to the force amplitude F according to Eq. (10.33):

$$E = \frac{6l}{Ewh^2}F \qquad (10.44)$$

Equations (10.43) and (10.44) are combined and result into

$$E = \frac{3hm\omega^2}{2l^2(k - m\omega^2)}U \qquad (10.45)$$

In terms of amplitudes, Eq. (10.35) can be written as

$$\Delta V_o = K_g v_i E \qquad (10.46)$$

Combining Eqs. (10.45) and (10.46) results in

$$\Delta V_o = \frac{3hm\omega^2 K_g v_i}{2l^2(k - m\omega^2)}U \qquad (10.47)$$

Since $\Delta V_o = KU$, the system's sensitivity K is therefore

$$K = \frac{3hm\omega^2 K_g v_i}{2l^2(k - m\omega^2)} \qquad (10.48)$$

Numerically, the equivalent mass is determined according to Table 3.1 as $m = 9.8 \times 10^{-13}$ kg, the equivalent stiffness is calculated based on Table 3.2 as $k = 2 \times 10^{-4}$ N/m; therefore, the natural frequency is $\omega_n = (k/m)^{1/2} = 14{,}282$ rad/s. Figure 10.15 shows the variation of K with ω when ω is subresonant, $\omega < \omega_n$. As indicated in Eq. (10.48), the microdevice sensitivity increases to infinity when the input frequency approaches the natural frequency.

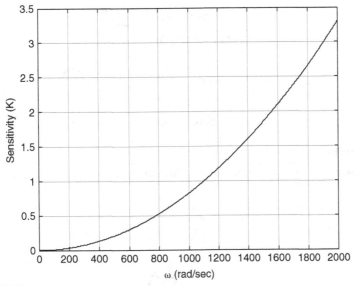

FIGURE 10.15

Variation of Microdevice Sensitivity with Input Frequency.

10.3.2 **Electromagnetomechanical Coupling**

Systems containing magnetic components in addition to mechanical and electrical elements result in coupled electromagnetomechanical systems, as illustrated by the following examples.

Example 10.6

Derive the mathematical model of the electromechanical system of Figure 10.16. The mechanical subsystem is formed of an arm (rotor) of mass moment of inertia J, which can rotate in two end bearings, and a rotary spring of stiffness k. This mobile arm is metallic and closes an electrical circuit, which has a voltage source (excitation) v_e, a resistance R, and an inductance L. An external constant magnetic field B is applied perpendicularly to the electrical circuit plane.

Solution

The system analyzed here is very similar to the dc motor examined at the beginning of this chapter. As shown next, a differential equation governs the dynamics of the mechanical system, another one defines the dynamic behavior of the electrical circuit, and two coupling equations connect mechanical to electrical variables. Let us discuss first the coupling equations based on Figure 10.17, which shows the top view of the mechanical rotary arm.

As known from electromagnetics, the interaction between the magnetic field B and the current i passing through the element of length l (which is part of both the rotary

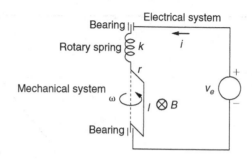

FIGURE 10.16

Schematic of an Electromagnetomechanical System with Rotary Mechanical Elements.

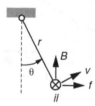

FIGURE 10.17

Top View of Mechanical Rotary Arm.

mechanical system and the electrical system) generates the following force f (which is shown in Figure 10.17):

$$\bar{f} = l\bar{i} \times \bar{B} \tag{10.49}$$

Because there is a right angle between the two product vectors of Eq. (10.49), the magnitude of f is the product of the magnitudes of the two vectors. The force f generates a torque about the pivot point that is

$$m_t = fr\cos\theta = lrBi\cos\theta = (AB\cos\theta)i \tag{10.50}$$

where $A = lr$ is the area of the rotating surface, see Figure 10.16. Equation (10.50) is a coupling equation, as it connects the mechanical torque m_t to the current i.

It is also known from electromagnetics that, due to the *Hall effect*, a *back electromotive force* (actually voltage) v_m is generated by the motion of a current-carrying conductor in a magnetic field B, which is perpendicular to the conductor; this voltage, which opposes the source voltage v_e, is calculated as

$$v_m = |\bar{v} \times \bar{B}|l = vBl\sin[90° - \theta(t)] = vBl\cos\theta(t)$$

$$= r\frac{d\theta(t)}{dt}Bl\cos\theta(t) = [AB\cos\theta(t)]\frac{d\theta(t)}{dt} \tag{10.51}$$

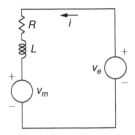

FIGURE 10.18

Electrical Subsystem of the Electromechanical System.

where v denotes velocity and $v = r\omega = rd\theta(t)/dt$. Equation (10.51) relates the voltage v_m (which is an electrical amount) to the angular velocity of the rotor $\omega = d\theta(t)/dt$ (which is a mechanical amount); therefore, this is another coupling equation.

The time-domain differential equation of the mechanical subsystem motion is

$$J\frac{d^2\theta(t)}{dt^2} = m_t(t) - k\theta(t) \tag{10.52}$$

where $k\theta$ is the opposing elastic torque produced by the spring. The differential equation defining the dynamic response of the electrical system is obtained based on Figure 10.18 and Kirchhoff's second law:

$$L\frac{di(t)}{dt} + Ri(t) = v_e(t) - v_m(t) \tag{10.53}$$

By combining Eqs. (10.50) and (10.52), as well as Eqs. (10.51) and (10.53), the following equations are obtained:

$$\begin{cases} J\dfrac{d^2\theta(t)}{dt^2} + k\theta(t) - ABi(t)\cos\theta(t) = 0 \\[2mm] L\dfrac{di(t)}{dt} + Ri(t) + AB\dfrac{d\theta(t)}{dt}\cos\theta(t) = v_e(t) \end{cases} \tag{10.54}$$

where the input is the voltage v_e, for instance, and the output is formed of the unknown rotation angle θ and the current i. The two Eqs. (10.54) can be combined to express one of the amounts i or θ in terms of v. ■

Example 10.7

A magnet is attached to a bridge to measure velocity and can slide inside a fixed coil, which is electrically connected to a voltmeter, as sketched in Figure 10.19. The bridge is bent out of its equilibrium position and allowed to undergo free vibrations. Determine the overall (equivalent) viscous-type damping (loss) coefficient of this mechanical system

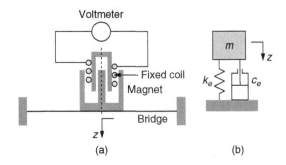

FIGURE 10.19

Electromechanical System with Fixed Coil and Magnet Attached to a Bending Bridge:
(a) Physical Model; (b) Lumped-Parameter Mechanical Model.

knowing the bridge characteristics: length $l = 0.4$ m, width $w = 0.02$ m, thickness $h = 0.001$ m, Young's modulus $E = 2.1 \times 10^{11}$ N/m², mass density $\rho = 7800$ kg/m³, the mass of the magnet $m_m = 0.02$ kg, as well as two consecutive voltage amplitude readings of 1 V and 0.98 V.

Solution

By *Faraday's law*, a motional electromotive force (voltage) v_m is induced in the electrical circuit containing the voltmeter, and its value is

$$v_m(t) = Bl_c \frac{dz(t)}{dt} \tag{10.55}$$

where l_c is the coil length and B is the magnetic field. The center bridge velocity is $dz(t)/dt$.

The actual mechanical system is equivalent to a lumped-parameter one formed of a mass m, a viscous damper with coefficient c_e, and a linear-motion spring of stiffness k_e, all placed at the bridge midpoint. The mass is the sum of the magnet mass m_m and the equivalent bridge mass m_e, which is given in Table 3.1, $m = m_m + m_e$. The midpoint stiffness of the bridge is provided in Table 3.2. As shown in Chapter 2, the equivalent damping coefficient is calculated as

$$c_e = 2m\omega_n \xi_e \tag{10.56}$$

where ω_n is the natural frequency, $\omega_n = \sqrt{k_e/m}$. In Chapter 2 as well, it is shown, Eq. (2.78), that the logarithmic decrement (the ratio of two consecutive motion amplitudes) is

$$\delta = \ln \frac{Z_{k-1}}{Z_k} = \frac{2\pi \xi_e}{\sqrt{1 - \xi_e^2}} \tag{10.57}$$

It is also shown that Eq. (2.72), $z(t) = Ze^{-\xi_e \omega_n t} \sin(\sqrt{1 - \xi_e^2}\,\omega_n t + \varphi)$, describes the free damped vibrations of a single-DOF mechanical system, which applies here by using z instead of x. The time derivative of z is therefore

$$\frac{dz(t)}{dt} = Z\omega_n e^{-\xi_e \omega_n t}\left[\sqrt{1 - \xi_e^2}\cos\left(\sqrt{1 - \xi_e^2}\,\omega_n t + \varphi\right)\right.$$

$$\left. - \xi_e \sin\left(\sqrt{1 - \xi_e^2}\,\omega_n t + \varphi\right)\right] \tag{10.58}$$

which can be written as

$$\frac{dz(t)}{dt} = Z_1 Z\omega_n e^{-\xi_e \omega_n t}\left[\sin\left(\sqrt{1 - \xi_e^2}\,\omega_n t + \varphi_1\right)\right] \tag{10.59}$$

where the new amplitude is a constant amount depending only on ξ_e and ω_n. Combining Eqs. (10.55) and (10.59) indicates that the voltage v_m is a sinusoidal function of time, whose amplitude is

$$V_m = Bl_c Z_1 Z\omega_n e^{-\xi_e \omega_n t} \tag{10.60}$$

As a consequence, the ratio of two consecutive voltage amplitudes is equal to the ratio of the consecutive deflection amplitudes at the bridge midpoint; therefore,

$$\ln\frac{V_{m,k-1}}{V_{m,k}} = \ln\frac{Z_{k-1}}{Z_k} = \frac{2\pi\xi_e}{\sqrt{1 - \xi_e^2}} \tag{10.61}$$

The equivalent damping ratio is found from Eq. (10.61):

$$\xi_e = \frac{\ln\dfrac{V_{m,k-1}}{V_{m,k}}}{\sqrt{4\pi^2 + \left(\ln\dfrac{V_{m,k-1}}{V_{m,k}}\right)^2}} \tag{10.62}$$

For the numerical values of this example, Eq. (10.62) yields an equivalent damping ratio $\xi_e = 0.0032$. The mass, equivalent midpoint bridge stiffness, and natural frequency are determined to be $m = 0.043$ kg, $k_e = 1050$ N/m, and $\omega_n = 156$ rad/s, so that Eq. (10.56) gives an equivalent viscous damping coefficient $c_e = 0.0433$ N-s/m. ■

10.3.3 Electromagnetomechanical Coupling with Optical Detection in MEMS

Optical capturing of the reflected beam from a tilting surface provides information about the amount of tilt by using a photodiode that can detect the deflection of a reflected ray. This principle is implemented in MEMS applications to measure angular position.

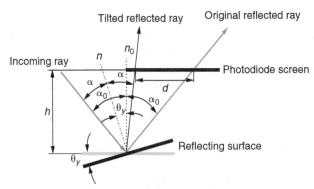

FIGURE 10.20

Optical Detection by Means of a Tilting Surface and Photodiode.

The distance d characterizing the deflection of a reflected ray after the reflecting surface has tilted an angle θ_y can be calculated simply based on Figure 10.20:

$$d = h[\tan \alpha_0 - \tan(\alpha_0 - 2\theta_y)] \qquad (10.63)$$

which allows determining the tilt angle θ_y as

$$\theta_y = \frac{\alpha_0 - \tan^{-1}(\tan \alpha_0 - d/h)}{2} \qquad (10.64)$$

Example 10.8

A sinusoidal current $i = I \sin(\omega t)$ passes through the circular-loop electrical circuit of radius r, which is printed on a rigid plate of thickness d_f and mass density ρ attached to a flexible microrod of length l, diameter d_r, and shear modulus G. The microdevice is placed under the action of an unknown constant magnetic field B, as shown in Figure 10.21. Determine the magnetic field B when $I = 0.002$ A, $\omega = 40$ rad/s, $r = 100$ μm, $d_f = 1$ μm, $l = 50$ μm, $\rho = 3600$ kg/m³, $G = 50 \times 10^{10}$ N/m². The tilt of the rigid plate is monitored optically by a photodiode, which is placed at a distance $h = 2$ mm from the plate, the incident laser beam angle being $\alpha_0 = 30°$, see Figure 10.20. It is also known that the maximum deviation of the reflected ray is $d = 4$ μm.

Solution

It can be shown using Faraday's law that the interaction between a current passing through a circular loop and an external magnetic field, which is parallel to the loop plane, results in a moment (torque) applied to the loop plane; the moment is on an axis that passes through the circular loop center and perpendicular to the magnetic field direction, see Lobontiu (2007) or Bueche and Jerde (1995) for more details. Moreover, the torque value is equal to the product of the loop area, magnetic field, and current:

$$m_t = ABi = \pi r^2 BI \sin(\omega t) = M_t \sin(\omega t) \qquad (10.65)$$

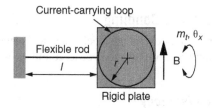

FIGURE 10.21

Top View of Electromagnetomechanical Sensor.

where the torque amplitude is $M_t = \pi r^2 BI$. The direction of the torque m_t being on the axis of the flexible rod, its effect is alternating torsion. For small tilt angles, the lumped-parameter mathematical model of the rigid plate mechanical motion is

$$J\ddot{\theta}_x = m_t - k_t\theta_x \qquad (10.66)$$

where J is the plate mass moment of inertia, k_t is the torsional stiffness of the rod (see Table 3.2), and θ_x is the tilt angle of the plate (identical to the rotation angle of the rod end). The mass moment of inertia for thin plates (where $2r \gg h = d_r$, see Appendix D) and torsional stiffness are calculated as

$$J = \frac{m(2r)^2}{12} = \frac{4\rho r^2 d_r (2r)^2}{12} = \frac{4\rho r^4 d_r}{3}; \quad k_t = \frac{GI_p}{l} = \frac{\pi Gd_r^4}{32l} \qquad (10.67)$$

With the numerical data of this example, $J = 48 \times 10^{-20}$ kg-m² and $k_t = 9.81 \times 10^{-10}$ N-m.

By applying the Laplace transform to Eq. (10.66), the following transfer function is obtained:

$$G(s) = \frac{\Theta_x(s)}{M_t(s)} = \frac{1}{Js^2 + k_t} \qquad (10.68)$$

The complex transfer function is therefore

$$G(j\omega) = \frac{1}{k_t - J\omega^2} = |G(j\omega)| = \frac{\Theta_x}{M_t} \qquad (10.69)$$

where Θ_x and M_t are the amplitudes of the tilt angle θ_x and torsional moment m_t, respectively. Combination of Eqs. (10.65) and (10.69) results in

$$\Theta_x = |G(j\omega)| M_t = \frac{\pi r^2 BI}{k_t - J\omega^2} \qquad (10.70)$$

Considering now that the angle Θ_x is θ_y of Eq. (10.64), Eq. (10.70) yields

$$B = \frac{(k_t - J\omega^2)\left[\alpha_0 - \tan^{-1}(\tan\alpha_0 - d/h)\right]}{2\pi r^2 I} \qquad (10.71)$$

and the numerical value of B is 0.0117 T.

10.3.4 Piezoelectric Coupling

A special type of interaction between the mechanical and electrical fields with applications in actuation and sensing is piezoelectric coupling, which is studied in this section.

Introduction to Piezoelectricity

In *piezoelectric transduction* (actuation or sensing) coupling between the mechanical and electrical fields is realized by means of the *direct* and *converse piezoelectric effects*, which are constitutive features of *piezoelectric materials*. In the direct piezoelectric effect, external application of a mechanical strain (deformation) to a piezoelectric material generates a voltage, which is proportional to the mechanical action. A device based on the direct piezoelectric effect can be used as a *sensor* or *generator*. The converse piezoelectric effect generates mechanical strain (deformation) from a piezoelectric material connected to an external voltage source (with the strain being proportional to the voltage); thus, this principle can be used for *actuation* resulting in *motors*.

Some natural materials, such as quartz crystals, the Rochelle salt or tourmaline, as well as ceramics, such as PZTs (based on lead [plumbum], zirconium, and titanium or barium and titanium) display piezoelectric behavior. Piezoelectric materials are formed of electric dipoles oriented randomly in the material structure, see Figure 10.22(a). Through a *poling process*, a strong external voltage (approximately 2000 V/m) is applied to the material, which has been heated to an elevated temperature, that aligns the dipoles as sketched in Figure 10.22(b); this structure is permanently impressed on the material by a fast cooling process. Through poling, the final dimension about the poling direction becomes larger, as illustrated in Figure 10.22(b). Based on the poling direction, three reference axes are used to define various properties and deformation motions of a piezoelectric block, as shown in Figure 10.23.

The numbers 1, 2, and 3 indicate translations about x, y, and z, respectively, whereas the numbers 4, 5, and 6 indicate the θ_x, θ_y, and θ_z rotations about the corresponding

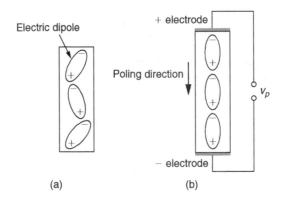

FIGURE 10.22

Piezoelectric Block: (a) Unpoled; (b) Poled.

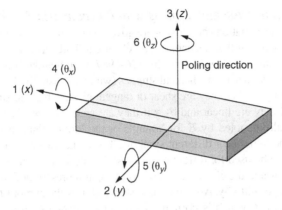

FIGURE 10.23

Reference Axes of a Piezoelectric Block.

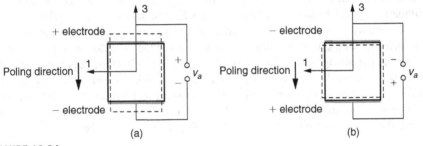

FIGURE 10.24

Piezoelectric Block as a Longitudinal Actuator Realizing: (a) Expansion; (b) Compression (Original Block Is shown with Solid Lines and Deformed Block Is Indicated with Dashed Lines).

axes. Axis 3 always indicates the poling direction. Let us consider the actuation situation where an external voltage v_a is applied to a piezoelectric block about the poling direction (which was generated by the poling voltage v_p), as illustrated in Figure 10.24(a). The result is the extension of the block dimension parallel to poling direction 3 and compressions in direction 1 (shown in the figure) and direction 2 (not shown, but perpendicular on the figure plane).

Conversely, when an external voltage is applied that has a polarity inverse to the one of the poling voltage, as indicated in Figure 10.24(b), the dimension parallel with direction 3 shrinks, whereas the dimensions about axes 1 and 2 extend, due to volume conservation. In either of the two situations, linear translatory actuation is possible about any of the axes, 1, 2, or 3. Actuation based on deformation about axis 3 is known as *longitudinal* actuation, whereas *transverse* actuation generates mechanical motion about either axis 1 or 2. A third operation mode, *shearing*, is presented in the companion website Chapter 10, together with detailed presentation of the *constitutive piezoelectric laws*, which are summarized in the following.

Longitudinal Actuation and Sensing with Piezoelectric Block

Both piezoelectric actuation and sensing are governed by *constitutive laws* that quantify the interaction between the mechanical and electrical fields, as briefly discussed next. Piezoelectricity standards (particularly, the 176-1987 IEEE Standard on Piezoelectricity) use special symbols, where mechanical stresses (which are usually denoted by σ if they are normal and by τ when they are shear or tangential), mechanical strains (which are denoted by ε if they are linear and by γ if they are generated by shear), and elasticity modulii (normally denoted by E for Young's or the longitudinal modulus and by G for the shear modulus) bear different symbols. The main reason, probably, is that the same symbol can be used for different amounts: For instance, E can stand for both the Young's modulus and the electric field, whereas ε can represent both the normal strain and the electric permittivity. Also, the same standard uses the symbol D for the charge density instead of σ, which is normally used in electrostatics. However, in this text, the classical physics notations are used. To avoid confusion between mechanical and electrical amounts that have the same symbol, the subscript m is used for mechanical amounts, such that ε means electrical permittivity and ε_m denotes mechanical strain. Similarly, E points to an electric field, whereas E_m designates Young's modulus.

We briefly present the actuator (motor) and sensor (generator) behaviors of a piezoelectric block with longitudinal action. The block's deformation and charge variation pertaining to the longitudinal direction result from superimposing the actions produced by the mechanical and electrical fields. Consider a piezoelectric block whose dimension about the poling axis (3) is l and whose cross-sectional area is A, as sketched in Figure 10.25. The Young's modulus of the piezoelectric material about the direction of interest is E_m and the electric permittivity is ε.

Actuation

The aim is to evaluate the total mechanical strain ε_m resulting from the actuator (motor) behavior of a piezoelectric block under the action of an external force f and an external (actuation) voltage v_a that has the same direction as the poling direction. The actuation voltage determines an increase in l through the converse piezoelectric effect, and the external force adds to the increase in l. The total deformation therefore is the sum of the piezoelectric (electrically generated) deformation and the mechanical one:

$$\Delta l = \Delta l_e + \Delta l_m \tag{10.72}$$

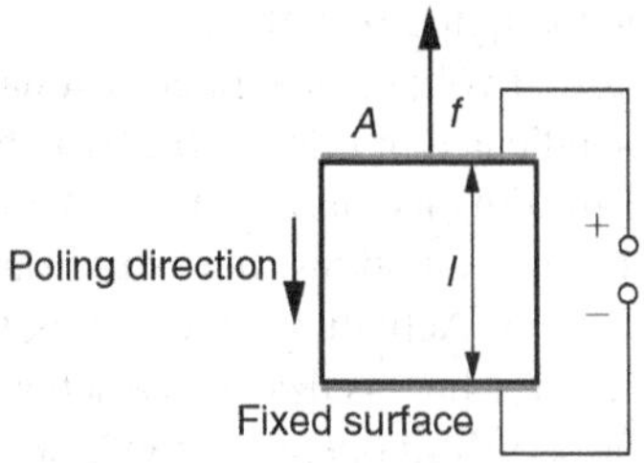

FIGURE 10.25

Piezoelectric Block Functioning as an Actuator.

According to the *converse piezoelectric effect principle*, the deformation is proportional to the applied voltage; therefore,

$$\Delta l_e = dv_a \tag{10.73}$$

with *d* being known as the *piezoelectric charge* (or *strain*) *constant*. In the literature, it is usually denoted by d_{33}, the first subscript indicating the poling direction and the second one pointing at the stress or strain direction. The constant is measured in m/V or, alternatively, in C/N. Equation (10.73) illustrates the coupling between the mechanical field (represented by Δl_e) and the electrical field (through v_a). As known from *Mechanics of Materials*, the mechanically generated deformation is

$$\Delta l_m = \frac{fl}{E_m A} = \frac{\sigma_m l}{E_m} \tag{10.74}$$

where $\sigma_m = f/A$ is the mechanical strain. As a consequence, the total deformation of Eq. (10.73) becomes

$$\Delta l = dv_a + \frac{\sigma_m l}{E_m} \tag{10.75}$$

Division by *l* in Eq. (10.75) results in the total mechanical strain:

$$\varepsilon_m = \frac{\Delta l}{l} = dE + \frac{\sigma_m}{E_m} \tag{10.76}$$

where $E = v_a/l$ is the electrical field corresponding to the voltage v_a set between the two electrodes spaced at *l*.

Example 10.9

A longitudinal piezoelectric actuator of length *l* = 0.03 m, cross-sectional area *A* = 1 cm², mass density ρ_{PZT} = 7500 kg/m³, Young's modulus E_m = 50 GPa, and piezoelectric charge constant *d* = 4 × 10⁻¹⁰ m/V is attached to a mass *m* = 0.020 kg, which further connects to a spring *k* = 200 N/m, as illustrated in Figure 10.26(a).

a. Evaluate the total system mechanical losses as an equivalent damping coefficient c_{eq} if experimental frequency-domain testing indicates the system's bandwidth is ω_c = 1.2 ω_n (where ω_n is the mechanical system's natural frequency). Consider the mass and stiffness contributions of the piezoelectric actuator.
b. Plot the system's response *y* as a function of time when a unit impulse voltage is applied to the piezoelectric actuator. Note: the electrical circuit is not shown in the figure.

Solution

a. Equation (9.57) gives the relationship between the bandwidth ω_c (which is the frequency corresponding to the state where the power becomes half the original power in a damped mechanical system under sinusoidal excitation), the natural frequency

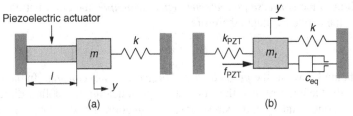

FIGURE 10.26

Longitudinal Piezoelectric Actuator: (a) Schematic of Actual System; (b) Lumped-Parameter Model.

ω_n, and in our case, the equivalent damping ratio ξ_{eq}. By using this example's specific requirement, Eq. (9.57) changes to

$$1.2\omega_n = \omega_n \sqrt{1 - 2\xi_{eq}^2 + \sqrt{4\xi_{eq}^4 - 4\xi_{eq}^2 + 2}} \qquad (10.77)$$

Equation (10.77) is solved for the equivalent damping ratio, which results in $\xi_{eq} = 0.56$. The mechanical lumped-parameter model of the piezoelectrically actuated system of Figure 10.26(a) is shown in Figure 10.26(b). The total mass m_t is formed of the mass m and the lumped mass fraction of the piezoelectric actuator, $m_{eq,PZT}$. The companion website Chapter 3 demonstrates that this latter mass fraction is one third of the total axially vibrating actuator mass; as a consequence,

$$m_t = m + m_{eq,PZT} = m + \frac{1}{3}m_{PZT} = m + \frac{1}{3}\rho_{PZT}lA \qquad (10.78)$$

Numerically, the total mass is $m_t = 0.0275$ kg. The piezoelectric block stiffness k_{PZT} and the one of the spring are connected in parallel; therefore, a total stiffness k_t is calculated as

$$k_t = k + k_{PZT} = k + \frac{E_m A}{l} \qquad (10.79)$$

with a numeric value of $k_t = 1.667 \times 10^8$ N/m. The equivalent damping coefficient can be determined from the equivalent damping ratio, the total mass, and total stiffness as

$$c_{eq} = 2\xi_{eq}\sqrt{m_t k_t} \qquad (10.80)$$

and its numerical value is $c_{eq} = 2397.8$ N-s/m.

b. Newton's second law is applied to the mechanical system of Fig. 10.26(b):

$$m_t \ddot{y} = f_{PZT} - c_{eq}\dot{y} - k_t y \qquad (10.81)$$

The force generated by the piezoelectric block is

$$f_{PZT} = k_{PZT} y_0 = \frac{E_m A}{l} y_0 \qquad (10.82)$$

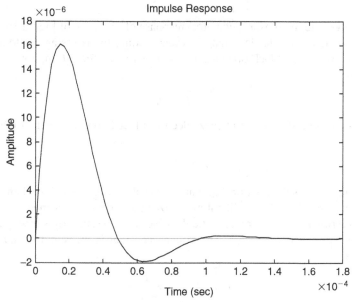

FIGURE 10.27

Displacement Response of Piezoelectrically-Actuated Mechanical System under Unit Impulse Voltage Input.

where y_0 is the free deformation of the piezoelectric block; therefore, in accordance with Eq. (10.73),

$$y_0 = dv_a \qquad (10.83)$$

Equations (10.82) and (10.83) are combined and substituted into Eq. (10.81), which becomes

$$m_t\ddot{y} + c_{eq}\dot{y} + k_t y = Cv_a \qquad (10.84)$$

where the constant C is

$$C = \frac{E_m A d}{l} \qquad (10.85)$$

with a numerical value of $C = 0.0667$ N/V. Laplace transforming Eq. (10.84) yields the transfer function

$$G(s) = \frac{Y(s)}{V_a(s)} = \frac{C}{m_t s^2 + c_{eq}s + k_t} \qquad (10.86)$$

By using the `tf` and `impulse` MATLAB® commands, the plot of Figure 10.27 is obtained. ∎

Sensing

We now evaluate the sensor (generator) response of the block of Figure 10.25 under the same external mechanical and electrical loading by assessing the total charge density σ. When the block operates as a sensor for the force f, the total generated charge is

$$q = q_m + q_e \qquad (10.87)$$

where the charge due to direct piezoelectric effect is proportional to the applied force:

$$q_m = d_m f \qquad (10.88)$$

Equation (10.88) is the relationship that couples the electrical and mechanical fields in sensing. The piezoelectric literature utilizes the g constant (named *piezoelectric voltage constant*), or g_{33}, for this particular case (with the subscripts having the same meaning with the ones of d_{33}). The voltage constant is measured in m²/C and is defined as

$$g = \frac{\varepsilon_m}{\sigma} \text{ or } g = \frac{E}{\sigma_m} \qquad (10.89)$$

where σ_m and ε_m are the mechanical stress and strain, and σ is the charge density ($\sigma = q/A$). Because $\sigma = \varepsilon_m/g = f/(E_m Ag)$, it follows by comparison to Eq. (10.88) that:

$$d_m = \frac{1}{gE_m} \qquad (10.90)$$

It can be shown that d_m is measured in C/N so the units of d (the piezoelectric charge constant) and d_m are identical. Equation (10.89) can also be written in the form

$$g = \frac{v/l}{f/A} = \frac{v}{\dfrac{fl}{E_m A} E_m} = \frac{v}{\Delta l E_m} \qquad (10.91)$$

which is

$$v = (gE_m)\Delta l \text{ or } v = \frac{1}{d_m}\Delta l \qquad (10.92)$$

Equation (10.92) indicates that the voltage generated through mechanical action is proportional to the mechanical deformation. The charge generated by the action of the voltage v_a is

$$q_e = Cv_a = \frac{\varepsilon A}{l} v_a = \varepsilon AE \qquad (10.93)$$

where C is the electrical capacitance of the piezoelectric block taken along its length l. Adding up the charges of Eqs. (10.88) and (10.93) gives the total charge of Eq. (10.87):

$$q = d_m f + \varepsilon AE \qquad (10.94)$$

Division by A in Eq. (10.94) produces the charge density $\sigma = q/A$ in the left-hand side:

$$\sigma = d_m \sigma_m + \varepsilon E \qquad (10.95)$$

where $\sigma_m = f/A$ is the mechanical stress.

Example 10.10

An accelerometer uses a piezoelectric plate to sense the acceleration of an input sinusoidal displacement using the device shown in Figure 10.28. Use a lumped-parameter model to express the acceleration amplitude by means of the sensing voltage amplitude. Known are the dimensions of the piezoelectric plate: thickness $h = 1$ mm, side of square cross-sectional area $a = 1$ cm, as well as Young's modulus $E_m = 5 \times 10^{10}$ N/m², piezoelectric mass density $\rho_{PZT} = 7500$ kg/m³, piezoelectric voltage constant $g = 0.074$ m²/C, mass $m = 0.05$ kg, and spring stiffness $k = 100$ N/m. Assume that the total system losses are lumped into the viscous damping coefficient $c_t = 136.821$ N-s/m. The voltage amplitude read by the sensor is $V = 0.05$ V. Consider that the spring and damper are precompressed to maintain permanent contact between the vibrating mass m and the piezoelectric sensor. Also, draw the Bode plot magnitude ratio to verify the results.

Solution

By taking into account that the piezoelectric plate can be represented as a mass-damper-spring system, Figure 10.29(a) shows the lumped-parameter model of the original system of Figure 10.28, which includes the effective mass, damping coefficient, and spring stiffness of the piezoelectric sensor.

Because the actual mass m and the piezoelectric plate equivalent (efficient) mass move together, the lumped-parameter model can be simplified further to the one of Figure 10.29(b), where the subscript t indicates total and adds up contributions from the mechanical system to the corresponding equivalent amounts of the piezoelectric sensor. By using the lumped-parameter model of Figure 10.29(b), the equation of motion of the system is

$$m_t \ddot{z} = -c_t(\dot{z} - \dot{u}) - k_t(z - u) \qquad (10.96)$$

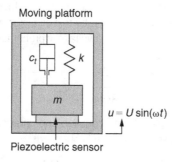

Moving platform

$u = U \sin(\omega t)$

Piezoelectric sensor

FIGURE 10.28

Piezoelectric Sensing of Acceleration.

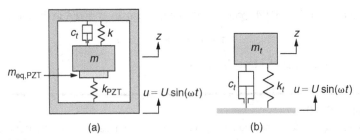

FIGURE 10.29

Lumped-Parameter Model of (a) Original System; (b) Equivalent System.

where z monitors the absolute motion of the mass m about the equilibrium position (the one obtained by the precompression of the dashpot). The total mass, damping coefficient, and spring rate are

$$\begin{cases} m_t = m + m_{eq,PZT} = m + \dfrac{1}{3}m_{PZT} = m + \dfrac{1}{3}\rho_{PZT}ha^2 \\[2ex] k_t = k + k_{PZT} = k + \dfrac{E_m a^2}{h} \end{cases} \qquad (10.97)$$

The total mass of Eq. (10.97) has been calculated by taking into account that the equivalent mass of the axially vibrating piezoelectric plate (fixed at one end and mobile at the other) is one third of the actual mass; see the companion website Chapter 3 for details. A similar accelerometer problem is studied in Chapter 9, and it is demonstrated that Eq. (10.96) can be brought to the form

$$m_t\ddot{y} + c_t\dot{y} + k_t y = -m_t\ddot{u} \qquad (10.98)$$

where $y = z - u$ is the coordinate indicating the relative motion of the mass with respect to the moving platform. It can be shown that, if the reference frame of z is placed at the point of static equilibrium, where the compression spring and damper forces balance out the reactions from the corresponding piezoelectric forces, then Eq. (10.96) and therefore (10.98) are valid; the interested reader can try this as a side exercise.

The piezoelectric actuator generates a voltage v proportional to the relative displacement y, as shown in Eq. (10.92):

$$y = \frac{1}{gE_m}v = d_m v \qquad (10.99)$$

The Laplace transform is applied to the equation and, after substituting Eq. (10.99) into Eq. (10.98), the following transfer function is obtained:

$$G(s) = \frac{V(s)}{U(s)} = -\frac{m_t s^2}{d_m(m_t s^2 + c_t s + k_t)} \qquad (10.100)$$

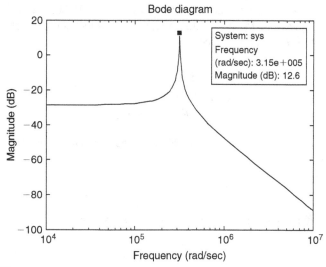

FIGURE 10.30

Magnitude-Ratio Bode Plot of Piezoelectric System's Transfer Function.

Figure 10.30 is the MATLAB® Bode plot of magnitude ratio corresponding to the transfer function of Eq. (10.100), obtained for the following parameter values: $m_t = 0.0503$ kg, $k_t = 5 \times 10^9$ N/m², and $d_m = 1/(gE_m) = 27 \times 10^{-11}$ C/N.

Our interest lies on transmissibility, the amplitude ratio of the voltage to the input displacement, which is the modulus of the complex function $G(j\omega)$, as discussed in Chapter 9; that is,

$$|G(j\omega)| = \frac{|V(j\omega)|}{|U(j\omega)|} = \frac{V}{U} = \frac{m_t \omega^2}{d_m \sqrt{(k_t - m_t \omega^2)^2 + (c_t \omega)^2}} \tag{10.101}$$

Equation (10.101) can also be written as

$$V = \frac{\dfrac{\omega^2}{\omega_{n,t}^2} U}{d_m \sqrt{\left(1 - \dfrac{\omega^2}{\omega_{n,t}^2}\right)^2 + \left(2\xi_t \dfrac{\omega}{\omega_{n,t}}\right)^2}} \tag{10.102}$$

where the total natural frequency $\omega_{n,t}$ and total viscous damping ratio ξ_t are

$$\omega_{n,t} = \sqrt{\frac{k_t}{m_t}}; \; 2\xi_t \omega_{n,t} = \frac{c_t}{m_t} \tag{10.103}$$

The numerical value of the natural frequency of Eq. (10.103) is $\omega_{n,t} = 3.154 \times 10^5$ rad/s; this value can also be retrieved by right clicking on the peak of the plot obtained using

MATLAB®'s `bodemag` command, as illustrated in Figure 10.30. An accelerometer is defined by a high natural frequency; therefore, the square root of Eq. (10.102) is approximately equal to 1. As a consequence,

$$V \approx \frac{\dfrac{\omega^2}{\omega_{n,t}^2}U}{d_m} = \frac{A}{d_m \omega_{n,t}^2} \tag{10.104}$$

where $A = \omega^2 U$ is the acceleration amplitude of the external motion. The acceleration amplitude is therefore expressed as

$$A = d_m \omega_{n,t}^2 V \tag{10.105}$$

By using the numerical values of the example, the acceleration amplitude is found to be $A = 1.345$ m/s^2. ■

Piezoelectric and Strain Gauge Sensory-Actuation

A piezoelectric stack can be equipped with two strain gauges, as shown in Figure 10.31, which is the photograph of an actuator realized by Thorlabs, that are connected in a half Wheatstone bridge. In this arrangement, the deformation of the piezoelectric stack can be monitored, and the block functions as both actuator and sensor.

An example is solved in the companion website Chapter 10 of the piezoelectric block of Figure 10.32, which has two strain gauges connected in a half Wheatstone bridge, as shown in Figure 10.33 (the strain gauges are denoted by R_1, R_2, whereas the other two resistances R are of constant value).

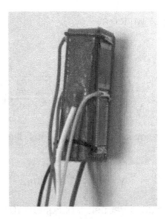

FIGURE 10.31

Piezoelectric Stack with Two Strain Gauges in a Half Wheatstone Bridge Connection.

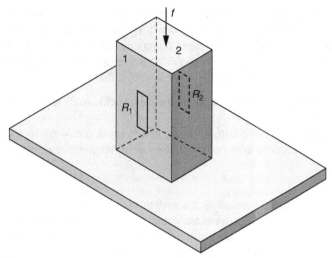

FIGURE 10.32

Piezoelectric Block with Two Strain Gauges.

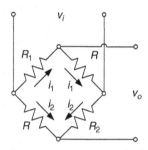

FIGURE 10.33

Half Wheatstone Bridge with Two Variable Resistors and Two "Dummy" Resistors.

It is also demonstrated that the output voltage variation is expressed as

$$\Delta v_o = \frac{1}{2} K_g \varepsilon_a v_i = \frac{1}{2} K_g \frac{f}{E_m A} v_i = \frac{K_g v_i}{2 E_m A} f \tag{10.106}$$

and that the *device sensitivity* (resulting from $\Delta v_o = Kf$) is therefore

$$K = \frac{K_g v_i}{2 E_m A} \tag{10.107}$$

Example 10.11

A two strain-gauge piezoelectric block is used to eliminate the effects of the force $f = 5 \sin(4t)$ N, which is applied at the end of the rigid rod shown in Figure 10.34. The rod is supported by an elastic beam (flexure hinge), both being originally horizontal. Known are L = 20 mm, the rod mass m = 0.05 kg, the flexure length l_f = 6 mm, the flexure cross-sectional dimensions, which are thickness h_f = 1 mm (dimension in the plane of the figure) and width w_f = 2 mm, as well as the beam Young's modulus E_f = 2.1 × 10¹¹ N/m². For the piezoelectric sensory-actuator known are l = 20 mm, cross-sectional area A = 6.25² mm², E_m = 5.2 × 10¹⁰ N/m², d = 390 × 10⁻¹² m/V, v_i = 15 V, and K_g = 2.

a. What actuation voltage is necessary for the piezoelectric block to counteract the external force f?

b. The readout indicates an output voltage amplitude of ΔV_o = 67.16 μV, which indicates there is an error in the actuator placement. What is the actual position of the actuator?

Solution

a. The lumped-parameter model of the system sketched in Figure 10.34 is shown in Figure 10.35, where the piezoelectric actuator has been replaced by a point force f_a, and the flexure hinge has been substituted by a spiral (torsional) spring of stiffness k_t. Static balancing of the horizontal rod requires that

$$f_a \frac{L}{2} = fL \tag{10.108}$$

In this case, there is no rotation of the rod about the pivot point and therefore no elastic torque generated by the torsional spring. By taking into account that $\Delta l = f_a l/(E_m A) = dv_a$, Eq. (10.108) allows expressing the actuation voltage as

$$v_a = \frac{2l}{E_m A d} f = \frac{2l}{E_m A d} 50 \sin(4t) \tag{10.109}$$

Numerically, the actuation voltage is v_a = 252.46 sin(4t) V.

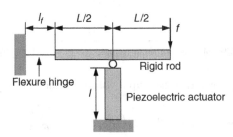

FIGURE 10.34

Lever-Type Mechanical System with External Force and Counteracting Piezoelectric Block.

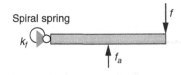

FIGURE 10.35

Equivalent Lumped-Parameter Model.

b. For a perfectly balanced mechanical system (with no rotation about the pivot point), the tip displacement of the piezoelectric actuator should be zero and this should translate into an output voltage of zero. The output voltage not being zero means that one possible cause for this discrepancy is the position of the actuator with respect to the pivot point. Assume the actual position parameter is *a* instead of the design parameter *L*/2. In this case, the dynamic equation of motion of the rod around the pivot point is

$$J\ddot{\theta} = f_a a - fL - k_f \theta \tag{10.110}$$

It is known that the elastic torque is equal to the stiffness times the deformation (rotation angle); thus, it is known from *Mechanics of Materials* that the spring rotary stiffness is

$$k_f = \frac{E_f I}{l_f} = \frac{w_f h_f^3 E_f}{12 l_f} \tag{10.111}$$

which is $k_f = 5.833$ N-m. According to Eq. (10.106), the following relationship can be written between the output voltage variation Δv_o and the actuation force f_a:

$$f_a = c_1 \Delta v_o; \; c_1 = \frac{2E_m A}{K_g v_i} \tag{10.112}$$

Assuming that the rod of Figure 10.35 rotates by an angle θ in the direction of action of f_a, this angle can be expressed for small deformations as

$$\theta = \frac{\Delta l}{a} = \frac{f_a l}{E_m A a} \tag{10.113}$$

Substituting f_a of Eq. (10.112) into Eq. (10.113) results in

$$\theta = \frac{c_2}{a} \Delta v_o; \; c_2 = \frac{2l}{K_g v_i} \tag{10.114}$$

Combining Eqs. (10.110), (10.112), and (10.114) produces the following differential equation:

$$\frac{Jc_2}{a} \times \frac{d^2}{dt^2}(\Delta v_o) + \left(\frac{k_f c_2}{a} - c_1 a \right) \Delta v_o = -fL \tag{10.115}$$

The Laplace transform applied to Eq. (10.115) yields the transfer function that connects $\Delta V_o(s)$, the Laplace transform of $\Delta v_o(t)$, to $F(s)$, the Laplace transform of $f(t)$. Because there is no damping in the system, the transfer function is also the ratio of the amplitudes of the two signals V_o and F:

$$\frac{\Delta V_o}{F} = \frac{L}{\frac{c_2}{a}(J\omega^2 - k_f) + c_1 a} \tag{10.116}$$

As a consequence, the following equation in a is produced:

$$c_1 a^2 - \frac{LF}{\Delta V_o} \times a + c_2 \left(J\omega^2 - k_f \right) = 0 \tag{10.117}$$

The mass moment of inertia of the rigid rod is determined as $J = mL^2/12$. By using the numerical values of this example, the value of $a = 11$ mm is obtained, which indicates that, for the given sensing voltage, the piezoelectric actuator needs to be moved 1 mm to the left to cancel the error. ■

10.4 THERMOMECHANICAL COUPLING: THE BIMETALLIC STRIP

The *bimetallic strip* is a cantilever beam consisting of two layers of different materials, as shown in Figure 10.36(a). Exposed to a temperature increase, the two materials expand differently along the length, but because they are attached, the compound beam bends, as illustrated in Figure 10.36(b) because the top layer material has a linear coefficient of thermal expansion α_1 that is larger than the coefficient α_2 of the bottom layer and therefore tends to expand more in the dimension l.

The radius of curvature R, shown in Figure 10.36(b), is related to the temperature increase $\Delta\theta$ as

$$R = \frac{h}{(\alpha_1 - \alpha_2)\Delta\theta} \tag{10.118}$$

where $h = h_1 + h_2$ (h_1 and h_2 are the two strips' thicknesses).

Mechanics of Materials shows that the curvature (the reciprocal of the radius of curvature) depends on the second derivative of the free end deflection $z(x)$, Fig. 10.36(b):

$$\frac{1}{R} = \frac{d^2 z(x)}{dx^2} \tag{10.119}$$

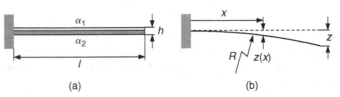

(a)   (b)

FIGURE 10.36

Bimetallic Strip in (a) Original (Undeformed) Condition; (b) Deformed Condition.

where x is measured from the fixed end of the cantilever of Figure 10.36(b). Combining Eqs. (10.118) and (10.119) followed by two integrations yields the slope and the deflection:

$$\begin{cases} \dfrac{dz(x)}{dx} = \dfrac{(\alpha_1 - \alpha_2)\Delta\theta}{h}x + C_1 \\ z(x) = \dfrac{(\alpha_1 - \alpha_2)\Delta\theta}{2h}x^2 + C_1 x + C_2 \end{cases} \qquad (10.120)$$

Both the slope and deflection are zero at the root, for $x = 0$, which results in the two integration constants, $C_1 = C_2 = 0$. The maximum deflection is at the free end and is found from the second Eq. (10.120) for $x = l$:

$$z = \frac{(\alpha_1 - \alpha_2)l^2}{2h} \times \Delta\theta \qquad (10.121)$$

Equation (10.121) indicates that the bimetallic strip functions as a thermal sensor when the free-end deflection can be measured.

Example 10.12

A two-material cantilever is calibrated by exposing it to a temperature increase of $\Delta\theta = 20°$ and evaluating the resulting free-tip displacement as a function of time. Known are the following parameters: $\alpha_1 = 22.2 \times 10^{-6}$ deg^{-1}, $\alpha_2 = 3.2 \times 10^{-6}$ deg^{-1}, $E_1 = 69$ GPa, $E_2 = 400$ GPa, $\rho_1 = 2700$ kg/m^3, $\rho_2 = 2400$ kg/m^3, common width $w = 50$ μm, common length $l = 350$ μm, $h_1 = 100$ nm, and $h_2 = 1$ μm. Plot the free-tip displacement using a lumped-parameter model and considering that the total viscous damping coefficient is $c = 2.3 \times 10^{-7}$ N-s/m.

Solution

When the inertia, damping, and stiffness properties of the bi-metallic strip are considered, the quasi-static model, which predicted the tip deflection as a function of temperature in Eq. (10.121), needs to be altered. The lumped-parameter model of the two-material cantilever consists of two collocated masses, m_1 and m_2, two springs in parallel (of stiffnesses k_1 and k_2) and an equivalent point force that produces the same tip displacement as the quasi-static thermal effect, see Figure 10.37.

The total mass is calculated by means of Table 3.1 as

$$m = m_1 + m_2 = \frac{33}{140} \times wl(\rho_1 h_1 + \rho_2 h_2) \qquad (10.122)$$

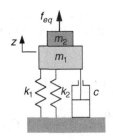

FIGURE 10.37

Lumped-Parameter Model of a Two-Material Cantilever under Thermal Deformation.

Similarly, the stiffness is calculated as indicated in Table 3.2:

$$k = k_1 + k_2 = \frac{3}{l^3} \times (E_1 I_1 + E_2 I_2) = \frac{w}{4l^3} \times (E_1 h_1^3 + E_2 h_2^3) \qquad (10.123)$$

With the numerical values of this example, $m = 1.1 \times 10^{-11}$ kg and $k = 0.1166$ N/m, which yields a natural frequency of 102,910 rad/s. The equivalent point force is calculated by means of Eqs. (10.121) and (10.123):

$$f_{eq} = kz = \frac{w(E_1 h_1^3 + E_2 h_2^3)(\alpha_1 - \alpha_2)}{8l(h_1 + h_2)} \times \Delta\theta = a\Delta\theta \qquad (10.124)$$

which shows the force is proportional to the temperature variation. The proportionality constant is $a = 1.234 \times 10^{-7}$ N-deg^{-1}. Based on Figure 10.37, the equation of motion of the mass is

$$m\frac{d^2 z(t)}{dt^2} = f_{eq} - c\frac{dz(t)}{dt} - kz(t) \qquad (10.125)$$

Combining Eqs. (10.124) and (10.125) results in

$$m\frac{d^2 z(t)}{dt^2} + c\frac{dz(t)}{dt} + kz(t) = a\Delta\theta \qquad (10.126)$$

The transfer function corresponding to Eq. (10.126) is

$$G(s) = \frac{Z(s)}{\Delta\Theta(s)} = \frac{a}{ms^2 + cs + k} \qquad (10.127)$$

By using the MATLAB® tf and step commands and shortening the time interval, the plot of Figure 10.38 is obtained; it can be seen that the steady state tip displacement is $z(\infty) = \lim_{s \to 0} sG(s)\Delta\Theta(s) = \Delta\theta a/k$, namely, $z(\infty) = 21.16$ μm.

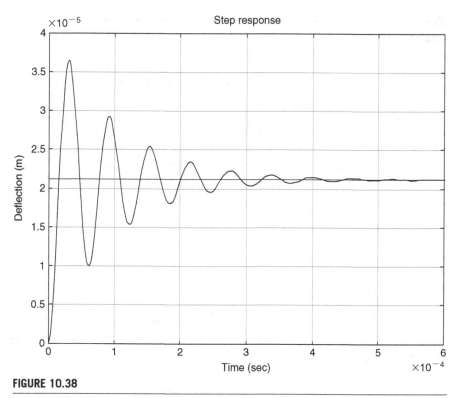

FIGURE 10.38

Free-End Displacement of Two-Material Cantilever under Thermal Deformation.

10.5 NONLINEAR ELECTROTHERMOMECHANICAL COUPLED-FIELD SYSTEMS

Obtaining the time-domain solution of nonlinear coupled-field systems is usually performed using time-stepping schemes, such as the finite difference method, the Runge-Kutta procedure, or other dedicated algorithms; the companion website Chapter 10 presents application of the forward and central finite difference methods to some nonlinear coupled-field mathematical models, whereas the last part of this section presents the use of Simulink® in numerically solving such a model.

The following example derives the nonlinear model of a dynamic system with coupling between the electrical, thermal, and mechanical fields. One equation connecting temperature variation $\Delta\theta$ to electrical resistance at the final temperature R is

$$R = R_0(1 + \alpha_R \Delta\theta) \tag{10.128}$$

where R_0 is the electrical resistance at the original (initial) temperature and α_R is the *thermal coefficient of resistance*.

Example 10.13

A voltage v is applied to a deformable bar by means of an external electrical circuit, as sketched in Figure 10.39. The voltage generates heating of the bar through the Joule effect and its subsequent axial deformation in the x direction. Derive a mathematical model of this coupled-field dynamic system with the output being the bar surface temperature θ_s and the axial deformation x of its guided end. Known are the linear coefficient of thermal expansion α, the specific heat c, the thermal coefficient of resistance α_R, as well as the electrical resistivity of the bar ρ_{el}, its original length at the air temperature l_0, length of square cross-sectional side a, and mass density ρ. The bar loses heat through convection in the air whose constant temperature is θ_∞ while the average convection coefficient is h. Assume that the variations of the cross-sectional dimensions and mass density with the temperature are negligible. Also assume that the electrical-to-mechanical energy conversion is instantaneous.

Solution

As seen in Chapter 5, the thermal capacity of the bar C_{th} is connected to its temperature and heat flow rates

$$C_{th}\dot{\theta}_s = q_i - q_0 \text{ or } mc\dot{\theta}_s = q_i - q_0 \tag{10.129}$$

where m is the bar mass, c is the specific heat, and q_i is the input heat flow rate gained through Joule heating defined as

$$q_i = \frac{v^2}{R} = \frac{v^2}{R_0[1 + \alpha_R(\theta_s - \theta_\infty)]} \tag{10.130}$$

and q_0 is the heat flow rate lost through convection

$$q_0 = hA_l(\theta_s - \theta_\infty) \tag{10.131}$$

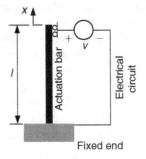

FIGURE 10.39

MEMS with Electrothermal (Joule Effect) Actuation.

The lateral area of the bar A_l is

$$A_l = 4al = 4al_0[1 + \alpha(\theta_s - \theta_\infty)] \tag{10.132}$$

while the constant mass is $m = \rho a^2 l_0$ and the original resistance is $R_0 = \rho_{el} l_0/a^2$. Using these parameters and substituting Eqs. (10.130), (10.131), and (10.132) in Eq. (10.129) results in

$$\rho a^2 l_0 c \dot{\theta}_s = \frac{a^2 v^2}{\rho_{el} l_0 [1 + \alpha_R(\theta_s - \theta_\infty)]}$$
$$- 4hal_0[1 + \alpha(\theta_s - \theta_\infty)](\theta_s - \theta_\infty) \tag{10.133}$$

Considering that the air temperature θ_∞ is constant, the notation $\theta = \theta_s - \theta_\infty$ is used, which allows reformulating Eq. (10.133) as

$$\rho\rho_{el}\alpha_R cal_0^2 \dot{\theta}\theta + \rho\rho_{el} cal_0^2 \dot{\theta} + 4\rho_{el}\alpha\alpha_R hl_0^2\theta^3$$
$$+ 4\rho_{el}(\alpha + \alpha_R)hl_0^2\theta^2 + 4\rho_{el}hl_0^2\theta = av^2 \tag{10.134}$$

The displacement x of the bar's movable end is expressed as

$$x(t) = \alpha l_0 \theta(t) \tag{10.135}$$

Equations (10.134) and (10.135) form the mathematical model of the *electrothermomechanical* system of Figure 10.39. The nonlinear Eq. (10.134), which expresses the *electrothermal* system coupling, can be solved for θ, whereas Eq. (10.135), which illustrates the *thermomechanical* system connection, is used to determine x, as shown in the solved Example 10.15.

10.6 SIMULINK® MODELING OF NONLINEAR COUPLED-FIELD SYSTEMS

This section comprises a few examples of coupled-field systems being described by nonlinear mathematical models and which are solved by means of Simulink®.

Example 10.14
Use Simulink® to plot the time variation of the arm rotation angle $\theta(t)$ and the current $i(t)$ corresponding to the electromagnetomechanical system of Example 10.6 and shown in Figure 10.16. Known are $r = 0.01$ m, $l = 0.04$ m, $k = 0.1$ N-m, $J = 2 \times 10^{-6}$ kg-m², $v_e = 60$ V, $R = 80$ Ω, $L = 0.04$ H, $B = 1$ T, as well as the nonzero initial condition $\theta(0) = 0.05$ rad.

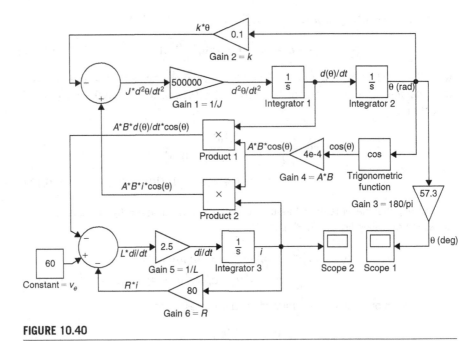

FIGURE 10.40

Simulink® Diagram for Solving the Mathematical Model of the Electromagnetomechanical System of Example 10.6.

Solution

The differential Eqs. (10.54) can be written as

$$\begin{cases} J\dfrac{d^2\theta(t)}{dt^2} = -k\theta(t) + ABi(t)\cos\theta(t) \\ L\dfrac{di(t)}{dt} = v_e - Ri(t) - AB\dfrac{d\theta(t)}{dt}\cos\theta(t) \end{cases} \qquad (10.136)$$

where $A = rl = 4 \times 10^{-4}$ m² is the rotating loop area. It can be seen that the differential equations are nonlinear, because the variable θ enters as a trigonometric function's argument and also because it multiplies the other variable i. Figure 10.40 contains the Simulink® block diagram, allowing us to solve the two differential Eqs. (10.136). The upper part of the diagram solves the first Eq. (10.136), while the lower part corresponds to the second Eq. (10.136). Blocks used in previous Simulink®-solved examples are seen in Figure 10.40; a new block here is the Trigonometric Function, which allows calculation of cosθ; this block is dragged from the Math Operations library. The nonzero initial condition of this example is specified by clicking the second integration block on the upper part of the block diagram (which gives the angle θ) and specifying the value of θ(0). Figure 10.41 illustrates the plots of θ and i as functions of time. It can be seen that the current reaches a constant value of less than 0.8 A very fast, whereas the arm has an oscillatory motion with an amplitude of approximately 3 degrees.

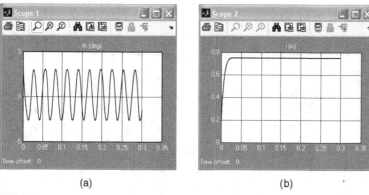

FIGURE 10.41

Plot of Solution in Terms of Time: (a) Rotation Angle θ (rad); (b) current I (A). Time Is Measured in Seconds.

Example 10.15

Solve Example 10.13 using Simulink® for a copper bar and the following numerical values: $l_0 = 0.1$ m, $a = 1$ mm (side of square cross-section actuator), $\rho = 8900$ kg/m^3, $\alpha = 17 \times 10^{-6}$ deg^{-1}, $\rho_{el} = 17 \times 10^{-9}$ Ω-m, $\alpha_R = 0.0039$ deg^{-1}, $c = 385$ J/(kg-deg), $h = 30$ J/(m^2-s-deg), and $v = 0.017$ V. Specifically, plot the time variations of the temperature change θ and the axial deformation of the bar x.

Solution

Equations (10.134) and (10.135) can be rewritten as

$$\begin{cases} \dot{\theta} = -a_1\theta - a_2\dot{\theta}\theta - a_3\theta^2 - a_4\theta^3 + a_5 \\ x = a_6\theta \end{cases} \tag{10.137}$$

with the coefficients a_1 through a_6 easily identifiable in Eq. (10.135) as

$$a_1 = \frac{4h}{\rho ca}; a_2 = \alpha_R; a_3 = \frac{4h(\alpha + \alpha_R)}{\rho ca}; a_4 = \frac{4h\alpha\alpha_R}{\rho ca};$$

$$a_5 = \frac{v^2}{\rho\rho_{el}cl_0^2}; a_6 = \alpha l_0 \tag{10.138}$$

The numerical values of the constants of Eq. (10.138) are $a_1 = 0.035$, $a_2 = 0.0039$, $a_3 = 1.3718 \times 10^{-4}$, $a_4 = 2.3219 \times 10^{-9}$, $a_5 = 0.4961$, and $a_6 = 1.7 \times 10^{-6}$. Figure 10.42 illustrates the Simulink® block diagram solving the nonlinear Eq. (10.138). The Add block is dragged from the Math Operations library and used to add the five terms on the right-hand side of Eq. (10.138); the Math Function block is also obtainable from the Math Operations library and is utilized to calculate the square of θ. The second

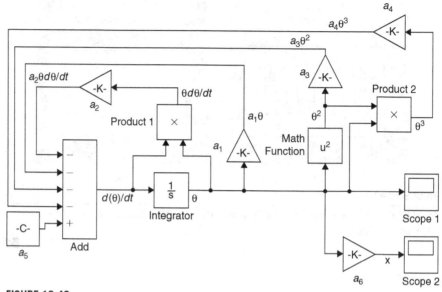

FIGURE 10.42

Simulink® Diagram for Solving the Mathematical Model of the Electrothermomechanical
System of Example 10.13.

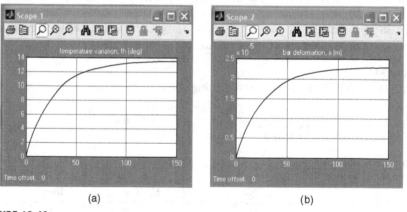

(a)   (b)

FIGURE 10.43

Plot of Displacement in Terms of Time, Time Is Measured in Seconds.

term in the right-hand side of Eq. (10.138) is generated as the product between θ and
$\dot{\theta}$ by means of the Product block from the Math Operations library. Figure 10.43
shows the rising and stabilizing portions of the system response over a time period of
150 s. The maximum bar temperature change is less than 14 deg and the maximum bar
deformation is approximately 23 μm.

SUMMARY

This chapter studies the interaction among systems containing elements of different physical types by formulating mathematical models of these coupled-field systems. As a preamble to system coupling, system analogies are studied using differential equations and transfer functions. Several examples of coupled-field systems are analyzed, including electromechanical and electromagnetomechanical ones, piezoelectrically-coupled applications, and bimetallic strips as thermomechanically coupled problems and electrothermomechanical systems. Examples of using Simulink® to model the dynamics of nonlinear coupled-field systems are included.

PROBLEMS

10.1 Formulate a state-space model for the dc motor of Example 10.1, shown in Figure 10.4, using two inputs (the armature voltage v_a and load moment m_l) and one output (the shaft angular velocity ω), then plot the system's time response (output) by means of Simulink®. Use the numerical values of Example 10.1 except for $v_a = 100 - 10/(t + 1)$ V and $m_l = 0.15 \cos(5t)$ V.

10.2 Consider that the mechanical system of Example 3.11, whose sketch is Figure 10.44, is actuated by a dc motor and its corresponding electrical circuit. Demonstrate the electromechanical coupling of this system deriving the transfer function matrix connecting the armature voltage v_a and load moment m_l as input variables to the midshaft cylinder rotation angle θ and armature current i_a as output variables.

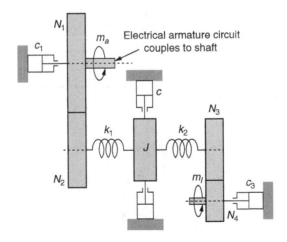

FIGURE 10.44

Gear-Shaft Rotary Mechanical System Driven by a dc Motor.

10.3 Calculate and plot the angular velocity of the midshaft cylinder studied in Problem 10.2. All parameters furnished in Example 3.11 are applicable here except for the actuation moment m_a. The other parameters are $R_a = 10\ \Omega$, $L_a = 5$ H, $v_a = 100 \sin(60t)$ V, $K_e = 0.02$ N-m/A, $K_t = 0.4$ N-m/A.

10.4 Use time-domain modeling to determine an electrical system that is analogous to the liquid system of Figure 10.45, where p_i is the input pressure and p_o is the output pressure. The system is formed of two tanks having the capacitances C_{l1} and C_{l2}, and three resistances: R_{l1}, R_{l2}, and R_{l3}.

10.5 Find a mechanical system that is analogous to the electrical system of Figure 10.46 using the transfer function approach.

10.6 Determine a mechanical system and an electrical system that are analogous. Both systems are defined by the mathematical model

$$\begin{cases} \ddot{x}_1 + 20\dot{x}_1 - 20\dot{x}_2 + 300x_1 = 80 \\ \ddot{x}_2 + 30\dot{x}_2 - 20\dot{x}_1 + 200x_2 = 220 \end{cases}$$

10.7 Propose an electrical circuit that is analogous to the lumped-parameter model of the MEMS sketched in Figure 10.47, which is formed of two shuttle masses, m_1 and m_2, six supporting identical beams (of bending stiffness k_1), and a serpentine spring (of stiffness k). The system is acted upon capacitively by the

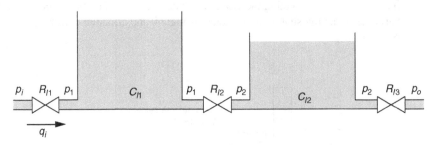

FIGURE 10.45

Liquid-Level System with Tanks and Valves.

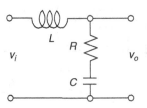

FIGURE 10.46

Electrical System for Mechanical System Analogy.

forces f_1 and f_2; viscous damping of coefficient c exists between the two masses and the underneath supporting base.

10.8 Identify an operational amplifier electrical system that is analogous to a mechanical system formed of a mass, a parallel spring-damper connection, and a force acting on the mass—see Figure 10.7(a). Use the transfer function approach.

10.9 A sinusoidal force f of unknown amplitude is applied to the mass-damper-spring mechanical system of Figure 10.48. Use the voltage-divider electrical circuit shown in the figure to evaluate the force amplitude F when known are $m = 1$ kg, $k = 200$ N/m, $c = 0.1$ N-s/m, $R = 30$ Ω, $R_s = 20$ Ω (the total sensing resistance), $L = 0.05$ H, $v = 120$ V (source voltage). The maximum voltmeter

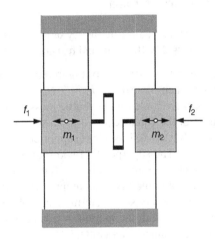

FIGURE 10.47

Microelectromechanical System with Capacitive Actuation.

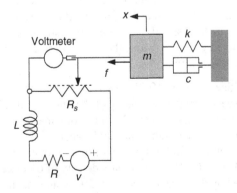

FIGURE 10.48

Mechanical Motion Sensing by a Voltage Divider Circuit.

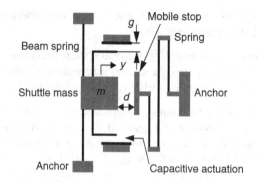

FIGURE 10.49

MEMS with Longitudinal Capacitive Actuation.

indication is 3 V at an excitation frequency $\omega = 20$ rad/s. Known also are the electric resistivity $\rho_{el} = 2 \times 10^{-7}$ Ωm and resistor wire area $A = 2 \times 10^{-7}$ m².

10.10 The shuttle mass $m = 1.4 \times 10^{-12}$ kg of Figure 10.49 is supported by two identical beam springs, each of stiffness $k_1 = 2$ N/m. Two identical longitudinal pairs create a capacitive force on the shuttle mass in the y direction, which is 1.4 times larger than the force required to close the gap $d = 6$ µm between the mass and a mobile, massless plate supported by a serpentine spring of stiffness $k_2 = 3$ N/m. The superposition width of the capacitor plates is $w = 250$ µm, the constant capacitive gap is $g = 5$ µm, the electric permittivity is $\varepsilon = 8.8 \times 10^{-12}$ F/m, and the viscous damping coefficient between the mass and substrate is $c = 0.001$ N-s/m. Determine and plot the mass displacement before and after contact with the mobile stop plate assuming that the capacitive actuation is discontinued after contact.

10.11 A voltage $v = 5\delta(t)$ µV (where $\delta(t)$ is the unit impulse) is supplied to the rotary capacitive MEMS actuator of Figure 10.50, which operates in air. Find and plot the time response of the mobile armature knowing its moment of inertia is $J = 10^{-18}$ kg-m², the viscous damping coefficient (not symbolized in Figure 10.50) is $c = 30$ N-m-s, and the spiral spring stiffness is $k = 2$ N-m. Known also are the fixed gap $g = 4$ µm, the radius $R = 220$ µm, and the capacitor armature width $w = 5$ µm.

10.12 The microbridge of Figure 10.51 is used to evaluate the dynamic viscosity of a fluid by transverse capacitive sensing of the midpoint displacement relative to the input displacement $u = 5\sin(200t)$ µm of a massless cage. Use a lumped-parameter model to determine the fluid dynamic viscosity knowing that the bridge has a length of 230 µm, width of 60 µm, thickness of 1.2 µm, Young's modulus of 1.6×10^{11} N/m², and mass density of 6100 kg/m³. The maximum voltage sensed by the capacitor is $\Delta V = 20$ µV and the mass of capacitor plate attached to the microbridge is $m_p = 2 \times 10^{-12}$ kg. For the capacitive sensor, the initial gap is $g_0 = 12$ µm and the bias voltage is $v_b = 10$ µV.

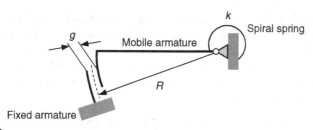

FIGURE 10.50

MEMS with Rotary Capacitive Actuation.

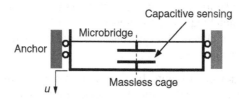

FIGURE 10.51

Microbridge with Displacement Input and Transverse Capacitive Sensing.

10.13 A microcantilever with a length $l = 200$ μm, width $w = 25$ μm, thickness $h = 2$ μm, Young's modulus $E = 160$ GPa, and mass density $\rho = 5800$ kg/m³ has two strain gauges ($K_g = 2$) attached at its root; the gauges are connected in a half Wheatstone bridge (with $v_i = 15$ μV). The cantilever is attached to a massless platform (base), which has a sinusoidal motion $u = 10 \sin(100,000t)$ μm. Determine the point mass that attaches at the cantilever free end knowing the output voltage amplitude of the Wheatstone bridge, which is $\Delta V_o = 5$ μV. Use a lumped-parameter model without damping.

10.14 A microcantilever is placed on a shaker that provides a variable-frequency input displacement motion $u = 5 \sin(\omega t)$ μm. The microcantilever has a length $l = 320$ μm, width $w = 50$ μm, thickness $h = 3$ μm, Young's modulus $E = 150$ GPa, mass density $\rho = 5400$ kg/m³ and is used in conjunction with a strain gauge ($K_g = 2.1$) attached to the cantilever root and connected in a Wheatstone bridge (with $v_i = 20$ μV). Knowing that the maximum voltage readout is $\Delta V_o = 8$ μV, determine the overall coefficient of viscous damping c.

10.15 The microcantilever of Problem 10.14 is used to sense the mass of a particle m_p that attaches to its free end. Calculate m_p knowing that the lumped-parameter viscous damping coefficient is $c = 4.27 \times 10^{-9}$ N-s/m, the maximum voltage readout is $\Delta V_o = 4.5$ μV, while all other parameters are the ones specified in Problem 10.14.

10.16 A straight deformable bar of cross-sectional area $A = 25$ mm², length $l = 100$ cm, mass density $\rho = 7800$ kg/m³, Young's modulus $E = 2.1 \times 10^{11}$ N/m², and

Poisson's ratio $\mu = 0.3$ is anchored at one end and subjected to a sinusoidal axial load of unknown amplitude and frequency $\omega = 45000$ rad/s at the other end. Disposing of three identical strain gauges ($K_g = 2.3$) that can be attached to the bar and connected in a Wheatstone bridge ($v_i = 30$ V), determine the force amplitude when the maximum output voltage of the bridge is 2.5×10^{-6} V. Consider the system is lossless.

10.17 The electromagnetomechanical system of Example 10.7 and shown in Figure 10.19 is used as a sensor for a sinusoidal displacement of frequency $\omega = 80$ rad/s and unknown amplitude. The external magnetic field is $B = 0.5$ T, the coil has $N = 50$ turns, and the coil diameter is $d_c = 0.01$ m. Use a lumped-parameter model together with the bridge and magnet parameters of Example 10.7 and the amplitude of the sensed voltage of 0.03 V to determine the input displacement amplitude. Also viscous damping affects the mechanical motion with a coefficient $c = 32$ N-s/m.

10.18 The MEMS device of Figure 10.52 is used to measure the external magnetic field B. A current $i = 0.005 \sin(60t)$ A passes through the circular-loop electrical circuit of radius $r = 160$ μm, which is printed on a rigid plate of thickness $h_p = 10$ μm and mass density $\rho = 6550$ kg/m³ attached to a flexible rod of length $l = 100$ μm, diameter $d_r = 2$ μm, and Young's modulus $E = 150$ GPa. The tilting of the rigid plate is monitored optically by a photodiode, which is placed at a distance $h = 3$ mm from the plate; the incident laser beam angle is $\alpha_0 = 25°$, as in Figure 10.20 and the maximum deviation of the reflected beam is measured to be $d = 6$ μm. Evaluate the magnetic field B.

10.19 An external magnetic field $B = 1$ T interacts with a sinusoidal current of unknown amplitude and rotates the rigid plate shown in Figure 10.53 by an angle θ. Optical detection with an incident beam at $\alpha_0 = 20°$ and a capture screen at a distance $h = 200$ μm from the microdevice plane indicates a maximum deflection of the reflected beam of $d = 50$ μm (see Figure 10.20). The three identical flexible members have a circular cross-section of diameter $d_r = 2$ μm, their length is $l = 300$ μm, while their material modulii are $E = 140$ GPA and $G = 112$ GPa. The rigid plate is defined by $r = 80$ μm, a thickness $h_p = 8$ μm, and a mass density $\rho = 6000$ kg/m³. Knowing that the

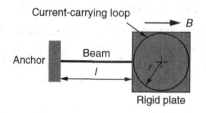

FIGURE 10.52

Top View of Electromagnetomechanical Beam Sensor.

current's frequency is one half the natural rotary frequency of the mechanical microsystem in the θ direction, determine the current amplitude.

10.20 An unknown actuation voltage is applied to the piezoelectric block of Figure 10.31 and generates an output (sensing) voltage variation of $\Delta v_o = 0.0001 \sin(20t)$ V in its half Wheatstone bridge sensing circuit. Determine the characteristics of the actuation voltage, as well as the maximum free-end displacement and the maximum force that can be generated by the piezoelectric block. Known are $K_g = 2$, $l = 20$ mm, $A = 42.25$ mm^2, $\rho_{PZT} = 5300$ kg/m^3, $v_i = 150$ V, $E_m = 5.2 \times 10^{10}$ N/m^2, $d = 390 \times 10^{-12}$ m/V. Consider the system is lossless.

10.21 A longitudinal piezoelectric actuator of length $l = 0.03$ m, cross-sectional area $A = 1$ cm^2, mass density $\rho_{PZT} = 7500$ kg/m^3, Young's modulus $E_m = 50$ GPa, and piezoelectric charge constant $d = 4 \times 10^{-10}$ m/V is attached to a mass $m = 0.20$ kg, which further connects to a spring $k = 200$ N/m, as illustrated in Figure 10.54.

    (a) Evaluate the total system mechanical losses as an equivalent damping coefficient c_{eq} if experimental frequency-domain testing indicates the

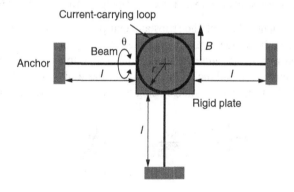

FIGURE 10.53

Top View of Compliant Microsystem for External Magnetic Field Detection.

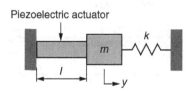

FIGURE 10.54

Longitudinal Piezoelectric Actuator.

system's bandwidth is $\omega_c = 1.2\ \omega_n$ (where ω_n is the mechanical system's natural frequency). Consider the mass and stiffness contributions of the piezoelectric actuator.

(b) Plot the system's response y as a function of time when a unit impulse voltage is applied to the piezoelectric actuator. Note: The electrical circuit is not shown in the figure.

10.22 Consider four strain gauges are available ($K_g = 2.2$) attached on the piezoelectric actuator of Problem 10.21; and they are connected in a full Wheatstone bridge with $v_i = 18$ V. Express the relationship between the actuation voltage of the piezoelectric block, $v_a = 50\sin(120t)$ V and the readout voltage ΔV_o for this device, then evaluate the viscous damping coefficient corresponding to the total (equivalent) system losses.

10.23 A circular piezoelectric plate with diameter $d = 20$ mm, thickness $h = 3$ mm, Young's modulus of 1.7×10^{11} N/m^2, mass density of $7,000$ kg/m^3, and $d_m = 3 \times 10^{-2}$ m/V is attached at the midpoint of a bridge, as shown in Figure 10.55. A displacement $u = 0.008\delta(t)$ m is provided to the bridge externally with $\delta(t)$ being the unit impulse. Determine and plot the sensing voltage variation with time. Known are the bridge data: length of 180 mm, width of 20 mm, thickness of 2 mm, Young's modulus of 2.1×10^{11} N/m^2, mass density of 7800 kg/m^3. The piezoelectric losses are equivalent to a viscous damping coefficient $c = 12$ N-s/m.

10.24 The two identical bimetallic strips and end-point mass of Figure 10.56, in conjunction with two identical strain gauges ($K_g = 2$) applied at the cantilever root, are used as a temperature sensor. Formulate the transfer function between the input temperature and the output voltage of the strain-gauge Wheatstone bridge ($v_i = 2$). Plot the output voltage using Simulink® when the sensor is exposed to

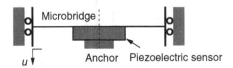

FIGURE 10.55

Microbridge with Piezoelectric Sensor.

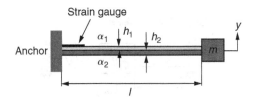

FIGURE 10.56

Bimetallic Strip with End-Point Mass and Strain Gauge.

a temperature increase of $\Delta\theta = 50°$. Known are $l = 50$ mm, $h_1 = 1$ mm, $h_2 = 2$ mm, $w = 8$ mm (the strips width), $\alpha_1 = 30 \times 10^{-6}$ deg^{-1}, $\alpha_2 = 4 \times 10^{-6}$ deg^{-1}, $E_1 = 80$ GPa, $E_2 = 200$ GPa, $\rho_1 = 5200$ kg/m^3, $\rho_2 = 8000$ kg/m^3, and $m = 0.05$ kg. Include the inertia contributions from the two beams in a lumped-parameter model.

10.25 The bimetallic strip of Problem 10.24 and Figure 10.56 is anchored at both ends after removing the mass m, such that it becomes a bridge used as an actuator. Evaluate the relationship between the maximum deflection at the midpoint and the temperature variation. Using a lumped-parameter model with viscous damping coefficient $c = 40$ N-s/m, determine and plot the system response to a temperature variation of 60 deg. The strain gauges are not utilized in this experiment.

10.26 Obtain a mathematical model for the MEMS sketched in Figure 10.57 to connect the input voltage to the output mechanical motion x. Consider electrical, thermal, and mechanical field interaction in the model.

10.27 Use Simulink® to plot x as a function of time and the input voltage v for the MEMS of Problem 10.26. Known are $l = 120$ μm, diameter of flexible beams cross-section $d = 3$ μm, $l_a = 200$ μm, thermal actuator side of square cross-section $a = 10$ μm, $\rho_a = 7600$ kg/m^3 (actuator mass density), coefficient of thermal expansion $\alpha = 2.5 \times 10^{-6}$ deg^{-1}, $R_1 = 0.1\Omega$, $v = 5$ μV, $\rho_{el} = 15 \times 10^{-9}$ Ω-m, $\alpha_R = 0.004$ deg^{-1}, $c = 380$ J/(kg-deg), $h = 32$ J/(m^2-s-deg), and $E = E_a = 200$ GPa.

10.28 Use a lumped-parameter model for the system of Figure 10.58, which is formed of two bimetallic strips and a connecting spring, to derive a time-domain mathematical model describing the mechanical motion in the spring direction when a temperature increase of $\Delta\theta$ is applied. Obtain a state-space model corresponding to this system.

10.29 Use Simulink® and the state space model of Problem 10.28 to plot the time response of the system sketched in Figure 10.58. Known are $l_1 = 60$ mm,

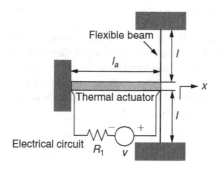

FIGURE 10.57

MEMS with Electrothermal (Joule Effect) Actuation.

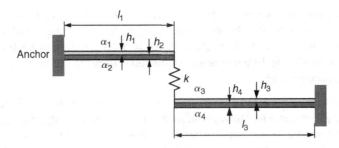

FIGURE 10.58

Two Bimetallic Strips with Connecting Spring.

$l_3 = 80$ mm, $h_1 = h_3 = 1.2$ mm, $h_2 = h_4 = 2$ mm, $w = 8$ mm (all strips' width), $\alpha_1 = 30 \times 10^{-6}$ deg^{-1}, $\alpha_2 = 4 \times 10^{-6}$ deg^{-1}, $\alpha_3 = 40 \times 10^{-6}$ deg^{-1}, $\alpha_4 = 8 \times 10^{-6}$ deg^{-1}, $E_1 = 80$ GPa, $E_2 = 200$ GPa, $E_3 = 90$ GPa, $E_4 = 180$ GPa, $\rho = 7000$ kg/m^3 (for all strips), $k = 30$ N/m, and $\Delta\theta = 65°$.

10.30 Use Simulink® to plot the time variation of the plate's tilting angle for the system sketched in Figure 10.52 of Problem 10.18 when $B = 0.2/(1 + 0.1t)$ T. The optical detection system is not employed here.

Suggested Reading

R. S. Figliola and D. E. Beasley, *Theory and Design for Mechanical Measurements*, 5th Ed. John Wiley & Sons, New York, 2007.

P. L. Reece (Editor), *Smart Materials and Structures*, Nova Science Publishers, New York, 2006.

R. H. Bishop (Editor), *The Mechatronics Handbook*, 2nd Ed. CRC Press, Boca Raton, FL, 2008.

N. Lobontiu and E. Garcia, *Mechanics of Microelectromechanical Systems*, Kluwer, New York, 2004.

MORGAN Electro Ceramic Piezoelectric Tutorials, available at: www.morganmatroc. commec_tutorials.

K. Ogata, *System Dynamics*, 4th Ed. Pearson Prentice Hall, New York, 2004.

W. J. Karplus and W. W. Soroka, *Analog Methods*, McGraw-Hill, New York, 1962.

N. Lobontiu, *Dynamics of Microelectromechanical Systems*, Springer, New York, 2007.

F. J. Bueche and D. A. Jerde, *Principles of Physics*, 6th Ed. McGraw-Hill, New York, 1995.

O. Buecher and M. Weeks, *Introduction to MATLAB® & Simulink®—A Project Approach*, 3rd Ed. Infinity Science Press, Boston, 2008.

Solution to Linear Ordinary Homogeneous Differential Equations with Constant Coefficients

Consider the following differential equations with the function x that depends on time, $x(t)$ being the unknown:

$$L_n[x(t)] = a_n \frac{d^n x(t)}{dt^n} + a_{n-1} \frac{d^{n-1} x(t)}{dt^{n-1}} + \cdots + a_1 \frac{dx(t)}{dt} + a_0 x(t) = 0 \qquad (A.1)$$

where $a_0, a_1, \ldots, a_{n-1}, a_n$ are constant coefficients. This equation is a homogeneous linear ordinary differential equation with constant coefficients, and the operator $L[x(t)]$ is a symbol notation that enables shortcut calculations. Searching for particular solutions of the form

$$x(t) = e^{\lambda t} \qquad (A.2)$$

in Eq. (A.1) results in

$$L[x(t)] = e^{\lambda t} \left(a_n \lambda^n + a_{n-1} \lambda^{n-1} + \cdots + a_1 \lambda + a_0 \right) = 0 \qquad (A.3)$$

The specific function of Eq. (A.2) is a solution to the differential Eq. (A.1) only when

$$a_n \lambda^n + a_{n-1} \lambda^{n-1} + \cdots + a_1 \lambda + a_0 = 0 \qquad (A.4)$$

Equation (A.4) is the *characteristic equation* associated with Eq. (A.1) and the roots λ are the *eigenvalues*. Solving Eq. (A.4) for the eigenvalues $\lambda_1, \lambda_2, \ldots, \lambda_{n-1}, \lambda_n$ results in the following solution to Eq. (A.1):

$$x_1(t) = e^{\lambda_1 t}, \, x_2(t) = e^{\lambda_2 t}, \ldots, x_{n-1}(t) = e^{\lambda_{n-1} t}, \, x_n(t) = e^{\lambda_n t} \qquad (A.5)$$

which are all independent functions. As a consequence, the general solution of Eq. (A.1) is a linear combination of the individual solutions of Eq. (A.5):

$$\begin{aligned} x(t) &= c_1 x_1(t) + c_2 x_2(t) + \cdots + c_{n-1} x_{n-1}(t) + c_n x_n(t) \\ &= c_1 e^{\lambda_1 t} + c_2 e^{\lambda_2 t} + \cdots + c_{n-1} e^{\lambda_{n-1} t} + c_n e^{\lambda_n t} \end{aligned} \qquad (A.6)$$

where $c_1, c_2, \ldots, c_n$ are constant coefficients.

There are four different cases, depending on the possible types of roots of the characteristic equation, as the eigenvalues can be (a) real and distinct (simple), (b) complex and simple, (c) real of order of multiplicity l, and (d) complex of order of multiplicity m. As a consequence, the general solution contains contributions from all four categories and therefore can formally be written as

$$x(t) = x_a(t) + x_b(t) + x_c(t) + x_d(t) \tag{A.7}$$

The solutions $x_a(t)$, $x_b(t)$, $x_c(t)$, and $x_d(t)$ are discussed next.

REAL DISTINCT ROOTS

Assuming that the characteristic Eq. (A.4) has n_1 simple real roots, their contribution into the solution to the homogeneous Eq. (A.7) is

$$x_a(t) = a_1 e^{\lambda_1 t} + a_2 e^{\lambda_2 t} + \cdots + a_{n_1 - 1} e^{\lambda_{n_1 - 1} t} + a_{n_1} e^{\lambda_{n_1} t} \tag{A.8}$$

where $a_1, a_2, \ldots, a_{n1}$ are constant coefficients.

COMPLEX DISTINCT ROOTS

If a root of the characteristic equation is complex, then its complex conjugate is also a root, and the corresponding eigenvalue pair is

$$\lambda = \sigma \pm j\omega \tag{A.9}$$

with $j = \sqrt{-1}$. The corresponding root of the homogeneous differential equation can be written as

$$e^{\lambda t} = e^{(\sigma + j\omega)t} = e^{\sigma t} e^{j\omega t} = e^{\sigma t} \left[\cos(\omega t) + j \sin(\omega t) \right] \tag{A.10}$$

Equation (A.10) indicates that if the final expression included there is a root to Eq. (A.1), the following are also roots to the same equation:

$$x_1(t) = e^{\sigma t} \cos(\omega t); \; x_2(t) = e^{\sigma t} \sin(\omega t) \tag{A.10}$$

Assuming the general solution of the homogeneous Eq. (A.1) has n_2 simple complex roots, the corresponding to the general solution is

$$x_b(t) = b_1 e^{\sigma_1 t} \cos(\omega_1 t) + b'_1 e^{\sigma_1 t} \sin(\omega_1 t) + \cdots + b_{n_2} e^{\sigma_{n_2} t} \cos(\omega_{n_2} t)$$
$$+ b'_{n_2} e^{\sigma_{n_2} t} \sin(\omega_{n_2} t) \tag{A.11}$$

with $b_1, b'_1, \ldots, b_{n_2}, b'_{n_2}$ being constant coefficients.

REAL MULTIPLE ROOTS

A root of the characteristic equation that has an order of multiplicity of n_3 generates n_3 roots of the homogeneous Eq. (A.1):

$$e^{\lambda t};\ te^{\lambda t};\ \ldots;\ t^{n_3-1}e^{\lambda t} \tag{A.12}$$

The contribution to the general solution of Eq. (A.1) of a total number n_3 of multiple real roots can be written as

$$x_c(t) = c_1 e^{\lambda t} + c_2 t e^{\lambda t} + \cdots + c_{n_3-1} t^{n_3-2} e^{\lambda t} + c_{n_3} t^{n_3-1} e^{\lambda t} \tag{A.13}$$

where $c_1, c_2, \ldots, c_{n_3}$ are constant coefficients.

COMPLEX MULTIPLE ROOTS

If a root of multiplicity n_4 is complex, then the following are also roots of the homogeneous Eq. (A.1):

$$e^{\sigma t}\cos(\omega t); e^{\sigma t}\sin(\omega t); te^{\sigma t}\cos(\omega t); te^{\sigma t}\sin(\omega t); \ldots; t^{n_4-1}e^{\sigma t}\cos(\omega t);$$
$$t^{n_4-1}e^{\sigma t}\sin(\omega t) \tag{A.14}$$

The contribution to the general solution of Eq. (A.1) of a total number n_4 of multiple complex roots is

$$x_d(t) = d_1 e^{\sigma t}\cos(\omega t) + d_1' e^{\sigma t}\sin(\omega t) + d_2 t e^{\sigma t}\cos(\omega t) + d_2' t e^{\sigma t}\sin(\omega t)$$
$$+ \cdots + d_{n_4} t^{n_4-1} e^{\sigma t}\cos(\omega t) + d_{n_4}' t^{n_4-1} e^{\sigma t}\sin(\omega t) \tag{A.15}$$

with $d_1, d_1', \ldots, d_{n_4}, d_{n_4}'$, being constant coefficients.

The constant coefficients of Eqs. (A.8), (A.11), (A.13), and (A.15) are determined by means of initial conditions that are applied to the assembled general solution of Eq. (A.7).

Review of Matrix Algebra

SPECIAL-FORM MATRICES

A *column vector* having n components is defined as

$$\{x\} = \begin{Bmatrix} x_1 \\ x_2 \\ \ldots \\ x_n \end{Bmatrix} \tag{B.1}$$

whereas a *row vector* is

$$\{x\} = \{x_1 \quad x_2 \quad \ldots \quad x_n\} \tag{B.2}$$

A *square matrix* of the n dimension (or an $n \times n$ matrix) comprises an equal number of rows and columns:

$$[A] = \begin{bmatrix} a_{11} & a_{12} & \ldots & a_{1n} \\ a_{21} & a_{22} & \ldots & a_{2n} \\ \ldots & \ldots & \ldots & \ldots \\ a_{n1} & a_{n2} & \ldots & a_{nn} \end{bmatrix} \tag{B.3}$$

A *symmetric matrix* is a square matrix for which $a_{ij} = a_{ji}$:

$$[A] = \begin{bmatrix} a_{11} & a_{12} & \ldots & a_{1n} \\ a_{12} & a_{22} & \ldots & a_{2n} \\ \ldots & \ldots & \ldots & \ldots \\ a_{1n} & a_{2n} & \ldots & a_{nn} \end{bmatrix} \tag{B.4}$$

A *diagonal matrix* is a square matrix with nonzero elements placed only on the main diagonal:

$$[D] = \begin{bmatrix} d_{11} & 0 & \ldots & 0 \\ 0 & d_{22} & \ldots & 0 \\ \ldots & \ldots & \ldots & \ldots \\ 0 & 0 & \ldots & d_{nn} \end{bmatrix} \tag{B.5}$$

The *identity matrix* (or *unity*) is a diagonal matrix with the nonzero elements being equal to 1:

$$[I] = \begin{bmatrix} 1 & 0 & \cdots & 0 \\ 0 & 1 & \cdots & 0 \\ \cdots & \cdots & \cdots & \cdots \\ 0 & 0 & \cdots & 1 \end{bmatrix} \tag{B.6}$$

A zero matrix, zero column vector, or zero row vector has all elements equal to 0.

The *transpose* of an $n \times m$ matrix $[A]$ is an $m \times n$ matrix denoted by $[B] = [A]^t$, whose elements are defined as $b_{ij} = a_{ji}$, where a_{ij} are the elements of the original matrix:

$$[A] = \begin{bmatrix} a_{11} & a_{12} & \cdots & a_{1m} \\ a_{21} & a_{22} & \cdots & a_{2m} \\ \cdots & \cdots & \cdots & \cdots \\ a_{n1} & a_{n2} & \cdots & a_{nm} \end{bmatrix}; \quad [B] = [A]^t = \begin{bmatrix} a_{11} & a_{21} & \cdots & a_{n1} \\ a_{12} & a_{22} & \cdots & a_{n2} \\ \cdots & \cdots & \cdots & \cdots \\ a_{1m} & a_{2m} & \cdots & a_{nm} \end{bmatrix} \tag{B.7}$$

BASIC MATRIX OPERATIONS

The result of multiplying a scalar a by an $n \times m$ matrix $[A]$ is

$$a[A] = a \begin{bmatrix} a_{11} & a_{12} & \cdots & a_{1m} \\ a_{21} & a_{22} & \cdots & a_{2m} \\ \cdots & \cdots & \cdots & \cdots \\ a_{n1} & a_{n2} & \cdots & a_{nm} \end{bmatrix} = \begin{bmatrix} aa_{11} & aa_{12} & \cdots & aa_{1m} \\ aa_{21} & aa_{22} & \cdots & aa_{2m} \\ \cdots & \cdots & \cdots & \cdots \\ aa_{n1} & aa_{n2} & \cdots & aa_{nm} \end{bmatrix} \tag{B.8}$$

Two matrices need the same dimensions in order to add algebraically; adding up the matrices $[A]$ and $[B]$, both of an $n \times m$ dimension, results in

$$[A] + [B] = \begin{bmatrix} a_{11} & a_{12} & \cdots & a_{1m} \\ a_{21} & a_{22} & \cdots & a_{2m} \\ \cdots & \cdots & \cdots & \cdots \\ a_{n1} & a_{n2} & \cdots & a_{nm} \end{bmatrix} + \begin{bmatrix} b_{11} & b_{12} & \cdots & b_{1m} \\ b_{21} & b_{22} & \cdots & b_{2m} \\ \cdots & \cdots & \cdots & \cdots \\ b_{n1} & b_{n2} & \cdots & b_{nm} \end{bmatrix}$$

$$= \begin{bmatrix} a_{11} + b_{11} & a_{12} + b_{12} & \cdots & a_{1m} + b_{1m} \\ a_{21} + b_{21} & a_{22} + b_{22} & \cdots & a_{2m} + b_{2m} \\ \cdots & \cdots & \cdots & \cdots \\ a_{n1} + b_{n1} & a_{n2} + b_{n2} & \cdots & a_{nm} + b_{nm} \end{bmatrix} \tag{B.9}$$

Two matrices $[A]$ and $[B]$ can be multiplied only if the number of columns of $[A]$ is equal to the number of rows of $[B]$; in other words, $[A]$ is of the $n \times m$ dimension and $[B]$ is of the $m \times p$ dimension. A generic element of the product matrix $[C]$, say c_{ij} (located in the ith row and the jth column of $[C]$) is calculated as

$$c_{ij} = \sum_{k=1}^{m} (a_{ik} b_{kj}) \tag{B.10}$$

for $i = 1, 2, ..., n$ and $j = 1, 2, ..., p$. Except for special cases, the product of two matrices $[A]$ and $[B]$ is noncommutative; therefore,

$$[A][B] \neq [B][A] \tag{B.11}$$

Multiplying a square matrix by the identity matrix leaves the original matrix unchanged:

$$[A][I] = [I][A] = [A] \tag{B.12}$$

For matrix addition and multiplication, the *associative* and *distributive laws* are applicable:

$$\begin{aligned}([A][B])[C] &= [A]([B][C]) \\ [A]([B] + [C]) &= [A][B] + [A][C]\end{aligned} \tag{B.13}$$

The *inverse* of a square matrix $[A]$ is denoted by $[A]^{-1}$, and the following property applies to the original matrix and its inverse:

$$[A][A]^{-1} = [A]^{-1}[A] = [I] \tag{B.14}$$

The inverse of a square nonsingular matrix $[A]$ is calculated as

$$[A]^{-1} = \frac{\text{adj}\,[A]^t}{\det\,[A]} \tag{B.15}$$

A nonsingular square matrix $[A]$ has a nonzero determinant: $\det[A] \neq 0$. The *determinant* of a square matrix $[A]$ is a number that can be calculated based on the elements of the first row, for instance,

$$\det[A] = a_{11}b_{11} + a_{12}b_{12} + \cdots + a_{1n}b_{1n} \tag{B.16}$$

where

$$b_{1j} = (-1)^{1+j}\det\left[A_{1j}\right] \tag{B.17}$$

(with $j = 1, 2, ..., n$) is the 1*j*th *cofactor* of $[A]$ and $[A_{1j}]$ is the 1*j*th *minor* of $[A]$ obtained by deleting row number 1 and column number j of the original matrix $[A]$. The *adjoint matrix* of $[A]^t$, which is denoted by adj$[A]^t$ in the numerator of Eq. (B.15), is a matrix that contains the cofactors of the elements of the transposed matrix $[A]^t$.

The solution to a system of n linear algebraic equations with constant coefficients of the form

$$\begin{cases} a_{11}x_1 + a_{12}x_2 + \cdots + a_{1m}x_m = b_1 \\ a_{21}x_1 + a_{22}x_2 + \cdots + a_{2m}x_m = b_2 \\ \cdots \\ a_{n1}x_1 + a_{n2}x_2 + \cdots + a_{nm}x_m = b_n \end{cases} \tag{B.18}$$

where x_1, x_2, ..., x_m are the unknowns can be obtained using matrix calculus. The system of Eq. (B.18) can be written as

$$[A]\{x\} = \{b\} \tag{B.19}$$

where

$$[A] = \begin{bmatrix} a_{11} & a_{12} & \cdots & a_{1m} \\ a_{21} & a_{22} & \cdots & a_{2m} \\ \cdots & \cdots & \cdots & \cdots \\ a_{n1} & a_{n2} & \cdots & a_{nm} \end{bmatrix}; \{x\} = \begin{Bmatrix} x_1 \\ x_2 \\ \cdots \\ x_m \end{Bmatrix}; \{b\} = \begin{Bmatrix} b_1 \\ b_2 \\ \cdots \\ b_n \end{Bmatrix} \tag{B.20}$$

The solution to Eq. (B.18) is found by left multiplying Eq. (B.18) by $[A]^{-1}$, which results in

$$\begin{aligned} [A]^{-1}[A]\{x\} &= [A]^{-1}\{b\} \\ [I]\{x\} &= [A]^{-1}\{b\} \\ \{x\} &= [A]^{-1}\{b\} \end{aligned} \tag{B.21}$$

The *eigenvalue problem* is defined by the following equation:

$$[A]\{X\} = \lambda\{X\} \tag{B.22}$$

where $[A]$ is an $n \times n$ square matrix, $\{X\}$ is a vector known as *eigenvector*, and λ is an *eigenvalue*. Equation (B.22) can be written as

$$([A] - \lambda[I])\{X\} = \{0\} \tag{B.23}$$

which represents a system of n homogeneous algebraic equations, the unknown being X_1, X_2, ..., X_n, the eigenvector components. For $\{X\}$ to be nontrivial (nonzero), the following condition needs to be satisfied:

$$\det([A] - \lambda[I]) = 0 \tag{B.24}$$

which is the *characteristic equation* attached to the eigenvalue problem of Eq. (B.22). The roots of the characteristic equation are the eigenvalues λ_1, λ_2, ..., λ_n. Each of the eigenvalues is substituted back into Eq. (B.23), which can be solved for a corresponding eigenvector. It should be noted that only $n - 1$ components of any eigenvector are independent; each of those components can be expressed in terms of the nth component, which can be chosen arbitrarily. Many eigenvalue applications use a value of 1 for the arbitrary component of each eigenvector. In other instances, the condition of *unit norm* is used, which requires that the norm of each eigenvector be 1:

$$\sqrt{X_1^2 + X_2^2 + \cdots + X_n^2} = 0 \tag{B.25}$$

Essentials of MATLAB® and System Dynamics-Related Toolboxes

This Appendix reviews the main MATLAB® based commands that are utilized in the book, as well as a summary of the main linear time invariant (LTI) models that are available in the Control System Toolbox™.

USEFUL MATLAB® COMMANDS

Presented here is a list with some useful commands from MATLAB®'s® Control System Toolbox™ and Symbolic Math Toolbox™. They have been organized under the following functional categories: mathematical calculations, visualization and graphics, linear systems modeling, time domain analysis, frequency domain analysis, and controls for quick reference. A complete function list with documentation (including Simulink® information) can be found at www.mathworks.com/access/helpdesk/help/helpdesk.html.

Comprehensive information on these functions and related functions can be accessed directly in MATLAB® by means of the doc command. For instance, if you need more information on Bode plots, just type >> doc bode at the MATLAB® prompt and you will be directed on the Help page to the pertinent information on the bode command. An additional informative source, where tutorials can be found, is www.mathworks.com.academia.

Mathematical Calculations

collect(p)

Collects all the coefficients of the same power in a symbolic polynomial previously defined as *p*.

det(m)

Calculates the determinant of a previously defined matrix *m*.

diag(v)

Generates a diagonal matrix from the elements of a previously defined vector *v*; the elements of *v* are placed on the main diagonal of the square matrix.

diff(f, v, n)

Calculates symbolically the *n*th derivative of a function *f* in terms of the variable *v*.

DOI: 10.1016/B978-0-240-81128-4.00016-7

`dsolve('eqn.1','eqn.2',…,'cond.1','cond.2',…,'x')`

Solves symbolically the differential equations `eqn.1`, `eqn.2`,… with the initial conditions `cond.1`, `cond.2`,… by considering the independent variable is x. If no symbol x is used, the default independent variable is time (t). If a differential equation contains, for instance, the second time derivative of an unknown function y, the MATLAB® notation of that is `D2y`. Similarly, the first time derivative of y is denoted as `Dy`. An initial condition that specifies the value of the first time derivative as being equal to a symbolic value of b at the initial time moment 0 is denoted as `'Dy(0) = b'`.

`expand(f)`

Performs symbolic expansion of functions (previously specified as f) such as polynomial multiplication and distribution of products over sums.

`eye(m)`

Returns the $n \times n$ identity matrix. When this command is used as `eye(m, n)`, the result is a matrix with m rows and n columns where $a_{ii} = 0$ (for $i = 1$ to n if $n < m$, and for $i = 1$ to m if $m < n$).

`factor(f)`

Transforms a rational-coefficient polynomial f into a product of irreducible, rational-coefficient, lower-degree polynomials.

`fzero([f, x0])`

Finds a root of the function f near the point $x0$. Instead of a point $x0$, an interval `[xmin, xmax]` can be specified. Both $x0$ and `[xmin, xmax]` are part of a vector x, and f is a function handle (denoted with the symbol @), which can be called in a MATLAB® session regardless of where that function has been introduced – see Example 11.16 in the companion website Chapter 11.

`ilaplace(f)`

Returns the inverse Laplace transform of the previously defined symbolic function f that depends on the variable s.

`int(f,x)`

Evaluates symbolically the indefinite integral of the function f that depends on the variable x. The following command evaluates the definite integral of f as a function of x between the integration limits a and b: `int(f,x,a,b)`.

`inv(m)`

Calculates numerically or symbolically the inverse of a previously defined matrix m. A matrix m that has two rows defined as first row: 1, 2, 3, 4; second row: -3, -2, -1, 0 is entered in MATLAB® as `m = [1,2,3,4;-3,-2,-1,0]`.

`laplace(f)`

Returns the Laplace transform of the previously defined symbolic function f that depends on the time variable t.

`limit(f,x,a)`

Calculates the symbolic limit of the function f when the variable x reaches the value of a. It can also be written as `limit(f,x,a,'right')`, in the case where $x > a$, or as `limit(f,x,a,'left')`, in the case where $x < a$. The value of a can be infinity, in which case where a is substituted by `inf`.

`poly(f)`

Returns the coefficients of a polynomial f, where f is a vector containing the roots of that polynomial; the returned coefficients are ordered in descending powers in a row vector. When f was defined as a matrix, the same command returns a row vector with the coefficients (also ordered in descending powers) of the characteristic polynomial, which is $\det(s[I] - [f])$, with $[I]$ being the identity matrix.

`polyval(f, v)`

Returns the value of polynomial f for the value v of the polynomial variable

`pretty(f)`

Returns the symbolic object f in a mathematical typeset format instead of the regular MATLAB® return.

`roots(n)`

Returns the roots of an s polynomial that was previously defined as the row vector n containing the coefficients of the s polynomial in descending order.

`simple(f)`

Returns the shortest algebraic form of the previously defined symbolic object (function) f by using `collect`, `expand`, `factor` or `simplify` (see explanation of this command next).

`simplify(f)`

Applies algebraic identities and functional identities (trigonometric, exponential and logarithmic) in order to simplify more complex symbolic functions.

`solve(eq1, eq2, ..., eqn, v1, v2, ..., vn)`

Solves the system formed of the equations `eq1`, `eq2`, ..., `eqn` for the variables `v1`, `v2`, ..., `vn`. In the form `solve(eq, v)`, it solves the equation defined as eq for the variable v.

`syms a real`

Specifies that the amount a is treated as a symbolic (algebraic) object, so no numerical value is expected of a, and also that a is real. Instead of the "real" qualifier, "unreal," or "positive" can be used. Also, if the nature of the amount a is not of interest, the command `syms a` can be used. The alternate command `a = syms('a','real')` has the same effect.

`taylor(f, n, v, a)`

Returns an $n - 1$th order Taylor polynomial approximation of a function f depending on the variable v and about the value a.

`zeros(m, n)`

Returns an $m \times n$ zero matrix.

Visualization and Graphics

`axis([xmin, xmax, ymin, ymax])`

Sets limits to axes to better visualize a portion of interest from a two-dimensional plot. In the configuration `axis ([xmin, xmax, ymin, ymax, zmin, zmax]`, the command operates in a similar manner with three-dimensional plots.

`ezplot('f')`

Plots a function f depending on a variable t over its default domain using a simplified syntax (where array operations such as multiplication or division are bypassed); for example, the function $f = 2t^2$ can be plotted using `ezplot('2t^2')`. When utilized in the variant `ezplot ('f', [a, b])`, the command realizes plotting of f between the domain limits of a and b.

`ltiview`

Opens an existing linear time invariant (LTI) object plot and enables changing the type of input and visualization of the (new) plot characteristics.

`mesh(z)`

Generates a three-dimensional plot as a wireframe mesh of the function z, which depends on two variables x and y; these are vectors having the dimensions m and n, respectively. The function works in connection with the `meshgrid` command. Instead of the `mesh(z)` command, the `surf(z)` command can be used, which generates a three-dimensional surface of the same shape as the wireframe.

`meshgrid(x, y)`

Creates two-dimensional arrays to enable three-dimensional plots of analytical functions of two variables. In the configuration `[X,Y] = meshgrid(x, y)`, where x is a vector of dimension m and y is another vector of dimension n, the command generates two arrays: One has the vector x as a row that repeats n times and the other array has the vector y as a column that repeats itself m times. With another variable, z, defined in terms of x and y, the command mesh(z) generates a three dimensional plot if used after the meshgrid command.

plot(t,y,'LineSpec')

Returns a two-dimensional line plot of the pairs (t, y), where t is the independent variable vector and y is the vector function that depends on t. LineSpec is a string of symbols that enable customizing the line used in the plot (its style, width, and color) as well as the marker used (its type, size, and color). Several function pairs, (t, y_1), (t, y_2), ..., can be graphically represented on the same plot, each pair with its own line specifications using the command plot(t,y1, 'LineSpec1',t,y2,'LineSpec1',…). Several spatially separated plots can be generated using the subplot command. For instance, if subplot(3,2,1) is used before an actual plot command, MATLAB® positions the respective plot in the first row and first column of a matrix with three rows and two columns. If the command subplot(3,2,3) is used, MATLAB® places the plot in the second row and first column, which is the third position in a sequence that runs from left to right and from top to bottom over the matrix elements.

plot3(x,y,z,'LineSpec')

Returns a three-dimensional line plot of the pairs (x, y, z), where x, y, and z can be defined as vectors. For instance, if x is considered to be a previously defined time vector between the limits of 0 and 10 s with a time increment of 0.01 s (e.g., as $t = 0{:}0.01{:}10$), the other two variables can be defined in terms of t. LineSpec specifies graphic details of the plot line. Several function pairs, (x_1, y_1, z_1), (x_2, y_2), ..., can be graphically represented on the same three-dimensional plot, each pair with its own line specifications using the command plot3(x1,y1,z1, 'LineSpec1',x2,y2,z2,'LineSpec1',…). Related MATLAB® functions are xlabel('text1'), ylabel('text2'), zlabel('text3'), title('text'), and legend('text'), which enable using customized text for the reference axes, a title for the plot and a legend.

surf(z)

Generates a three-dimensional surface plot; see meshgrid.

Linear System Modeling

dss(a,b,c,d,e)

Creates a descriptor state space LTI model based on the previously defined matrices a, b, c, d, and e.

frd(r,f)

Creates a frequency response data LTI object from a vector r formed of the values of the $G(j\omega)$, the complex transfer function and a vector f formed of the frequencies ω (in rad/s) corresponding to the values of $G(j\omega)$. The adjacent command freqresp(frd(r,f),om0) where om0 is a frequency value that was, for instance, not specified in the vector f, generates the value of $G(j\omega)$ that corresponds to om0. Mention should be made that om0 can also be a vector.

frd (sys, f)

Converts a previously defined LTI object (such as zero-pole-gain, transfer function, or state space) into a frequency response data object (model) that corresponds to a vector f containing specified frequencies.

[n,d] = ss2tf(a,b,c,d,ui)

Returns the transfer functions that correspond to the state space model defined by the a, b, c, d matrices and the ith input, denoted by ui. Specifically, d is a row with the coefficients of an s polynomial in descending order (the characteristic polynomial), whereas n is an array with as many rows as outputs, each row containing the coefficients of an s polynomial in descending order. If only the command ss2tf(a,b,c,d,ui) is used, MATLAB® returns the array denoted by n in the original command syntax.

ss(a,b,c,d)

Creates a state space LTI model based on the previously defined matrices a, b, c, and d. The command ss(sys) converts a previously defined zero-pole-gain, zpk, or transfer function, tf, model into a state space model. The command [a,b,c,d] = ssdata(sys) returns the a, b, c, d matrices of the state space model denoted here by sys.

tf(n,d)

Generates a transfer function LTI object (model) from a previously defined numerator (denoted here by n) and a previously defined denominator (denoted here by d). The function can also be used directly as tf([ ],[ ]), where the two one-row matrices denoted by [] contain the coefficients of the numerator (the first [] matrix) and of the denominator (the second [] matrix) in descending order (from the largest power of s to the constant term). The function tf(sys) converts another LTI model (such as zero-pole-gain, zpk, or state space, ss,) into a transfer function model.

tf2ss(sys)

Converts an existing transfer function model denoted by sys into a state space model. The command [a,b,c,d] = tf2ss(sys) works similarly and returns the a, b, c, d matrices of the state space model.

Time Domain Analysis

impulse(sys)

Produces the two-dimensional plot of a previously defined LTI object, denoted here as sys, such as a transfer function or a state space model under a unit impulse (delta Dirac function) input. All the features presented for the step command are also valid for the impulse command.

`initial(sys,x0)`

Generates a two-dimensional plot based on a previously defined state space model denoted here by `sys` and an initial-state vector denoted here by `x0`. Several state space models, `sys1`, `sys2`,... can be used in conjunction with the same initial condition `x0` with the command `initial(sys1,sys2,...,x0)`. Time can also be specified with this command using `initial(sys,x0,t)` where *t* can either be a value or a previously defined vector. The command `[y,x,t] = initial(sys,x0)` stores the state space trajectory vector *y*, the state vector *x*, and the time vector *t* of a state space model `sys` subjected to the initial-state vector `x0`. The same results are obtained using the function `initialplot` instead of `initial`, with the same syntax.

`lsim(sys,u,t)`

Generates the two-dimensional plot of the response of a previously defined LTI object `sys` under the action of a time-dependent input *u* over a time interval defined by the vector *t*.

`step(sys)`

Produces the two-dimensional plot of a previously defined LTI object denoted here as `sys`, such as a transfer function or a state space model, under a unit step input. Time is the independent variable and the time range for the plot is chosen by MATLAB®. If a time vector *t* has been defined, then the command `step(sys,t)` will generate the plot over the specified range of *t*. Another variant of this command is `x = step(sys)`, whereby the time response is not plotted, but the vector *x* (which is the time-domain function resulting from applying a unit step input to a system) is generated and stored for further manipulation.

`zpk(n,d,k)`

Creates a zero-pole-gain model in the form of a factored polynomial fraction. If the command `zpk([1,2],[3,3,5,6],8)` is used, MATLAB® returns

```
Zero/pole/gain:
8 (s-1) (s-2)
-------------------
(s-3)^2 (s-5) (s-6)
```

The command `zpk(sys)` converts a previously defined transfer function model or state space model denoted by `sys` into a zero-pole-gain model.

Frequency Domain Analysis

`bode(sys)`

Produces the Bode plots (magnitude in decibels and phase angle in degrees) based on a previously defined linear time invariant object denoted here by `sys`, such as transfer function, state space, zero-pole-gain, or frequency response data. The command bode(sys1,sys2,…) generates Bode plots for the LTI objects denoted by 'sys1', 'sys2', ….

damp(sys)

Returns three groups of data, namely eigenvalues, damping ratios, and natural frequencies pertaining to a previously defined LTI object, such as a transfer function or a state space model.

eig(m)

Returns the eigenvalues (the squares of the natural frequencies) of a previously defined matrix m. The use of the command [V,D] = eig(m) returns the modal matrix [V] (where each column is an eigenvector) and a diagonal matrix [D] with the eigenvalues located on the main diagonal.

eigs(m)

Calculates the six largest eigenvalues of the matrix m and locates them in a vector. The command [V,D] = eigs(m) operates similarly to the [V,D] = eig(m), but each of the two matrices [V] and [D] are 6×6 matrices.

pole(sys)

The command finds the poles of the LTI object previously defined as sys. For sys being a transfer function, the poles are the roots of the denominator (characteristic equation).

pzmap(sys) or pzplot(sys)

Return a plot showing the positions of poles and zeroes for a previously defined LTI object, such as transfer function or state space model, denoted by sys. The command [p,z] = pzmap(sys) returns the vector p containing the poles and the vector z containing the zeroes of the LTI object. The command pole(sys) plots only the pole positions, whereas the zero(sys) plots only the zeroes positions associated with the LTI object sys.

residue(n,d)

Generates the residues (which are the numerators in a partial fraction expansion where all fractions have first-degree polynomials in their denominators) of a previously defined fraction (such as a transfer function) that uses the polynomial n as its numerator and the polynomial d as its denominator. If instead of this command, the following command is used, [r,p,k] = residue(n,d), MATLAB® returns three groups of numbers: the residues, the poles (which are the roots of d, the denominator polynomial, with changed signs), and the direct term k (which is the quotient of dividing n by d and returns 0 when the degree of n is less than the degree of d).

`zero(sys)`

Finds the zeroes of the LTI object previously defined as `sys`. For a transfer function, the zeroes are the roots of the numerator.

Controls

`feedback(sys1, sys2)`

Generates the closed-loop transfer function for a basic negative feedback control system, where `sys1` is the open-loop transfer function $G(s)$ and `sys2` is the feedback transfer function $H(s)$.

`nyquist(sys)`

Plots the Nyquist chart for a Laplace-domain function defined as `sys`. The alternate command `nyquist(sys,w)` returns the Nyquist plot for a set of specified frequencies w.

`rlocus(sys)`

Calculates and plots the root locus for the open-loop transfer function of a SISO system; the function was previously defined as `sys`. In the format `rlocus(sys1, sys2,..)`, MATLAB® plots the root loci for the open-loop transfer functions previously defined as `sys1`, `sys2`, ...

CONTROL SYSTEM TOOLBOX™ LINEAR TIME INVARIANT MODELS

The Control System Toolbox™ of MATLAB® has the capability of creating single entities from zero-pole-gain, transfer function, state space, and frequency response data models (which are defined as collections of vectors and matrices), as exemplified in Chapters 7 through 10. These objects are known as *linear time invariant models* and are invoqued by the MATLAB® commands in Figure C.1.

Details on creating individual LTI objects have been given in the main text; a few additional details are included here on object manipulation (operations), conversion, and visualization.

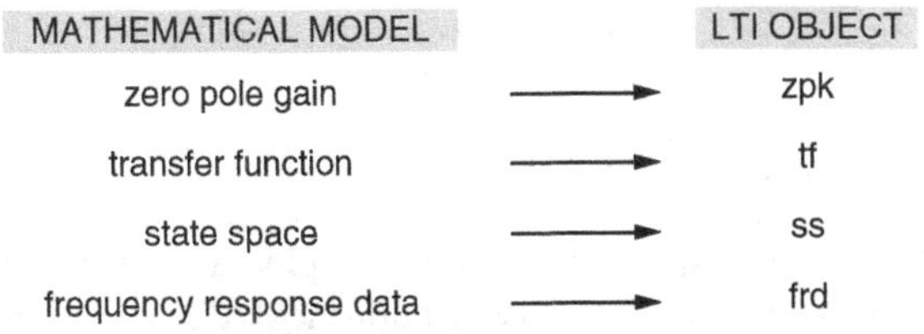

MATHEMATICAL MODEL		LTI OBJECT
zero pole gain	⟶	zpk
transfer function	⟶	tf
state space	⟶	ss
frequency response data	⟶	frd

FIGURE C.1

Linear Time Invariant Objects Corresponding to Various Mathematical Models.

Simple mathematical operations, such as addition and multiplication, can be applied between LTI objects. Consider for instance the following transfer function object, $G(s) = 1/(s^2 + 2s + 3)$, and assume the transformation $2G(s) + 5$ is needed. The following MATLAB® code realizes this task:

```
>> g = tf(1,[1,2,3]);
>> 2*g+5
```

which returns

```
Transfer function:

5 s^2 + 10 s + 17

------------------

s^2 + 2 s + 3
```

Converting between LTI objects, as seen in the book chapters, can be performed by calling a particular LTI object as a function of a different LTI object that has already been defined. To convert the transfer-function model used in the previous example into a state space model, for instance, the command `>> ss(g)` realizes this, and to obtain a zero-pole-gain model from the same transfer function model, the command `>> zpk(g)` is needed. As mentioned, it is not possible to transform any of the zpk, tf, or ss models into a frd model, but an existing frd model can be converted to any of the other three LTI models. There are no restrictions on converting zpk, tf, and ss models between themselves.

Another important rule, the *precedence rule*, frd > ss > zpk > tf, basically states that operations among various LTI objects are possible and arithmetical combinations among all four objects produce an frd model. Similarly, combination of a zpk model with a tf model (for instance), according to $2g - 3z$ (g is the transfer function defined previously and z is a zpk model described here), results in a zpk model:

```
>> z = zpk(0,[1,2],1);
>> 2*g-3*z
```

returns

```
Zero/pole/gain:

-3 (s-0.2473) (s^2 + 1.581s + 5.391)

-------------------------------------

(s-2) (s-1) (s^2 + 2s + 3)
```

LTI objects can subsequently be used to plot a system's time-domain or frequency-domain response by means of built-in MATLAB® commands, such as `step`, `impulse` or `bode`, to mention just a few. Once a response plot has been created, it can be visualized by means of the `ltiview` command. This command also enables changing the type of input and visualizing the new response of the same system. Once an LTI object plot is obtained, the command >> `ltiview` (typed at the MATLAB prompt) opens the LTI Viewer window. You can import an existing LTI object either from the Workspace or from a MAT-file. Assuming the transfer function that has been used thus far is imported into the LTI Viewer, the plot of Figure C.2 results, which illustrates the system response to a unit step (which is

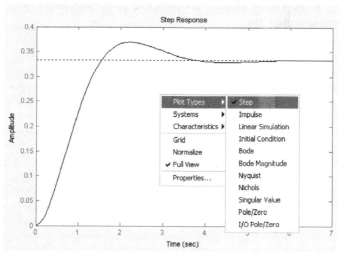

FIGURE C.2

LTI Viewer of a Second-Order System Unit Step Response.

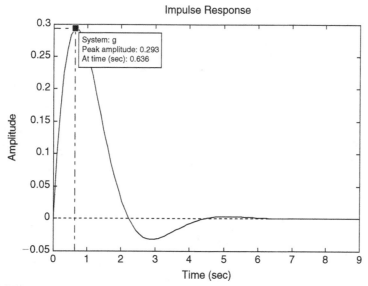

FIGURE C.3

LTI Viewer of a Second-Order System Unit Impulse Response with Peak Response.

the default input). By right clicking the plot space, you can change the plot type by selecting another input or response type altogether. Once an input has been selected and the corresponding plot has been obtained, important system response characteristics can easily be obtained from the plot. By changing the Plot Types to Impulse, for instance, the plot of Figure C.3 is produced, and after selecting Peak Response under Characteristics and clicking on the peak point, the explanatory box appears, which gives the value of the maximum (peak) response and the corresponding time.

Deformations, Strains, and Stresses of Flexible Mechanical Components

BARS UNDER AXIAL FORCE

A bar of length l, rectangular cross-sectional area A, and modulus of elasticity (or Young's modulus) E, which is acted upon by axial forces f_x, as sketched in Figure D.1, elongates (or compresses if the forces have opposite directions) by a quantity

$$u_x = \Delta l = \frac{f_x l}{EA} \tag{D.1}$$

By definition, the *axial* (or *normal*) *strain* is

$$\varepsilon_a = \varepsilon_x = \frac{u_x}{l} = \frac{\Delta l}{l} = \frac{f_x}{EA} \tag{D.2}$$

The *normal stress*, which is perpendicular to any cross-section of the bar of Figure D.1 (and therefore parallel to the forces f_x), is calculated as

$$\sigma_a = \sigma_x = \frac{f_x}{A} \tag{D.3}$$

Comparison of Eqs. (D.2) and (D.3) yields

$$\sigma_a = E\varepsilon_a \tag{D.4}$$

which is *Hooke's law*.

Extension about the axial (x) direction is accompanied by compressions in the y and z directions of the bar shown in Figure D.1. These compressive deformations

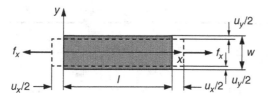

FIGURE D.1

Bar under Axial Point-Force Loading.

DOI: 10.1016/B978-0-240-81128-4.00017-9

are perpendicular to the force direction and result in *transverse* (or *lateral*) *strains*, which are defined as

$$\varepsilon_t = \varepsilon_y = \frac{u_y}{w} = \frac{\Delta w}{w}; \ \varepsilon_t = \varepsilon_z = \frac{u_z}{h} = \frac{\Delta h}{h} \tag{D.5}$$

where h is the bar thickness (dimension in the z direction, which is normal to the drawing plane of Figure D.1). The axial and transverse strains are related as

$$\varepsilon_t = -\mu\varepsilon_a \tag{D.6}$$

where υ is *Poisson's ratio*, a material constant. Equations (D.1) through (D.6) are also valid for a bar that is clamped at one end and free at the other end, where a force f_x is applied.

BARS UNDER AXIAL TORQUE

A point moment (torque) applied at the free end of a clamped-free bar, such as the one of Figure D.2(a), or at the midpoint of a clamped-clamped bar, such as the one of Figure D.2(b), produces an angular deformation (either at the free end of the clamped-free bar or at the midpoint of the clamped-clamped bar) that is equal to

$$\theta_x = \frac{m_x l}{GI_t} \tag{D.7}$$

where G is the shear modulus of elasticity and I_t is the torsion moment of inertia. The moduli of elasticity E and G are related by means of Poisson's ratio as

$$G = \frac{E}{2(1 + \upsilon)} \tag{D.8}$$

For a circular cross-section bar of diameter d, the torsion moment of inertia is actually the *polar moment of inertia*, which is equal to

$$I_t = I_p = \frac{\pi d^4}{32} \tag{D.9}$$

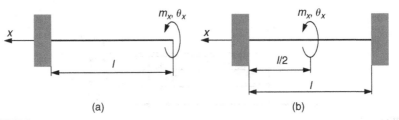

(a)   (b)

FIGURE D.2

Bars in Torsion under Point-Moment Loading: (a) Clamped-Free Bar; (b) Bridge (Clamped-Clamped Bar).

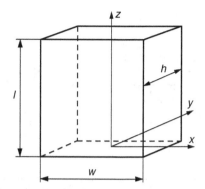

FIGURE D.3

Prismatic Block.

For a bar of rectangular cross-section of dimensions w (width) and h (thickness), where $w > h$, the torsion moment of inertia is calculated as

$$I_t = wh^3\left[0.33 - 0.21\frac{h}{w}\left(1 - \frac{h^4}{12w^4}\right)\right] \qquad (D.10)$$

The mass (mechanical) moment of inertia of a circular cross-section cylinder of mass m, radius R, length l, and mass density ρ with respect to its centroidal (symmetry) axis is

$$J_t = J_p = \frac{mR^2}{2} = \rho l\frac{\pi d^4}{32} = \rho l I_p \qquad (D.11)$$

For a prismatic block, such as the one of Figure D.3, the mass moments of inertia are

$$J_x = \frac{m(l^2 + h^2)}{12}; \; J_y = \frac{m(l^2 + w^2)}{12}; \; J_z = \frac{m(w^2 + h^2)}{12} \qquad (D.12)$$

BEAMS IN BENDING

For a cantilever (or clamped-free beam) such as the one of Figure D.4(a), the maximum *deflection* and *slope* (or rotation) produced at the free end by a point force f_z applied at the same point are

$$u_z = \frac{f_z l^3}{3EI_y}; \; \theta_y = \frac{f_z l^2}{2EI_y} \qquad (D.13)$$

with E being the elasticity modulus and I_y being the cross-sectional moment of inertia about the bending y axis.

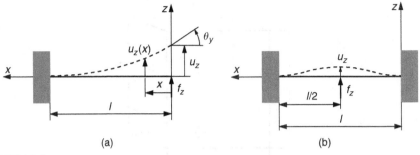

FIGURE D.4

Beams in Bending under Point-Force Loading: (a) Cantilever; (b) Bridge.

For a bridge (or clamped-clamped beam) such as the one of Figure D.4(b), the maximum deflection produced at the midpoint by a point force applied at the same point is

$$u_z = \frac{f_z l}{192EI_y} \tag{D.14}$$

Both the deflection and the slope are zero at the clamped ends of the beams of Figure D.4.

Figure D.5 shows a portion of a bent beam under the action of two end bending moments m_y. The upper area of the beam is compressed while the lower area is extended under the action of the two bending moments. As a consequence, a layer (or fiber) is undeformed, and this is the *neutral axis* (or *fiber*). Compressive stresses are set up on all the layers between the neutral axis and the upper fiber, and extensional stresses are applied to the fibers limited by the neutral axis and the lower fiber. As is the case was with bars under axial loading, the bending stresses are normal (i.e., perpendicular on any cross-section). According to *Navier's equation*, the maximum and minimum stresses corresponding to the lower and upper fibers are equal to

$$\sigma_{max} = |\sigma_{min}| = \frac{m_y \dfrac{h}{2}}{I_y} \tag{D.15}$$

where h is the thickness of the beam cross-section (dimension measured between the undeformed lower and upper fiber). The cross-section is assumed to be symmetric about the y axis.

Normal stresses are accompanied by *shear stresses* (denoted by τ), which are produced by a force f_z, such as the ones of Figure D.4, and are calculated as

$$\tau = \frac{f_z}{A} \tag{D.16}$$

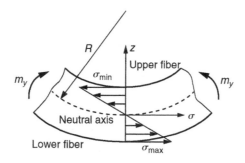

FIGURE D.5

Portion of a Bent Beam.

where A is the cross-sectional area. Similar to flexible mechanical members, shear stresses are generated between adjacent flowing fluid layers and Eq. (D.16) is therefore applicable to fluids as well.

For a circular cross-section beam of diameter d, the moment of inertia is

$$I_y = I_z = \frac{\pi d^4}{64} \tag{D.17}$$

For a rectangular cross-section beam of width w and thickness h, the moment of inertia is

$$I_y = \frac{wh^3}{12} \tag{D.18}$$

The *radius of curvature R* of the bent beam shown in Figure D.5 is calculated as

$$\frac{1}{R} = \kappa = \frac{d^2 u_z(x)}{dx^2} \tag{D.19}$$

where κ is the *curvature* and $u_z(x)$ is the deflection of a beam at a distance x; see Figure D.4(a).

Index